Biology
for Life

Edited by Kenneth J. Stein, Ph.D.

Seton Press
Front Royal, VA

© 2019-2021 Seton Home Study School
All rights reserved.
Printed in the United States of America

Seton Home Study School
1350 Progress Drive
Front Royal, VA 22630
Phone: (540) 636-9990
Fax: (540) 636-1602

ISBN: 978-1-60704-168-9

For more information, visit us on the web at www.setonhome.org
Contact us by e-mail at info@setonhome.org

Table of Contents
Biology for Life

Foreword v

Chapter 1: The Unborn Baby and the Miracle of Life 1

Chapter 2: Cells 35

Chapter 3: The Circulatory System 67

Chapter 4: The Endocrine System 93

Chapter 5: The Respiratory and Excretory Systems 113

Chapter 6: The Nervous System and Senses 131

Chapter 7: The Musculoskeletal and Integumentary Systems . 155

Chapter 8: The Digestive System 179

Chapter 9: Reproduction and Regulation of the Cell 207

Chapter 10: Single-celled Organisms and Fungi 231

Chapter 11: The Plant Kingdom 267

Chapter 12: Invertebrates: Part 1 301

Chapter 13: Invertebrates: Part 2 333

Chapter 14: Vertebrates: Part 1 361

Chapter 15: Vertebrates: Part 2 387

Chapter 16: Plants, Animals, and Man 435

Image Attributions 455

Glossary 461

Index 487

Foreword

Why Should Catholics Study Biology?

The picture above is a painting of the creation of man by Michelangelo. It is at the center of his famous painting on the ceiling of the Sistine Chapel in the Vatican.

> And He said: "Let us make man to our image and likeness: and let him have dominion over the fishes of the sea, and the fowls of the air, and the beasts, and the whole earth, and every creeping creature that moveth upon the earth." And God created man to His own image: to the image of God He created him: male and female He created them. And God blessed them, saying: "Increase and multiply, and fill the earth, and subdue it, and rule over the fishes of the sea, and the fowls of the air, and all living creatures that move upon the earth." And God said: "Behold I have given you every herb bearing seed upon the earth, and all trees that have in themselves seed of their own kind, to be your meat: And to all beasts of the earth, and to every fowl of the air, and to all that move upon the earth, and wherein there is life, that they may have to feed upon." And it was so done. And God saw all the things that he had made, and they were very good. And the evening and morning were the sixth day. *(Genesis 1:26-31)*

The preceding quote from the first chapter of the book of Genesis was the inspiration for Michelangelo's masterpiece. Here we see Adam, the man, having been made in glory at the right hand of God.

Genesis tells us that man, both male and female, are made in God's own image. Yet they look different, so how can that be? What does that mean? Genesis also teaches that "God blessed them" and told them to "increase and multiply," and finally, Genesis I concludes that what God has created is "very good."

In the preceding parts of Genesis, after God created the land, sea, and sky, and the creatures to fill the land, sea, and sky, each of which were given to man to "have dominion over," God declared each thing to be "good." However, it was only after creating man that God declared what He had made to be "very good."

Biology For Life

FOREWORD

But what is man? The writers of the Bible contemplated this great question, "What is man?"

Job 15:14: What is man that he should be without spot, and he that is born of a woman that he should appear just?

Psalms 8:5: What is man that thou art mindful of him? or the son of man that thou visitest him?

Psalms 143:3: Lord, what is man, that thou art made known to him? or the son of man, that thou makest account of him?

Ecclesiastes 2:12: I passed further to behold wisdom, and errors and folly, (What is man, said I, that he can follow the King his maker?)

Ecclesiasticus 18:7: What is man, and what is his grace? and what is his good, or what is his evil?

Hebrews 2:6-8: But one in a certain place hath testified, saying: What is man, that thou art mindful of him: or the son of man, that thou visitest him? Thou hast made him a little lower than the angels: thou hast crowned him with glory and honour, and hast set him over the works of thy hands: Thou hast subjected all things under his feet. For in that he hath subjected all things to him, he left nothing not subject to him.

The Garden of Eden, Thomas Cole

By looking carefully at the way these questions were asked, we can conclude that man, made in the image and likeness of God, should be perfect, "without spot," and that he was "crowned with glory and honour," so that God Himself would be mindful of him and visit him and that God would take account of him. God made man a little lower than the angels.

The part of man that was made in the image and likeness of God and that beholds wisdom and is just, would be studied one way. The part of man that has plants and animals for food, and that could increase and multiply, should be studied another way.

The first part is the spiritual part of man and is studied in philosophy and theology.

The second part is the physical part of man. This part is studied in the science of biology.

This is why we study Biology: to learn about man, and the other living things that God created for man.

Biology for Life

As Catholics, we should embrace the study of science. It is the teaching of the Church that God is the Author of all truth, and so there can be no conflict between revealed truth and scientific truth.

In fact, the Catholic Church, through her saints, doctors, and popes, has recognized this throughout the centuries. St. Augustine, St. Thomas Aquinas, and many popes have taught that science is compatible with faith. Many scientists agree. Francis Collins, for example, a modern genetic scientist and convert to Christianity who was appointed to the Pontifical Academy of Sciences by Pope Benedict XVI, has written, "Science is not threatened by God; it is enhanced." Collins reminds us that it is God Who makes science, and all things, possible.

In this textbook, we will study the parts that make up man, as well as the parts of animals and plants, from the smallest parts to entire groups of large parts that make complex systems. We will learn about some of the non-living components that are necessary for life, such as essential chemicals and water.

As this is a pro-life text, it is fitting that we begin with the study of the human body, starting with conception and fetal development in the womb. We will then progress through cells and cell structure, continuing through the various systems of the human body. After completing our study of human biology, we will study general biology, including biological classification, systems of plants and animals, and non-human organisms from single-celled through complex multi-celled organisms like fish, reptiles, birds, and mammals.

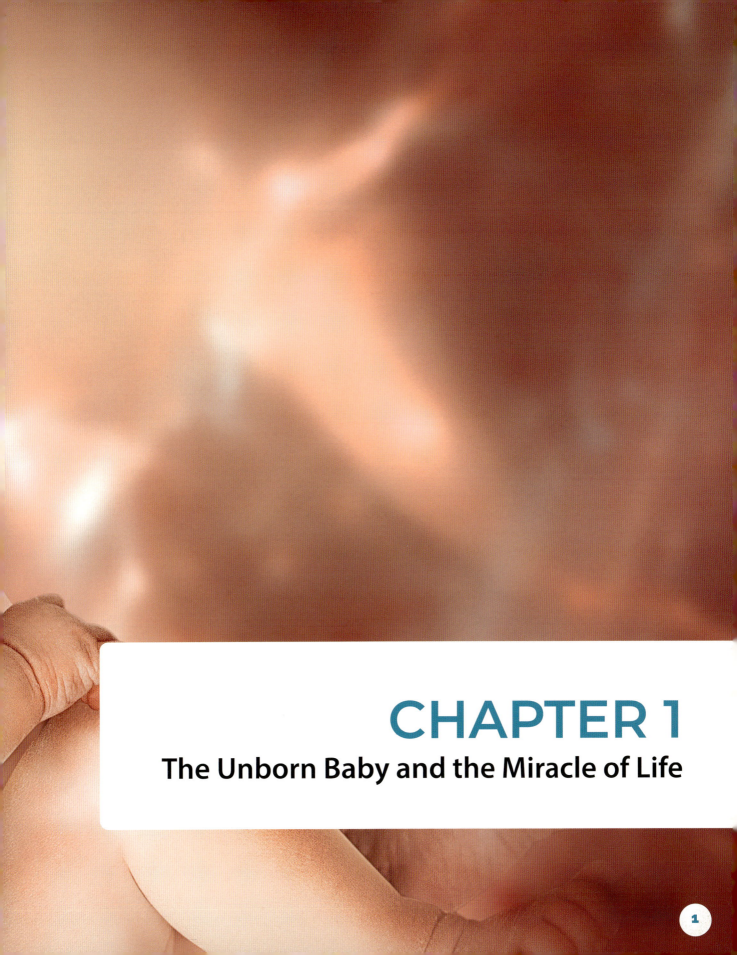

CHAPTER 1
The Unborn Baby and the Miracle of Life

Chapter 1

The Unborn Baby and the Miracle of Life

Chapter Outline

1.1 Introduction
   A. In God's Kingly Image
   B. Life, Death, and Consequences

1.2 The Mystery of Life
   A. Fundamental Genetics
   B. The Genetic Code: Blueprint of Life
   C. Conception

1.3 Growth of the Unborn Baby: Week One to Week Eight
   A. Week One
   B. Week Two
   C. Weeks Three and Four
   D. Weeks Five to Seven
   E. Week Eight

1.4 Growth of the Unborn Baby: Third Month to Ninth Month
   A. Month Three
   B. Month Four
   C. Month Five
   D. Month Six
   E. Month Seven
   F. Month Eight
   G. Month Nine

1.5 Labor

1.6 The Catholic Perspective of Life

1.1 Introduction

> **Section Objectives**
>
> - Give the Catholic Church's teachings about life

A. In God's Kingly Image

In the form of Adam and Eve, our first parents, man was created in the "image and likeness" of God (Gen. 1:26). The Church teaches that Adam and Eve sinned against God and fell from the state of grace. Although hampered by Original Sin, man would retain his inherent goodness, his primacy among earthly creatures, and his status as having "dominion over the fishes of the sea, and the fowls of the air, and the beasts, and the whole earth, and every creeping creature that moveth upon the earth" (Gen. 1:26).

Man, then, is a beautiful and holy creation who reigns upon earth, and is in no way "just another animal," as many modern scientists claim. This is reflected in the Biblical passages where it is written, "Then God said: 'Let us make man in Our image, after Our likeness" Gen. 1:26; and, "the Lord God formed man out of the clay of the ground and blew into his nostrils the breath of life, and so man became a living being" (Gen. 2:7).

The Church has repeatedly affirmed the indisputable value and sanctity of human life primarily based on the Bible, both in the Old Testament and the words of Jesus in the New Testament. In the past, few disputed this most obvious reality. Hence, popes preceding those of the second half of the twentieth century had little reason to explain the simple and apparent truths of human life and procreation. They did not need to write lengthy encyclicals explaining the great dignity of human life, or explain why attacks upon life are clearly evil, or insist upon man's unique headship amid earthly creatures.

The disease of our "modern age" is characterized by a certain darkness of both the intellect and will, and a severe sickness of the soul. Modern popes must teach the most basic principles of civil society, and address the fundamental aspects of what is now called "human rights." Perhaps this is a necessary task, but one must not overlook the profound significance of human life and the utter wickedness of attacks upon the human reproductive system, unborn babies, and the elderly.

The key aspects of man's existence and sanctity pertain not to man, but to God. Since God created the human body and soul, attacks upon the human body and soul are thus attacks upon God our Creator. This crucial point must never be minimized or obscured. Birth control, human cloning, embryonic stem-cell research, abortion, and euthanasia, are grievous affronts to God; they are rejections of God Himself, His primary Creation, His right to create life, His right to determine one's time of death, and His revealed Truth. These sins directly defy God who is essentially the Creator; His essential work is to create eternal beings.

Figure 1-1. Tertullian is considered to be one of the Fathers of the Church

B. Life, Death, and Consequences

The Church, with its Popes, Fathers, Doctors, and Saints, has steadfastly upheld the sanctity of human life, and, likewise, has always condemned any assault upon life. Here are some compelling examples of the Catholic Church's defense of human life, the first of which dates back to the first century:

- "You shall not kill by abortion the fruit of the womb and you shall not murder the infant already born" (from *The Didache,* an early work of Christian catechesis which attributes its teaching to the Twelve Apostles, circa 90 AD).

- "We say that women who induce abortions are murderers, and will have to give account of it to God. For the same person would not regard the child in the womb as a living being and therefore an object of God's care, and then kill it… But we are altogether consistent in our conduct. We obey reason and do not override it" (from the *Legatio* of Saint Athenagoras, a Father of the Church, circa 165 AD).

- "For us [Christians], we may not destroy even the fetus [Latin for "young one"] in the womb, while as yet the human being derives blood from other parts of the [mother's] body for its sustenance. To hinder a birth is merely speedier man killing; nor does it matter when you take away a life that is born, or destroy one that is coming to birth. That is a man which is going to be one: you have the fruit already in the seed" (from Tertullian's *Apology*, circa 160-240 AD—Tertullian has not been declared a saint, but he is considered to be one of the Fathers of the Church).

- "She who has deliberately destroyed a fetus has to pay the penalty of murder…here it is not only the child to be born that is vindicated, but also the woman herself who made an attempt against her own life, because usually the women die in such attempts. Furthermore, added to this is the destruction of the child, another murder" (from a letter written by Saint Basil the Great, a Father of the Church, circa 330-379 AD).

CHAPTER 1: **THE UNBORN BABY AND THE MIRACLE OF LIFE**

Figure 1-2. St. Jerome, a Father of the Church.

- "They drink potions to ensure sterility and are guilty of murdering a human being not yet conceived. Some, when they learn that they are with child through sin, practice abortion by the use of drugs. Frequently they die themselves and are brought before the rulers of the lower world guilty of three crimes: suicide, adultery against Christ, and murder of an unborn child" (from a letter written by Saint Jerome, a Father of the Church, circa 342-420 AD).

- "That person is a murderer who causes to perish by abortion what has been conceived" (from a writing of Pope Stephen V, circa 885-891 AD).

Section Review — 1.1 A & B

1. In whose likeness was man created?
2. What is the name of the earliest known work in which the Church proclaimed the sanctity of human life? When was this written?
3. Name two saints who were Fathers of the Church who defended the sanctity of human life.

Biology For Life

CHAPTER 1: THE UNBORN BABY AND THE MIRACLE OF LIFE

1.2 The Mystery of Life

With the teachings of the Catholic Church and the writings of various Popes, Fathers, and Doctors, one may begin a truthful Catholic exploration of the mystery of life. The principles in this chapter are designed to provide biological insights in defense of the constant proclamation of the Church that life begins at **conception**.

Conception occurs at the moment of **fertilization**, when a male sperm is united with a female egg. That is the moment at which an individual human being begins to exist, and thus when human life begins. At that moment, God creates a human soul (out of nothing), and unites this soul to the new physical entity that He forms through the union of the sperm and the egg.

Church teaching, which is always consistent with the Sacred Scriptures, affirms the sanctity of life in Jeremiah 1:5, "Before I formed you in the womb, I knew you; before you were born, I dedicated you; a prophet to all nations I appointed you." Even before Jeremiah was born, God knew and loved him and destined him to be a great prophet. Likewise, one observes in Psalm 139:14-15, "I praise you, so wonderfully you made me; wonderful are your works! My very self you knew; my bones were not hidden from you when I was being made in secret, fashioned as in the depths of the earth." The words "depths of the earth" are figurative language for "in the womb."

> **Section Objectives**
> - Define the terms: gene, trait, fetus
> - Describe genetics

Why must life begin at fertilization? In order for fertilization to occur, both the sperm and the egg must be **living**. Only living cells can give rise to new living cells. Thus, one must conclude that there is no point or time interval between fertilization and birth when the baby is not alive. It is impossible for a living sperm cell and a living egg cell to unite and not be alive for any period of time. That is why a new living being is formed the instant these unite. If at any time either the sperm cell or the egg cell were not living, they could not unite.

Consider your own life and proceed backwards in time. Before you were a teenager, you were a

DEFINITION OF TERMS

This section of the chapter provides some preliminary discussion of the conception and development of the unborn child. You may be unfamiliar with some of the terms used. While the textbook will go into more detail later, the following paragraph provides a brief summary that explains some terms that may be helpful in understanding this section of the chapter.

When an egg cell is fertilized, it is a single cell with a complete genetic code for a new human being. This single cell is called a zygote. Several hours after fertilization, the zygote divides into two cells, which are now called blastomeres. This is the beginning of the stage of the embryo, which lasts for eight weeks. The embryo grows as the blastomeres continue to divide until at the stage with 16 cells, which form a solid "ball" of cells, which is called the morula. The cells continue to divide until they number about one hundred, and the solid "ball" becomes a semi-hollow ball, called a blastocyst. On approximately the seventh day since fertilization, the blastocyst gets implanted in the uterus, the womb. After eight weeks the embryo has more recognizable human features and is then called a fetus, which means young one.

Biology For Life

child and before that, an infant. Before then, you were a baby fetus and before that a baby embryo. Before that, you were a blastocyst, before that a morula, and first of all a zygote. These various stages comprise all of a person's early development.

This chapter will examine each of the above living stages of a conceived baby inside the womb and discuss the one thing these stages all have in common: they are all comprised of active living cells. Cells are the basic whole units of life. In a human body there are many different types of cells. Each cell has a definite size and function. Individual cells require oxygen and food, and are capable of self-repair and reproduction. If we start with the very first stage, a fertilized egg, we are amazed that this one living cell is incredibly complex.

Scientists estimate that this one-celled fertilized egg, the result of the union of a living sperm cell and a living egg cell, each containing its own complex information data, has an informational content equal to 1000 volumes of the Encyclopedia Britannica! In another sense, the information in a single-celled fertilized egg would fill hundreds of CD-ROM discs. This one-celled fertilized egg is completely unique, and is more or less in charge of its own destiny. It is, however, dependent on other human beings for its survival.

A. Fundamental Genetics

All the characteristics of a particular organism are contained in the DNA of its chromosomes. The scientific study of how parents pass on their characteristics to their offspring is called **genetics**.

To understand life, one must have a basic understanding of our blueprint, known as our genetic code, and the mechanism of our genes. A **gene** is a specific part of a chromosome in a section of DNA. Each gene is responsible for a particular trait in the organism, for example, eye color. The gene is the basic, physical and active unit in heredity. Body cells have different functions, but all cells in a body contain the same genetic code. **Every cell in the baby differs from every cell in the mother.** There are two different

genetic codes and thus two different persons. Many times the blood type of the child is different from the mother, and half of the time, the sex of the baby is different from the mother. To claim that the mother should have the right to choose abortion "because it is her body," ignores the scientific fact that two bodies are involved.

The Church teaches that God provides a specifically created soul for each human child conceived by its parents. The fact that the child has different DNA from the mother's DNA proves that it is not part of her body, but is a different person who has a new soul from God. This life is sacred and was created for eternal happiness in Heaven with God. God wants this child to grow to know, love and serve Him. The child is a blessing, according to Genesis chapters I and II, so the parents must care for him, and educate him towards the goal of eternal life with God. Thus abortion is a violation of the Will of God and the natural order of human development.

Dr. A. W. Liley (1929 – 1983) is known as the Father of Fetology (the medical study and treatment of the unborn child). He was a professor in New Zealand and is credited with the first blood transfusion in an unborn baby. Dr. Liley makes the following remarks in *A Case Against Abortion*:

"Physiologically, we must accept that the conceptus, that is the conceived baby, is, in a very large measure, in charge of the pregnancy... Biologically, at no stage can we subscribe to the view that the fetus, or unborn baby, is a mere appendage of the mother... It is the embryo, another stage of unborn baby,

CHAPTER 1: THE UNBORN BABY AND THE MIRACLE OF LIFE

that stops his mother's periods and makes her womb habitable by developing a placenta and a protective capsule of fluid for him. He regulates his own amniotic fluid volume and, although women speak of *their* waters breaking or of *their* membranes rupturing, these structures belong to the fetus.

"And finally, it is the fetus, not the mother, who decides when labor should be initiated. Although the fetus is growing inside of the mother, it does not mean that he is not his own unique and genetically distinct living person…This, then, is the fetus we know and, indeed, we each once were. This is the fetus we look after in modern obstetrics [äb-ste-triks], the same baby we are caring for before and after birth, who before birth can be ill and needs diagnosis and treatment just like any other patient."

In fact, these days, doctors actually perform operations on an unborn baby who needs surgery!

> **Section Review — 1.2 A**
>
> 1. List the major way in which the baby is different from the mother. List two other ways the baby can be different from the mother.
> 2. Which person is in charge of the pregnancy of a mother?

B. The Genetic Code: the Blueprint of Life

Cells contain four major classes of molecules: carbohydrates, fats, proteins, and nucleic acids. For the purposes of genetics, the two fundamental molecules that we will discuss are proteins and nucleic [nu-klā-ik] acids. The term Protein comes from the Greek word for "primary," and nucleic acids are named for their location in the nucleus of the cell. **Proteins** control most of the functions of the cell, including growth and metabolism. The nucleic acids have an even higher importance because they control the formation of proteins. We can restate the above as follows: **nucleic acids provide the information, and proteins are the products of this information.**

Nucleic acids are made up of **nucleotides** [nu-klē-ō-tīdz], which are units composed of: (1) a nitrogenous (nitrogen-carrying) base, (2) a sugar molecule and (3) a phosphate group. Four basic nucleotides are strung together in long chains that form a double stranded helix, known as the Deoxyribonucleic [dē-äk-si-rī-bō-n(y)u-klā-ik] acid or DNA molecule. DNA is the "computer program" for life, and each person's DNA is unique and found in every cell within the body. In fact, it is more distinctive than fingerprints

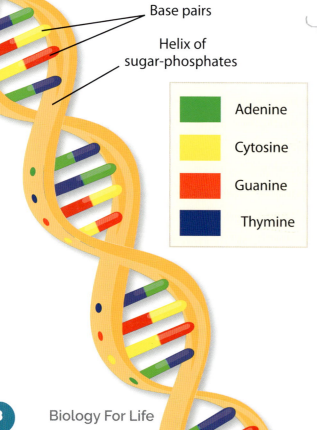

Figure 1-3. DNA is a nucleic acid that contains the genetic instructions used in the development and functioning of all known living organisms and some viruses.

Base pairs
Helix of sugar-phosphates

- Adenine
- Cytosine
- Guanine
- Thymine

Biology For Life

CHAPTER 1: **THE UNBORN BABY AND THE MIRACLE OF LIFE**

for every person, and like fingerprints, the police use DNA evidence in legal investigations. DNA contains the genes that determine eye color and hair color, height, and whether the being is a dog or a plant or a human.

The double helix shape of a DNA molecule resembles a twisted ladder with each of its sides composed of alternating phosphate groups and sugar molecules; the "rungs" on the staircase consist of pairs of nitrogenous bases. These bases are adenine (A), guanine [gwä-nēn] (G), cytosine [sī-tō-sēn] (C) and thymine [thī-mēn] (T). Each nitrogenous base within a DNA molecule can only bind to one other nitrogenous base. That is, adenine will only bind with thymine, and cytosine can only bind with guanine; A with T, C with G. This principle is referred to as complementary base pairing.

DNA exists within cells as structures called **chromosomes** [krō-mə-sōmz]. That is, the DNA molecules are attached to or chemically bound to proteins. Chromosomes contain all of the traits or characteristics for the child that he or she will ever have. The transmission of traits from the parent to the child is called **heredity**. Chromosomes reside in the nucleus

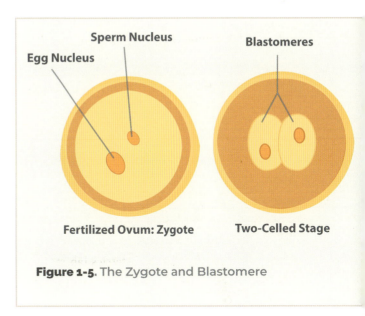

Figure 1-5. The Zygote and Blastomere

of egg and sperm cells, and provide the chemical instructions for human development.

Chromosomes are complex and contain all of an organism's genetic information for synthesizing many different kinds of proteins and enzymes. A specific region of the complex DNA molecule, which codes for one kind of protein, is called a **gene**. All of the genes contained in the chromosomes of the fertilized egg determine our distinctive, inherited traits. A child resembles its parents because its genes come from both parents. That is, one-half of a child's genes come from the mother and one-half come from the father. The genetic code that is stored in the nucleus of the fertilized egg is copied and transferred to every new cell that is formed. The complete genetic code will remain in each of the trillions of cells throughout an individual's life. Although the exact number of genes in each cell is not known, some scientists estimate that there are approximately 20,000 genes.

Human egg and sperm cells have the unique capability to create a new cell and eventual being with a new chromosomal makeup. Each egg and sperm cell contains only twenty-three chromosomes and are distinct from the somatic or other cells of the body. **Fertilization** is the process that results in

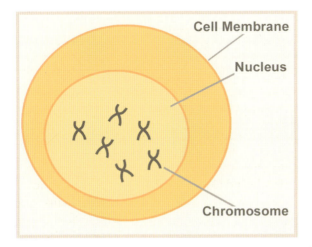

Figure 1-4. A fertilized egg contains 46 chromosomes, 23 from the father and 23 from the mother.

Biology For Life

Process of Cleavage

Figure 1-6.

Zygote
(formed when sperm and egg nuclei combine)

Two-Cell Cleavage

Four-Cell Cleavage

Eight-Cell Cleavage

the joining of the chromosomes of an egg cell and a sperm cell to produce a new unique **genome** [jē-nōm]. This is the term for the genetic code or the hereditary information that is contained within the DNA. The new genome will direct the future developments of the new person: hair color, skin texture and color, height, and all of the person's physical attributes. Think of the genome as an organic code system, a type of library that provides all of the information for a person, beginning with conception, through its fetal development, and throughout its life outside of the womb.

The number of chromosomes in the human species is forty-six. Thus, a human **zygote** [zī-gōt], or a fertilized egg, contains forty-six chromosomes, twenty-three of which were received from the female parent via the egg cell and twenty-three from the male parent by the sperm cell.

As the zygote begins to grow and divide in a process called **mitosis** [mī-tō-sis], one cell becomes two, and two cells become four, four become eight, and eight become sixteen. Each new cell contains a complete set of forty-six chromosomes. Eventually, all cells of the body will contain twenty-three pairs of chromosomes, except the egg and sperm cells, both of which contain only half the number of chromosomes.

Following fertilization, the process of cleavage begins to occur. **Cleavage** is a special kind of cell division in which the cells divide into smaller cells rather than same-sized cells. The large fertilized egg cell (the light-yellow area in Figure 1-6) divides into two smaller cells, and the two into four smaller cells, and so on. Thus, cleavage begins with one very large egg cell and ends with a greater number of smaller cells. These continue dividing by mitosis.

Regular mitosis produces two cells from one cell. Initially the cell grows about twice its original size before dividing into two daughter cells, each of which ends up the same size as the dividing, parent cell. These two daughter cells are identical to the parent cell in both chromosomal makeup and size.

After the first few cell divisions, the cells begin to differentiate. This is due to various genes directing the cells to assume diverse roles. Genes play a role in directing where cells migrate to and control what functions they will perform. Genes also determine the rate by which mitosis will occur, when cells cease to function, and when they will die.

The human body is constantly replacing dead cells with new ones by mitosis. In fact, the entire outer layer of our skin is really composed of dead cells.

Section Review — 1.2 B

1. Define the following terms:
   a. chromosome b. gene c. mitosis
   d. cleavage.

2. How are the human sperm and egg cells different from other body cells?

3. Explain how the traits of the parents are transmitted to their child. Your answer should include the terms chromosome, gene, egg, sperm cells.

CHAPTER 1: **THE UNBORN BABY AND THE MIRACLE OF LIFE**

C. Conception

The Catholic Church, inspired by the Holy Spirit, has consistently taught that a human life begins at conception, which occurs at the moment of fertilization, as science clearly proves.

Dr. Jerome Lejeune (1926-1994) was a French geneticist and professor of genetics at a university in Paris. Dr. Lejeune tells us the following: "It is now an experimentally demonstrated fact that at the three-celled stage, every individual is uniquely different from every other individual and the probability that the genetic information found in one cell would be identical to another person is less than one in one billion."

The single-celled zygote is the first cell of a person created by God with the help of the parents. The process by which this unborn, one-celled person grows into a baby ready for birth is a process that involves many steps. The development of the child can be broken down into four interdependent parts: growth, morphogenesis [mor-fō-jen-ĭ-sis], determination, and differentiation.

Figure 1-8. Dr. Jerome Lejeune was a French geneticist who discovered the chromosomal cause of Down Syndrome.

Figure 1-7. The moment of fertilization, that is, the living products of the male and female reproductive systems uniting to form a single-celled living person. The sperm is about 40 micrometers from head to tail. The egg is much larger, 100 micrometers in diameter, which is about the size of a pinhead.

- **Growth** refers to an increase in size.
- **Morphogenesis** is the origin of form or shape such as the appearance of fingers and toes.
- **Determination**, also called restriction, means a part of the embryo becomes restricted to a specific role or function.
- **Differentiation** is a process whereby cells become specialized and the appearance of a new property emerges, such as the development of muscle cells.

Professor Jerome Lejeune was one of the world's leading experts in the field of Fundamental Genetics, and was the discoverer of the chromosomal cause of Down Syndrome. Professor Lejeune wrote that the single-celled zygote is the "most specialized cell under the

Biology For Life

CHAPTER 1: **THE UNBORN BABY AND THE MIRACLE OF LIFE**

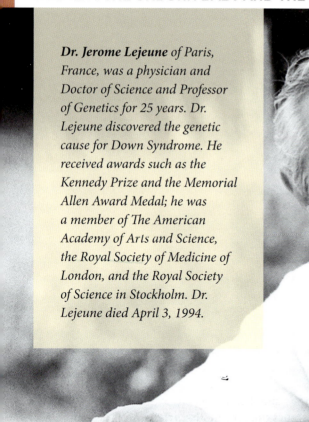

Dr. Jerome Lejeune of Paris, France, was a physician and Doctor of Science and Professor of Genetics for 25 years. Dr. Lejeune discovered the genetic cause for Down Syndrome. He received awards such as the Kennedy Prize and the Memorial Allen Award Medal; he was a member of The American Academy of Arts and Science, the Royal Society of Medicine of London, and the Royal Society of Science in Stockholm. Dr. Lejeune died April 3, 1994.

sun. After conception, much information is lost, but absolutely nothing is added or gained except nourishment."

Furthermore, Dr. Lejeune states, "I cannot see any difference between the human being you were and the late human being you are, because in both cases, you were and are a member of our species." Human cells cannot be classified with any other species, including monkeys, because **each cell is distinctly human**, having forty-six human chromosomes.

LIFE BEGINS AT CONCEPTION

In the words of Dr. Lejeune:

"Each of us has a very precise starting moment which is the time at which the whole necessary and sufficient genetic information is gathered inside one cell, the fertilized egg, and this is the moment of fertilization. There is not the slightest doubt about that and we know that this information is written on a kind of ribbon which we call the DNA…Nature has used the smallest possible language to carry the information from father to children, from mother to children, from generation to generation…At no time is the human being a blob of protoplasm. As far as your nature is concerned, I see no difference between the early person that you were at conception and the late person which you are now. You were and are a human being."

When Dr. Lejeune testified in the Louisiana legislature he stated:

"Recent discoveries by Dr. Alec Jeffreys of England demonstrate that the information

Biology For Life

on the DNA molecule] is stored by a system of bar codes not unlike those found on products at the supermarkets…it's not any longer a theory that each of us is unique."

In 1989, during testimony on the Seven Human Embryos in Tennessee, Dr. Lejeune made the following observation: "…as soon as he has been conceived, a man is a man."

> **Section Review — 1.2 C**
>
> 1. What is fertilization?
> 2. Define a.) growth b.) morphogenesis c.) determination d.) differentiation.

1.3 Growth of the Unborn Baby: Week One to Week Eight

> **Section Objectives**
>
> - Identify the steps in development from fertilization to implantation
> - Describe the Period of the Child at Embryo stage
> - Explain the role of the placenta

Process of Growth

As we keep in mind that each and every cell in our bodies contains a complete genome, let us look at the process of growth. The complete prenatal period (the time before birth) of development usually lasts around forty weeks. Two of these forty weeks occur before the egg is fertilized. During these two weeks, the mother's egg undergoes changes and matures to prepare it for fertilization. With this in mind, the actual number of weeks from fertilization to birth is thirty-eight.

Scientists divide these thirty-eight weeks into different stages or periods. Some scientists use the trimester system, with each trimester consisting of three calendar months. Other scientists use a different terminology to divide the thirty-eight weeks. This alternate system calls the first stage the **Period of the Embryo**, which lasts from day one until the eighth week. The second stage is called the **Period of the Fetus** and lasts from the ninth week until birth.

A. Week One

During fertilization of the **ovum** (egg cell) by the sperm, twenty-three chromosomes from the father's sperm unite and chemically bind with twenty-three chromosomes from the mother's ovum. This occurs in a period of about twelve hours. From this point onward, the fertilized ovum is called the **zygote** [zī-gōt]. In this stage, the cellular machinery begins to duplicate so that it can divide into two new cells. Approximately thirty hours after fertilization, the duplication process finishes and cleavage begins.

Cleavage [klē-vij] is the process in which the living fertilized egg divides, or cleaves, into two smaller cells. The process continues with the two new cells duplicating their cellular machinery so that they can begin dividing into more, equal-sized cells. The second cleavage is finished within forty-eight hours after fertilization.

At the end of three days, there are sixteen to thirty-two cells, and by the fourth day, there may be as many as sixty or seventy cells. The first few divisions have a slower rate of mitosis than the succeeding divisions. All of the cells that result from the early cleavage stage are referred to as **blastomeres** [blăs-tō-mēr].

After about three days, the zygote has become a solid ball of blastomeres that resemble a berry and is termed a **morula**, from the Latin word for

CHAPTER 1: THE UNBORN BABY AND THE MIRACLE OF LIFE

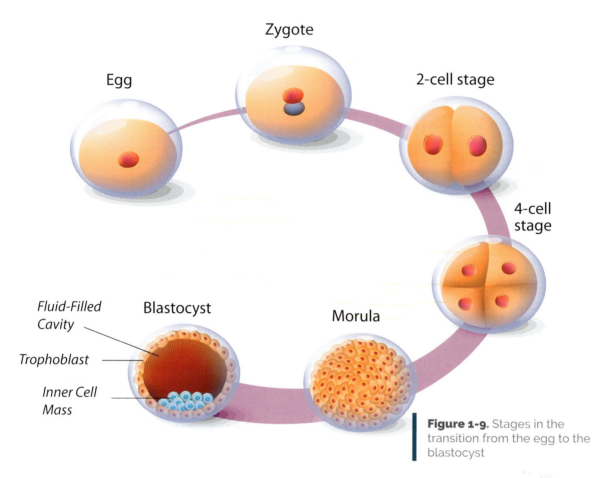

Figure 1-9. Stages in the transition from the egg to the blastocyst

mulberry. Over the next few days, the solid living morula will eventually hollow out in the center and fill with fluid. When the hollow center appears, it is called a **blastocyst** [blăs-tō-sist] and marks the end of cleavage.

As the cells continue to multiply, a group of cells begins to increase within one side of the hollow sphere. This **inner cell mass** lies within a shell-like layer of cells and is the size of a pinpoint. It marks the location of the living embryo. The outer shell-like layer of protective cells is called the **trophoblast** [trō-fə-blast] or **trophectoderm** [trō-fek-tə-dərm] and will be the future location of the fetal placenta.

Beginning at the time of fertilization and through the blastocyst stage, the yolk that was part of the mother's egg provides nutrition for the growing cells. Even though the mother's egg is one of the largest cells in her body, it only contains a limited amount of nutrients. The dividing cells will quickly use up this small food supply. When the yolk is depleted, the dividing cells of the newly conceived child must find another food source. They will do so when the blastocyst reaches its next destination.

The blastocyst slowly passes down the **fallopian tube** [fa-lō-pē-an] or **uterine tube** until it reaches its place of rest called the **uterus** or womb. In order to attach to the uterus, the outer layer of the blastocyst releases enzymes that digest part of the uterine lining. These enzymes cause some of the cells of the mother's womb to burst. This area will become the point of attachment for the blastocyst.

The implantation of the blastocyst within the uterus occurs around six days following fertilization.

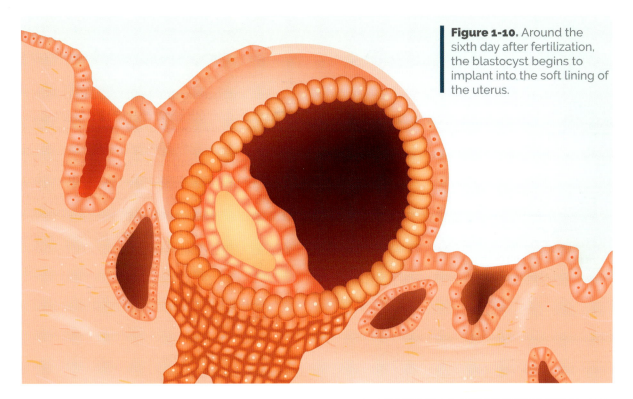

Figure 1-10. Around the sixth day after fertilization, the blastocyst begins to implant into the soft lining of the uterus.

It will become the soft, nourishing home of the child for the remainder of the pregnancy. Implantation of the blastocyst also causes its cells and the cells of the uterus to divide rapidly, forming the rudiments, or beginnings, of a placenta and various protective membranes.

During the same time, the ovaries of the mother respond by secreting hormones that stimulate the uterine cells to release nutrients, including: glycogen [glī-kō-jen], proteins, lipids, and even some essential minerals. These nutrients are referred to as "**uterine milk**" and serve to nourish the growing child. Furthermore, the uterine cells have the ability to increase in size, thereby allowing the storage of excess amounts of these vital nutrients.

For a time, the uterine cells continue to provide nourishment, but work in conjunction with the placenta. After the eighth week of pregnancy, the uterine cells are no longer needed to provide nourishment and the placenta takes over this important role for the growing baby.

Section Review — 1.3 A

1. How long is the Period of the Embryo and the Period of the Fetus?
2. What is the difference between the ovum and the zygote?
3. What is the meaning of cleavage? What is the meaning of blastomere?
4. What is a morula? When is it called a blastocyst?
5. What is a trophoblast?
6. Define a fallopian tube.
7. What are the nutrients in "uterine milk"?
8. How many days after fertilization is the unborn child implanted in the womb?

Biology For Life

CHAPTER 1: **THE UNBORN BABY AND THE MIRACLE OF LIFE**

B. Week Two

The period of the **embryo**, a Greek word meaning, "to swell," marks the beginning of the second week after fertilization and lasts through the eighth week. This period marks the development of the placenta, the formation of the primary internal organs of the unborn baby, and the appearance of the baby's major external body structures.

The **placenta [pla-sĕn-ta]** is the organ that attaches the unborn child to the uterine wall of the mother. The placenta has many functions and serves as a liver, kidney, and lung for the embryo. It is responsible for nourishment and the production of hormones. These hormones are important in maintaining the pregnancy. In addition, the placenta can be likened to a gate by selectively letting substances in and out of the mother and the baby. Although the placenta allows substances to pass through to the baby, it keeps the mother's circulatory system separate from that of the baby.

Following implantation in the uterus, some of the hormones that have been released into the mother's body have taken effect and the mother may experience discomfort. During this time, many of the child's 120 cells have already become specialized! These specialized cells have various functions, one of which is to cause the digestive enzymes to be released by the blastocyst. In addition, these specialized cells produce a hormone called **human chorionic [kŏr-ē-ä-nĭk] gonadotropin [gō-năd-ō-trō-pĭn]** (hCG). This hormone causes the other cells in the living embryo to produce additional enzymes and hormones. Both of these aid the growth and development of the growing child.

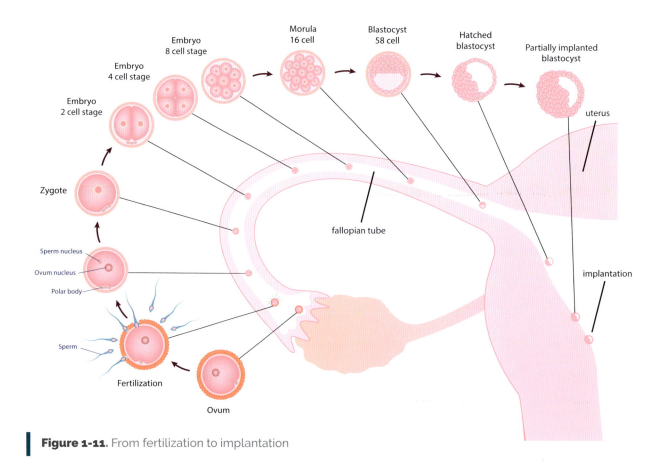

Figure 1-11. From fertilization to implantation

16 Biology For Life

In addition to the embryo, the placenta also produces hCG, and many other hormones. One of these is **progesterone [prō-jes-tĕ-rōn]**, which plays a role in maintaining the pregnancy. As this tiny ball of cells begins to multiply, a saga of life continues to unfold. The mother is affected and almost transformed by this little life inside of her. These hormones are powerful and affect and even change the mother's immunity, digestion, blood chemistry, emotions, lung capacity, and the tendons and ligaments (sinews) that hold her body together. For the last, her body becomes more elastic to prepare for stretching of tissues and the birth of her baby.

During the time when the blastocyst implants into the uterus, finger-like projections called **chorionic villi** begin to grow and branch off from the trophoblast (recall that this is the half of the blastocyst that forms the placenta). This process is comparable to tree roots growing and extending into soft soil. Fetal capillaries that contain fetal blood form in these chorionic villi. The trophoblast cells begin to release digestive enzymes. These continually erode the uterine tissue to increase the space necessary for the developing child.

The mother's tiny blood vessels, called **capillaries**, are broken during implantation. Blood flushes out from these and seeps into sacs or **lacunae [la-kü-nā]** that surround the chorionic villi. These lacunae contain nutrients and carry oxygen from the mother's blood that diffuse across the thin membrane of the chorionic villi to assist with nourishment. Waste products from the cells of the blastocyst diffuse across the chorionic villi and are removed from the mother's blood. These processes occur before full development of the placenta. Eventually the chorionic villi and other developing tissues will become the fetal placenta and will take over as the dominant means of oxygenation and nourishment.

The development of the placenta during the second week of pregnancy is a critical period for the well being of the child. The baby is connected to

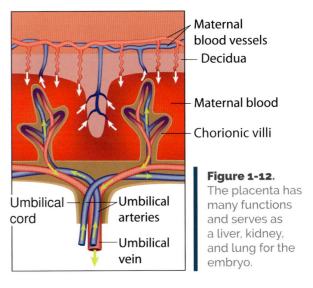

Figure 1-12. The placenta has many functions and serves as a liver, kidney, and lung for the embryo.

the placenta by the umbilical cord, which eventually grows to a length of twenty inches. The **umbilical cord** contains one vein that supplies the child with nutrition and oxygen. It also contains two arteries that remove waste and deoxygenated blood. In the placenta, the capillaries of the mother and the child intertwine without joining, like two vines wrapped around a tree.

A thin membrane located between the capillaries of the mother and child allows nutrients and wastes to cross both sets of capillaries by **diffusion** and **active transport**. This process is similar to water passing through a sponge. Diffusion is the random movement of molecules from an area of higher concentration of molecules to an area of lower concentration of molecules. When you see a clump of un-dissolved tea slowly dissolve at the bottom of a glass, you are watching diffusion. Active transport requires energy to move a substance across the cell membrane.

The placenta grows until the seventh month, at which time it remains static for one month, and then regresses during the last month. **Passive immunity** from some diseases, which is built up by the mother during pregnancy, can be passed through the placenta to the child and will remain with the baby for up to six months following birth. In the latter stages of development, should the child need to fight infection, which is quite rare, the placental tissue can swallow or engulf the

CHAPTER 1: THE UNBORN BABY AND THE MIRACLE OF LIFE

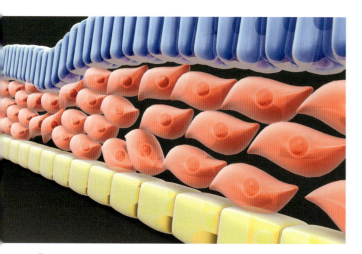

Figure 1-13. The three germ layers are called the Ectoderm, the Mesoderm, and the Endoderm.

mother's antibodies and carry them to the fetus to be released into the fetal bloodstream. This temporary resistance to disease helps protect the child from many infectious diseases during the first few months of its life.

The placenta is involved with bringing water to the growing baby as well. The early fetus is about 90% water by weight, and the percentage of water will decrease to about 70% by birth. The placenta has other functions and responsibilities. At about ten weeks after fertilization, the placenta is producing about 1.5 grams of protein per day, which makes the placenta the hardest-working organ in the body. This is a significant amount of material when considering that during this time, the unborn baby weighs about 1 gram.

Early in the embryonic stage, the cells of the inner cell mass become organized into a flattened embryonic disk. It begins with two layers of cells and shortly afterwards gains a third layer. These three layers are called **primary germ layers** and give rise to the future organs and other specialized parts of the baby's body. These three germ layers are called the Ectoderm, the Mesoderm and the Endoderm. The **ectoderm** [ek-tō-derm] is the first and **outer germ layer**, which becomes skin, hair, and nervous system. The **middle germ layer**, or **mesoderm** [me-zō-derm], becomes the skeletal system, muscular system, circulatory system, and urinary system. The **endoderm** [en-dō-derm] is the **inner germ layer**. It forms the digestive tract and some internal organs, including the liver, pancreas, and thyroid.

The embryonic disk begins to fold itself in the beginning of the third week by changing the shape of individual cells. At this time, cells begin multiplying at different rates. The flat disk soon becomes tubular, or cylindrical, as the body of the embryo becomes more distinct. The future location of the brain and spinal cord appears in a channel-like structure called a neural groove. Small, paired clumps of cells will begin to form on each side of the neural groove. These clumps of cells are the early stages of the vertebrae and back muscles.

A mass of cells appears where the heart tube is beginning to form, and blood vessels begin to join. These cells will soon become closed and functional as a circulatory system. At the end of the third week, the embryo is about as long as the thickness of a nickel and weighs about one gram (1/28th Oz.).

Section Review — 1.3 B

1. What happens during the period of the embryo?
2. What is the placenta? What are some of its functions?
3. What is progesterone?
4. What does the chorionic villi contain?
5. What are lacunae? What do they do?
6. What does diffusion mean?
7. What is passive immunity? Is it a good thing?
8. List three germ layers, and explain their function.
9. What is the size of the unborn child at the end of the third week?

C. Weeks Three and Four

Weeks three and four mark a period of rapid progression and dynamic change even though the baby is still smaller than an apple seed. The heart, which is not very complex at this stage, consists of two tubes that will begin to fuse together by day eighteen. One of the most remarkable features of the child in the womb is that by day twenty-one, the heart begins beating within a **closed system of blood vessels**. Only twenty-one days after the merging of egg and sperm, blood is circulating between the embryo and the placenta via the umbilical cord!

After the third week, the very delicate spinal cord begins to be covered over and protected from spina bifida [bif-i-da]. **Spina bifida** is a defect of the spinal cord where part of the spinal cord may stick out through an opening in the vertebra. It can be seen in the early stages of development. Spina bifida can be caused by a lack of B vitamins, especially folic acid. Therefore, pregnant mothers are often advised to take folic acid supplements during pregnancy.

The folds of the head and face are distinguishable in the fourth week, and spots appear where the eyes and ears are beginning to form. Tiny buds appear where the arms and legs are beginning to form.

The fourth week is characterized by the development of rudimentary organs in a process called **organogenesis [ȯr-ga-nō-je-ne-sis]**. The rudimentary parts of the thyroid gland, lungs, liver, pancreas, and kidneys are apparent just after three weeks! Among these organs, the liver appears very early in the development of the child. Specialized liver cells begin to clump together on the twenty-first day.

In another week, the liver cells are clearly grouped into a liver, which appears about the same time as the gall bladder. The upper and lower jaws are also discernible at this time. By the end of the first month, the human baby embryo has progressed from a one-celled state to millions of cells. It has grown from microscopic size to an intricately organized group of cells, with each cell having specific functions in the developing nervous, muscular, circulatory, digestive, and skeletal systems.

D. Weeks Five to Seven

The fifth week through the seventh week is a time of continued organogenesis and growth. The embryo has grown 10,000 times in size, a brain has developed, and blood is flowing through its vessels! The head grows rapidly and becomes round and erect. The eyes, nose, and mouth develop and form the face of the baby. This may seem like great progress, and it is, but the developing baby needs to progress more as it marches on toward the trillion cells that will be present in its body at birth.

At six and one-half weeks, all of the twenty milk tooth buds are present. The arms and legs are no longer buds but limbs. They begin to elongate, and fingers and toes emerge from the previous elbow-like buds.

All of the major organs have formed and continue to grow, giving the body a more human-like form. Glucagon [glü-ka-gon] has been found in the embryo's pancreas as early as the sixth week, and insulin by the seventh or eighth week. At the eighth

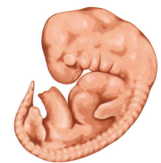

Figure 1-14. By week four, organogenesis begins where rudimentary organs become apparent.

week, the baby's stomach is secreting gastric juice. The presence of glucagon, insulin, and gastric juices are strong evidence for the complexity of life, even in these early stages of development.

Sometime during the end of the fifth and the beginning of the sixth week, brain waves become detectable with an instrument known as an **Electroencephalograph [ē-lĕk-trō-ĕn-sef-a-lō-graf] (EEG)**. This was first reported in the Journal of the American Medical Association in October

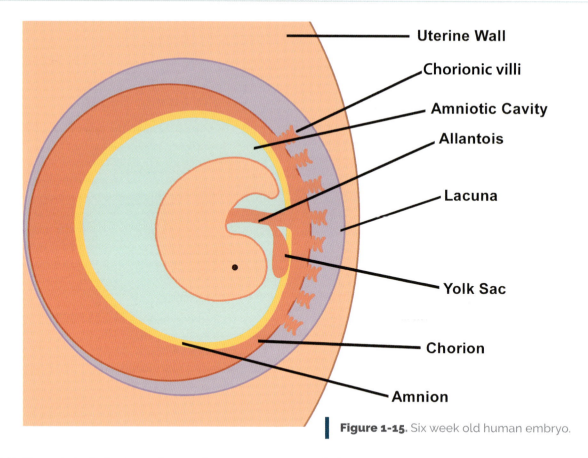

Figure 1-15. Six week old human embryo.

1964. Not only is the brain working at this stage, but also signals are transmitted along the nerves.

During these weeks, the movement of the embryo is quite extensive, and has been documented by many scientists and doctors. In 1986, Dr. Liley photographed a fascinating display of embryonic movement by capturing a baby sucking his thumb. Another scientist, L. B. Arey, described in his textbook the following displays of motion in the womb: "By the sixth week, if the area of the lips is stroked, the child responds by bending the upper body to one side and making a quick backward motion with the arms. In the eighth week, if you tickle the baby's nose, he will flex his head backwards away from the stimulus."

Likewise, an article by Dr. James Dobson gives the following account: "When a surgical technician opened the abdomen for a tubal pregnancy [the fertilized ovum developed within the Fallopian tube], the tube expelled a 1-inch long child, about 4-6 weeks old. It was still alive in the sac and the tiny baby was waving its little arms and kicking its legs and even turned its whole body over."

Section Review — 1.3 C & D.

1. What is spina bifida? What causes it? How can a pregnant mother try to avoid it for her baby?

2. During the process of organogenesis, which organs begin their development?

3. The letters EEG stand for electroencephalograph. What does this instrument detect in the unborn baby?

4. What features develop and can be seen during the fifth to seventh week of the unborn baby?

E. Week Eight

At the end of the eighth week (second month), almost all of the major structures of the body have been formed, though these are miniature. Embryonic **bone formation** has begun and the bones are very elastic, and remain cartilaginous until the sixth month. That is, they are flexible and not hard, or calcified, like adult bones.

By this time, though some of the chorionic villi have degenerated, the remaining chorionic villi continue to develop the placenta. Various protective membranes that have been forming since the second week are now more pronounced and functional. The trophoblast has developed into a membrane called the **chorion** [kŏr-ē-on]. It surrounds the embryo and will also form a part of the placenta. *Figure 1-15 shows the chorion.*

While the placenta is forming from the chorion, the chorionic villi, and the uterine cells, another membrane is developing around the embryo. This membrane is called the **amnion** [am-nē-on], which forms a protective layer around the embryo and forms the amniotic cavity. The amniotic cavity fills with a liquid, called the amniotic fluid, which serves several purposes, including protection, shock absorbency, and temperature regulation. In addition, it provides a watery environment in which the embryo can grow freely without being encumbered or compressed by surrounding tissues. When the baby advances from the embryonic stage into the fetal stage, he will continually drink the amniotic fluid. The amniotic fluid is recycled at a rapid rate; every three hours, the fetal baby gets new "bath water."

In addition to the chorion and amnion membranes, two other membranes function in the development of the child. The **yolk sac**, although containing no yolk, performs three important functions. It forms blood cells in the early stages of development, provides circulation until the circulatory system develops, and it gives rise to germ cells, which will eventually become the sperm or egg cells when a child matures into the stage known as adolescence. Finally, the yolk sac forms future immune cells. The second membrane is called the **allantois** [al-lan-tō-is]. Like the yolk sac, it forms blood cells; however the allantois plays a major role in the development of the placenta and the umbilical cord, especially the two umbilical arteries and the umbilical vein.

At the end of the first week, the embryo was barely one millimeter in length. Since then, in a matter of eight weeks, the embryo has become thirty times as long. An important time of initial development is now over. The embryonic stage has

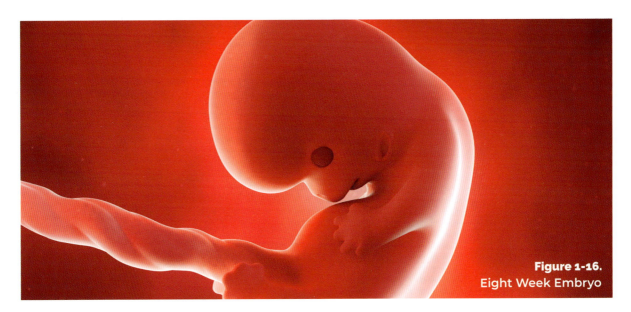

Figure 1-16.
Eight Week Embryo

been completed by the end of the eighth week. The embryo has progressed from a microscopic bundle of cells to having all of the essential internal and external body parts!

Any disturbances during these initial eight weeks of development will most likely result in major malformations or malfunctions. If the defects are too great, then the life of the unborn child usually ends naturally at this stage. The premature, natural ending of a pregnancy is often referred to as a miscarriage or spontaneous abortion. It is important that the mother abstain from smoking or consuming alcohol during the pregnancy. These two substances, tobacco smoke and alcohol, are among the leading causes of birth defects.

> **Section Review — 1.3 E.**
>
> 1. How do we describe the bones of the unborn in the second month?
> 2. What is the difference between the chorion and the amnion membranes in the eighth week?
> 3. What are the three important functions of the yolk sac?
> 4. How is the allantois different from the yolk sac?
> 5. What are some of the purposes of the amniotic fluid?

1.4 Growth of the Unborn Baby: Third Month to Ninth Month

Section Objectives

- Describe the Period of the Child at Fetus stage
- Describe the circulation in the womb of the child in the fetal stage

The Living Fetus

Beginning with the ninth week, the baby is called a fetus, a Latin word meaning "young one." The period of the fetus continues until birth, and the tiny young one remains immersed in a sea of amniotic fluid. The **fetus** is a completely formed baby in miniature, and is very lively and active. The unborn baby has the senses of touch and hearing, which enable it to respond to pain, pressure, and loud noises. In the fetal period, the baby has formed all of the systems needed to maintain life after birth. Over the next six months, these systems will mature and function as a unified system under a single command center, the brain.

A. Month Three

It is during the third month that the baby's neck elongates and straightens. The eyelids grow over the eyes, sealing them shut for most of the next six months. The ninth week marks the beginning of fingernail and toenail growth. Not only are the nails growing, but also the toes are able to curl. The baby may even scratch himself if the nails grow too much. Often, these nails will need to be clipped at birth, especially in post-term babies.

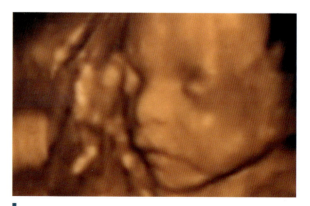

Figure 1-17. 3D ultrasound of an unborn child at the fetus stage.

Fetal Growth

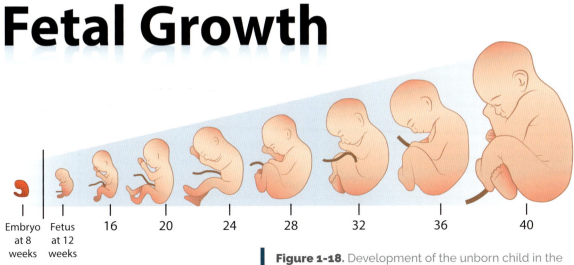

Figure 1-18. Development of the unborn child in the fetus stage from the 8th to the 40th week.

Although bones began to grow in the second month, these begin to calcify, or harden, in the third month. During this month, the physiological differences in the sexes become prominent. If the child is a boy, then male hormones are released. If the child is a girl, then nothing will happen for a little while. By week eleven, the baby is able to open his little mouth and may even place a finger or thumb inside the newly discovered mouth. Some babies are even born with calluses on their thumbs from thumb sucking.

The open mouth will begin swallowing the amniotic fluid. Since the amniotic fluid contains carbohydrates, proteins, and lipids, it undoubtedly plays a role in providing nutrition. This month reveals a time of rapid growth, as the baby doubles in size. He is now approximately the size of an adult finger. The amniotic sac has increased from the size of a golf ball to that of a baseball. By the end of the month, the fetus is moving vigorously, although he is still not large enough to wedge a shoulder and a foot on both sides of the womb, and so the mother does not yet feel the baby moving.

Fetal circulation is different than circulation outside of the womb. In the womb, there is very little fetal blood circulating to the lungs. Instead of the lungs exhaling carbon dioxide into the air as is done after birth, these functions are conducted by the placenta and umbilical cord. The infant receives oxygenated blood and nutrients from the placenta via the umbilical vein. Deoxygenated blood containing waste products leaves the fetus via umbilical arteries, and is carried to the bloodstream of the mother. Arteries take blood away from the body of the baby to the mother. Conversely, in both newborn babies and all of us, arteries take blood away from the heart. The fetal lungs remain collapsed until a few seconds after birth when blood is forced to the lungs with the baby's first breath.

> **Section Review — 1.4 A.**
>
> 1. When is the unborn baby often called a fetus? What does it mean in Latin?
> 2. Name some of the changes in the unborn during the third month.
> 3. From what does the unborn receive nutrients?
> 4. In what month do the bones of the unborn begin to harden?

CHAPTER 1: THE UNBORN BABY AND THE MIRACLE OF LIFE

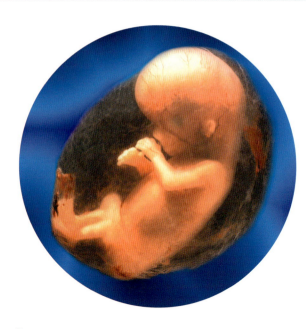

Figure 1-19. The unborn child as a fetus.

B. Month Four

The fourth month is very distinguishable because the baby begins developing a very fine hair-like coating over the entire body, beginning with the eyebrows and the upper lip. This hair is called **lanugo**, coming from the Latin word *lana* for "fine wool." These hairs are only temporary, and usually disappear before birth.

Up to this point, the baby's head was much larger in proportion to the rest of his body, but during this month, the body begins to equalize with the head. The baby continues to grow at a rapid rate. He has not gained weight because he has not started to accumulate the ordinary **white fat** that is present at birth. Around this time, the whorls, ridges, and loops that make up our unique fingerprints are formed and will remain the same throughout life. As the fingerprints develop, the baby develops a keener sense of touch.

The nerve cells in the brain have stopped dividing because they have reached their maximum number. From this point onward, the brain will continue to lose cells until death. From the twelfth to sixteenth week, the baby's diaphragm moves up and down as if he is breathing with a set of inflated lungs. However, he is not yet breathing with his lungs, and his lungs are deflated like empty balloons. During this time, the baby obtains oxygen from the mother's blood. The movement of the diaphragm will soon disappear and then reappear later in the third trimester. The baby's heart is quite strong and contracts regularly. His heartbeat is about twice as fast as his mother's. At the end of the fourth month, the baby weighs about a half a pound, and is about 10 inches (25 cm) in length from head to toe.

C. Month Five

The fifth month marks a major change in physical appearance.

The process of replacing dead skin cells with living cells begins in this month. The **sebaceous glands**, or oil-producing glands, begin working during this month. These are the same glands that may cause acne during one's teenage years. The oil that is produced by these glands is called **sebum**. The sebum combines with the dead skin cells to form a white cheese-like paste all over the baby's skin. This coating, called **vernix**, acts as a protective layer for the new tender skin that is continually exposed to the mineralized water of the amniotic cavity. The lanugo hairs that were formed earlier help the vernix to stick to the baby. The sweat glands, called **sudoriferous glands**, also start forming in this month, but do not open until the seventh month.

This month does not mark the beginning of movement, which began much earlier, but it usually marks the mother's first **sensation of the movement**. The baby's movement is often mistaken for indigestion or gas until it has happened multiple times and becomes more distinct. The first time that a mother feels the baby move is called **quickening**, and is one of many gauges that help the doctor determine if the baby is developing normally.

The baby also begins to fall into a routine of sleeping and waking. The mother is over halfway

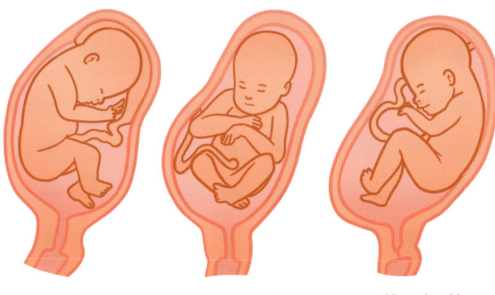

20 weeks old
8 to 12 inches long
weighs 1/2 to 1 pound

24 weeks old
About 14 inches long
weighs 1 to 1 1/2 pounds

28 weeks old
About 15 inches long
weighs 3 pounds

Figure 1-20. Development of the unborn child from the 20th week to the 28th week.

through the pregnancy by the end of this month, and the baby is approximately half of the size expected at birth. From head to foot, he is about twelve inches long, and he weighs about one pound. From this point onward, he will start to put on weight at a rapid rate.

From the beginning, the placenta has been connected to the fetus by the **umbilical cord**. This cord contains two arteries and one vein to assist in the delivery of nourishment and oxygen, and to carry away waste material. These three blood vessels in the umbilical cord are able to withstand considerable pressure because of blood circulating in them. In addition, the blood vessels are covered with a gelatinous substance, known as **Wharton's jelly**. It allows the blood vessels to withstand vigorous squeezing.

The covering of the umbilical cord is part of the amniotic membrane, and will seldom kink or knot, no matter how much the fetus turns or twists during his movements in the womb. The cord appears coiled at birth because the vessels inside have grown longer than the cord itself.

Section Review — 1.4 B & C.

1. What do we call the hair-like coating that covers the unborn at four months?
2. What unique human part develops in the fourth month?
3. From where does the baby obtain oxygen?
4. How fast does the baby's heart beat?
5. What are sebaceous glands?
6. What is vernix? What does it do?
7. What are the sudoriferous glands?
8. What do we call the first time the mother feels the baby move?
9. What do we call the jelly-like substance that covers the umbilical cord?

D. Month Six

The sixth month is considered a turning point in the pregnancy because the child is able to survive outside the mother's womb with the aid of modern medical interventions. At this time, the lungs are developing and preparing for life outside of the womb. A sticky molecule, called **pulmonary surfactant**, begins to coat the lungs. Following birth, this coating helps the lungs stay inflated when the baby gasps his first mouthful of air. In addition, the baby begins to develop alveoli during this time. **Alveoli** are tiny air sacs that are critical for gas exchange.

The baby begins to put on an important kind of fat called **brown fat**, which differs from white fat. **Brown fat** has special cellular structures that provide it with its characteristic brown color. These cellular structures *directly* convert the brown fat to heat energy. Brown fat is a necessary safety device should the baby be born prematurely. This fat provides heat to regulate the temperature of the baby, who otherwise has too little muscle mass, body hair, and insulating fat to prevent hypothermia. Conversely, heat production would not be immediate if the baby used white fat, which would have to be transported out of the cells and converted, which requires time and even more energy.

The brainwaves generated by the six-month old fetus are advanced and resemble the brainwaves of the newborn baby. There is evidence that the fetal baby may be startled by loud, unfamiliar noises. The baby is capable of remembering sounds, such as the mother's voice and her digestive noises. There is much evidence which shows that babies can react to and become familiar with music during this time.

A team of psychologists from Canada, working with Barbara Kisilevsky, a professor of nursing at Queens University, Ontario, did research on the hearing capabilities of infants in the womb. They discovered that the fetus recognized its mother's voice and could distinguish it from other voices. They concluded that even while still within the womb, the developing brains of babies are "learning speech patterns and laying the groundwork for language acquisition."

Dr. William Sears is an Associate Clinical Professor of Pediatrics at the University of California, Irvine, School of Medicine. He has eight children of his own, and has written many books on childcare. He reports that recent research shows that from the fifth or sixth month on, a fetus can hear what is going on in the world outside the womb. Studies show that unborn babies kick violently when they hear rock music, and they calm down when they hear classical music. Other studies indicated that that six-month-old fetuses will move in rhythm with their mothers' speech.

The unborn sixth-month-old baby puts on more than a pound this month and may even look old and wrinkled because the upper and lower layers of the skin have different growth rates. The fat needed to fill out the body under the skin is still lacking.

The bones continue calcifying in the direction of solid skeletal structures. The first place that this occurs is in the sternum. This change from cartilage to bone is called **ossification** [ŏs-i-fĭ-kā-shun]. In adults, bones have around 90% calcium,

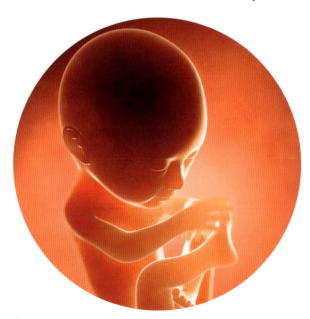

Figure 1-21. Rendering of an unborn child in the 6th month.

but in the six-month old baby fetus, the bones contain only about 12% calcium. The baby is about fifteen inches in length.

E. Month Seven

In the seventh month, the fetal baby is well adapted to living outside of the womb. One of the main reasons for a good survival rate during this month is that the nervous system, particularly the brain, has sufficiently developed to control breathing movements. The brain is also capable of controlling the muscles involved in swallowing if food is put into the mouth. The brain regulates the body temperature, and the lungs are capable of filling with air should they need to perform this function. There is considerable surface area in the lungs, which allows an effective exchange of gases in the lungs.

The baby's blood is able to absorb enough oxygen in order to supply the body tissues with the levels of oxygen required to maintain life. The central nervous system continues to mature and the expanding brain begins to wrinkle to fit an even larger amount of gray matter into the skull. The brain is the first part of the body to develop because it controls most functions of the body.

At this point, all babies are startled by loud noises, and are able to react to disturbing noises immediately, whereas previously, there was a delayed response. It is at this time that the baby becomes more rotund by adding white fat to fill in the wrinkled skin. The baby can weigh up to four and a half pounds in this month. Another significant change in this month is the loss of some fine lanugo hairs that were previously covering his whole body.

In the seventh month, scientists believe they have observed a Rapid Eye Movement (**REM**) sleep state in the fetal baby. There are four stages of sleep and the REM stage is the most restful deep sleep, and equivalent to our dream sleep. While in this REM state, the eyes, ears, and other organs are stimulated, and the baby's heart rate and blood pressure change. Scientists estimate that the baby spends 70-80% of its time in this REM state. How relaxing!

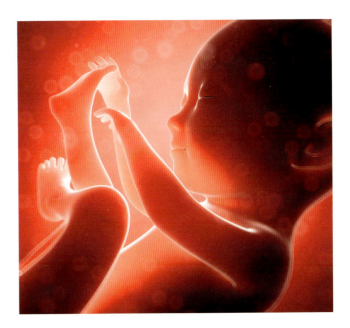

Figure 1-22. Rendering of an unborn child in the 7th month.

Section Review — 1.4 D & E.

1. What is pulmonary surfactant?
2. Why is brown fat important for the unborn baby?
3. Describe what evidence shows regarding how babies can hear sound at six months.
4. In the sixth month, what do we call the change from cartilage to bone?
5. How do we know the unborn baby can survive outside the womb in the seventh month? Mention the nervous system.
6. What does REM mean? What happens during REM for the unborn?

Biology For Life

CHAPTER 1: THE UNBORN BABY AND THE MIRACLE OF LIFE

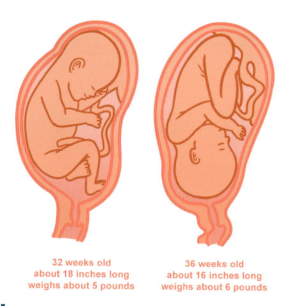

32 weeks old
about 18 inches long
weighs about 5 pounds

36 weeks old
about 16 inches long
weighs about 6 pounds

Figure 1-23. Development of the unborn child from the 32nd week to the 36th week.

limits of pregnancy. The placenta begins to degenerate; blood vessels clot, and cells calcify and become tough. This prevents the mother's blood from reaching the baby fetal capillaries. Labor is imminent. The baby becomes less active and has less room to move around. The baby continues to put on white fat at a rapid pace, and some of the brown fat becomes stored around the vital organs for eventual heat production.

After birth, the white fat will act like an insulator, conserving heat, and the brown fat will be used for making heat. The vernix covering the lanugo hairs is almost completely shed during this month. The vernix will remain in the amniotic fluid for the remaining days of the pregnancy. Some lanugo hairs that have fallen off the baby are swallowed and accumulate in his bowel, along with other secretions. This first bowel movement

F. Month Eight

The eighth month is a time of slow development, as the fetus begins to completely develop all major bodily systems. The baby's nervous system has matured enough so that it controls breathing movements and performs common **reflexes**, including pupil constriction in response to bright light. The lens of the eye is able to focus and the eyelids are able to blink. The skeletal muscles are continuously contracting and relaxing, giving the baby a strong and steady exercise routine.

Much time is spent "breathing" and recycling amniotic fluid. The baby's skin becomes smooth, and is covered with a layer of white fat. In these last two and a half months, the baby will double in weight again and is approximately eighteen inches long from head to toe. There is little obvious difference between an eight-month old unborn baby and a newborn baby.

G. Month Nine

In the ninth month, the rate of growth slows considerably and the placenta regresses, because both mother and child are nearing the physical

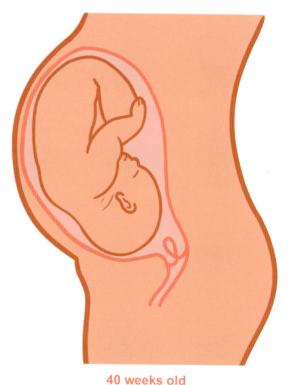

40 weeks old
20 inches long
weighs 7 to 8 pounds

Figure 1-24. The unborn child in its mother's womb in the 40th week, just before birth.

produces **meconium**. It is a greenish black stool that began collecting in the intestines in the sixth month, and is expelled shortly after birth.

Since pigmentation has not fully formed, the baby's eyes are usually blue at this time, even though they may change to brown or green. At birth, most babies are around twenty inches long, and weigh between six and eight pounds. Before delivery, the baby will usually assume a head down position, but this is not always the case.

Shortly before birth, 300 quarts of blood per day are transported back and forth from the placenta to the fetus.

> **Section Review — 1.4 F & G.**
>
> 1. What are some of the things that happen for the unborn in the eighth month?
> 2. After birth, how will the white fat help the baby? How will the brown fat help?
> 3. What is the first bowel movement called?
> 4. What color are the eyes of most babies? Why?
> 5. What is the length of most babies? What is the weight?

1.5 Labor

Section Objectives

- Describe the process of labor.

Labor

Labor is the term used to describe the process by which a mother gives birth to, or delivers, her unborn child. Science indicates that it is the baby who signals for delivery, not the mother. The mother responds to the baby's chemical signal to begin the process of delivering the child. These chemical signals are properly called hormones and are produced and released by specialized structures of the endocrine system called glands. A small area in the baby's brain, called the **hypothalamus** [hī-pō-thăl-a-mus], stimulates another part of the baby's brain, the **pituitary gland**. In turn, the **adrenal** [a-drē-nal] **glands** are stimulated to release stress hormones. These stress hormones trigger the necessary changes for labor to occur.

The First Stage of Labor and Delivery

Labor becomes initiated through signals from the baby. This starts the first stage for the three stages of labor. During the **Early Phase** of the first stage, the process starts out slowly with short, infrequent **contractions** of the uterus. A contraction is a flexing of the muscles of the uterus to pull open the cervix so that the baby can pass through. The uterus is very muscular, but these muscles are rarely used, so it can cause discomfort when they are first used for birth. Later the **Active Phase** starts. Here, the contractions become stronger and increase in intensity, causing the cervix to open and thin out. Next comes the **Transition Phase** where contractions grow in intensity and the cervix dilates, or opens, the last few centimeters. When the cervix is fully open and the baby is ready to move through the birth canal, the first stage is over.

CHAPTER 1: **THE UNBORN BABY AND THE MIRACLE OF LIFE**

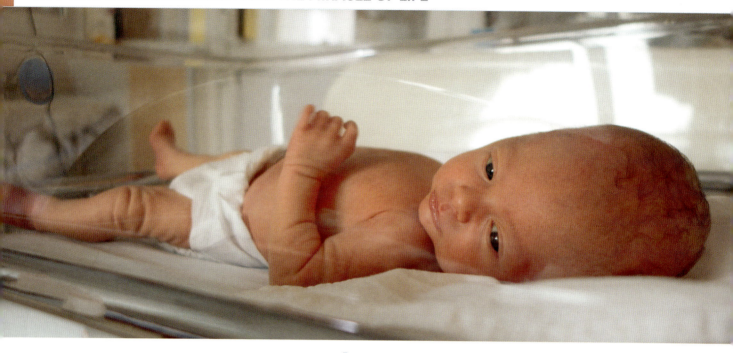

Figure 1-25. Because the skull of a newborn child is not fully developed, it has a soft spot called a fontanel.

The Second Stage of Labor and Delivery

In the second stage the contractions of the uterus grow even stronger and the baby is pushed through the birth canal until the baby is delivered. These contractions can be very powerful and may cause great discomfort.

The Third Stage of Labor and Delivery

In the third stage of labor the placenta is eliminated from the body of the mother. This is sometimes called the after birth. This usually happens within ten minutes to a half an hour after the birth of the baby. The full placenta must be "delivered" to make sure that the mother does not develop a hemorrhage (bleeding), or an infection.

After Delivery of the Baby

When the child is born, many things happen at once. A reduction in temperature starts a physiological process, which causes the interior of the umbilical cord to swell and collapse its interior blood vessels. This, in effect, creates a natural clamp, which would halt the flow of blood in about five minutes if left to progress naturally. However, the cord is quickly clamped and cut by a physician or midwife. The baby usually develops an oxygen debt coming through the birth canal and, when stimulated, will make a gasp for air. A good hearty cry will force air into the little sacs in the lungs, and the surfactant that has been produced will keep the sacs from collapsing.

Once the lungs are being worked, the heart must undergo a major change. Before birth, the heart was pumping very little blood to the lungs because oxygen was entering via the umbilical cord. After birth, the lungs take over the function of the umbilical cord and supply oxygen to the blood. At this time, the heart reroutes the blood to the lungs to pick up the required oxygen.

The skull of the newborn is not fully developed. The parts of the skull can flex so the baby can easily pass through the birth canal. Hence, a baby's head has a soft spot, called a fontanel [fän-ta-nel]. The fontanel usually closes sometime after the baby's first birthday, usually around eighteen months.

CHAPTER 1: **THE UNBORN BABY AND THE MIRACLE OF LIFE**

1.6 The Catholic Perspective of Life

> **Section Objectives**
>
> - Give the Catholic Church's theological view about life

Now that we have looked at life biologically, we need to consider it theologically. The Catholic Church, with her Popes, Saints, and honored Theologians, has consistently condemned abortion as a crime against both God and man. The Bible, the Sacred Word of God, has much to say on this subject. Genesis 1:26 teaches us that man was made in the image and likeness of God. God made man the best of His earthly creation, possessing a worth far greater than the plant or animal kingdom. God bestows personhood at the moment of conception. A person's age, maturity, social status, or mental capabilities do not diminish his status as a person.

In 2 Maccabees 7:22-23, a mother, seeing her seven sons perish in the same day, proclaims these prophetic, noble words: "I do not know how you were formed in my womb, for I gave you neither breath, nor soul, nor life, neither did I frame the limbs of every one of you. But the Creator of the universe who shapes each man's beginning, as He brings about the origin of everything, He in His mercy, will give you back both life and breath…"

In other words, she acknowledges that it is God who gives breath, life, and soul, and He does it in the womb. Likewise, now that her seven sons have died for God, He will give them eternal life. God is the Creator of all life.

Human life begins in the womb, and eternal life begins at Baptism. Solomon affirms this in the book of Wisdom, 7:1-2: "…and in my mother's womb, I was molded into flesh, in a ten-months period–body and blood, from the seed of man." (The ten-month period that Scripture is referring to is in lunar months, not calendar months. Lunar months are twenty-eight days long.)

Scripture is filled with references to the sacredness of life and the evil of murder. The sin of abortion is condemned under the Old Law, as it is a direct violation of the fifth commandment.

The New Testament, though, offers the faithful an even deeper perspective. The New Testament fulfills the Old Testament, and its dictates proceed well beyond the scope of the Old Law. In the New Testament, particularly in the Holy Gospels, one observes a new and unique understanding of the holiness and sanctity of the child. In Saint Luke's Gospel, when the disciples try to keep children away from Jesus, Our Lord says, "Let the children come to me, and do not hinder them, for the kingdom of God belongs to such as these." Our Lord loved all people at all stages of development, and commanded us to love our neighbor as ourselves.

More than anyone else, the innocent children in their mother's womb need to be loved. The

"Let the children come to me, and do not hinder them, for the kingdom of God belongs to such as these."

Biology For Life

Biblical viewpoint is that they are a precious gift and a most valuable blessing. Psalm 127:3-5 states, "Sons are a heritage from the Lord, the fruit of the womb a reward." Other than the worship of false gods, and other evils which are direct offenses against God Himself, abortion is perhaps the greatest of the remaining evils.

We need to stir our consciences and awaken the hardened hearts of those in this world who do not accept the truth, those who reject God and His commandments. There are many who are not willing to accept the reality that abortion is murder, a grave offense against God. Blessed with the gift of faith, we must proclaim our faith in Christ's one, true Church, and defend the Church's moral teachings. For members of the Body of Christ, promoting the sanctity of human life is a basic obligation, a fundamental requirement.

Hence, as a Catholic soldier, one must truly recognize and defend the wondrous gift of life. We must support what we say with our faith, the true Faith of Jesus Christ, the Son of God, and live what we believe, in both our words and in our example. Our Lord commands: "You are the light of the world. A city that is set on a mountain cannot be hid. Neither do men light a candle, and put it under a bushel (basket), but upon a candlestick, that it may give light to all that are in the house. Let your light so shine before men, that they may see your works, and glorify your Father who is in heaven" (Mt. 5:14-16).

Section Review — 1.5 & 1.6

1. What does labor mean?
2. What starts the process of labor?
3. What forces air into the newly-born baby?
4. What happens to the baby's lungs after birth?
5. What is the fontanel? When does it close?
6. When does God give breath, life, and the soul?
7. What did Our Lord say about children in St. Luke's Gospel?
8. What did Our Lord say in the Gospel of Matthew in Chapter 5?

CHAPTER 1: THE UNBORN BABY AND THE MIRACLE OF LIFE

Chapter 1 Supplemental Questions

Answer the following questions.

1. Define the terms: growth, morphogenesis, determination, and differentiation.
2. Describe genetics.
3. Identify the steps in development from fertilization to implantation.
4. What is the connection between a DNA molecule and a gene?
5. Describe the Period of the Child at Embryo stage.
6. Describe the Period of the Child at Fetus stage.
7. Describe cleavage.
8. Describe the circulation in the womb of the child in the fetal stage.
9. Summarize each month of development of the unborn child. Limit your answer to two-to-three sentences per month.
10. Explain the role of the placenta.
11. Describe the process of labor.
12. In 5-8 sentences, summarize the Catholic Church's view of life.
13. Study all diagrams in this chapter. Be able to draw and label all diagrams.

Biology For Life

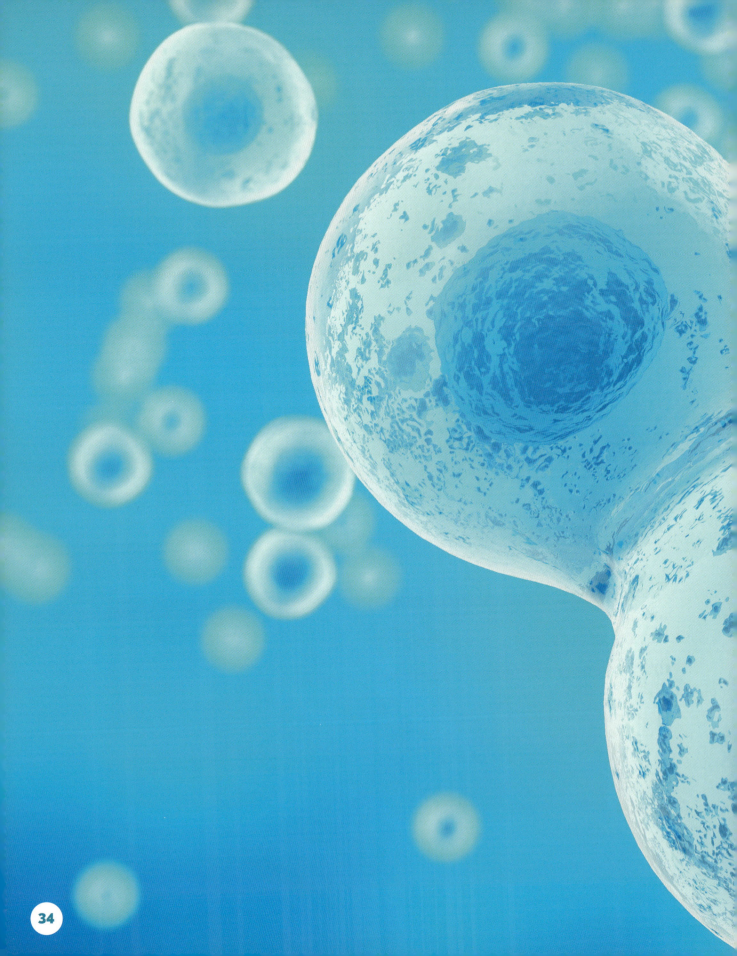

CHAPTER 2
Cells

Chapter 2
Cells

Chapter Outline

2.1 Introduction
- A. The Building Blocks of Our Spiritual Life
- B. What is Life?
- C. The Building Blocks of Our Physical Life

2.2 Cells: the Building Blocks of Physical Life
- A. Cell Theory
- B. Cell Functions

2.3 The Parts of a Cell
- A. The Cell Wall
- B. Membranes
- C. The Nucleus
- D. Cytoplasm and Organelles
- E. Plastids
- F. The Living Cell
- G. Differences in Cells

2.4 The Arrangement of Living Matter
- A. The Variety of Unicellular Organisms
- B. Tissues to Systems

2.5 The Chemistry of Carbon
- A. The Properties of Carbon
- B. Alcohols, Organic Acids, and Organic Bases

2.6 Macromolecules
- A. Carbohydrates
- B. Lipids

2.7 Proteins and Nucleic Acids
- A. Amino Acids and Proteins
- B. Protein Shape
- C. Enzymes
- D. Nucleic Acids
- E. ATP and ADP

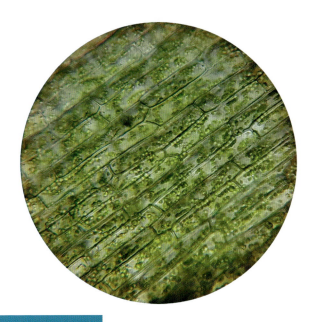

2.1 Introduction

> **Section Objectives**
>
> - Define what it means for a thing to be alive.
> - Name some differences between living things and machines.

A. The Building Blocks of Our Spiritual Life

Human beings have a physical and a spiritual nature, and both of these have unique building blocks. The cell is the building block for the physical nature. Grace and virtue are the building blocks of our spiritual nature. We will shortly begin our investigation into the cell. Before doing so, let us review how we grow as spiritual beings.

The sacraments enable us to develop in a manner that is pleasing to God. All seven sacraments bring us closer to Him and help us advance towards the perfection of charity. Baptism is absolutely essential for developing a spiritual life. When we are baptized we become new creations. Baptism gives a new character to our soul and makes us a member of the one, true Church. What are the effects of Baptism? The effects of Baptism are great, as *The Catechism of the Catholic Church (CCC)* notes: "…the two principal effects are purification from sins and new birth in the Holy Spirit" (CCC 1262). The Catechism goes on to say that "it makes us adopted sons of God, incorporates us into the Church, and seals us with an indelible spiritual mark of our belonging to Christ" (*CCC 1265-1274*). We also receive grace that enables us to believe in God, to hope in God, and to love God. Grace aids us in discerning the prompting of the Holy Spirit, and it also helps us to be of goodwill and avoid sin. Thus, Baptism is the first and fundamental building block of the spiritual life.

Without Baptism we are not spiritually reborn and cannot enter Heaven. Our Lord solemnly assures us of this very important truth in John 3:5: "Amen, amen, I say to you, unless a man be born again of water and the Holy Spirit, he cannot enter into the kingdom of God." Here Jesus reminds us of the importance of water for our spiritual life and salvation. Of course, we also know that water is necessary for our life on Earth. Although Baptism is a "gateway sacrament" for growing in virtue and character, the life of divine grace is not limited to the sacrament of Baptism. The Holy Eucharist nourishes our life, and Confirmation strengthens it.

B. What is Life?

Aristotle defined life as the capacity for self-change. All living things change and must obtain energy by making their own food or by obtaining it from the food that they eat. They grow, reproduce, and eventually die. In addition, all organisms can change themselves or adapt to their environment, thus ensuring their survival.

All things, living and non-living, are composed of atoms, which are the smallest whole units of elements having the same chemical properties of the element. Atoms themselves are made up of protons, neutrons, and electrons. Molecules are the smallest units of a substance that contain two or more atoms, which may be of the same or different kinds of atoms. These atoms are held together by chemical bonds.

Sometimes we refer to living organisms as machines. No harm results from doing so, as long as we realize that we are using a metaphor. For example, the heart of an animal resembles a pump in many ways. It is helpful to describe its functions as an engineer might by using descriptions of pressure, flow, energy, etc. The heart differs from any mechanical pump, however, because it grows, repairs itself, and adjusts to the needs of an organism. In a word, it is alive!

There are great differences between living things and machines:

- Machines cannot build or rebuild their own parts. All organisms, however, constantly renew their tissues and cells.
- Machines do not grow from seeds or eggs but are composed of unchanging parts that are assembled by people or robots. Living things develop from single cells, grow their own parts, and develop functions specific for these parts.
- A living cell possesses characteristics that are specific to an organism. These characteristics are studied by specialized scientists called cytologists. By looking at a cell, it is possible for cytologists to identify its job and what organism the cell came from. However, by examining a ball bearing or length of copper wire, an engineer cannot determine the kind of machine it belonged to.

Figure 2-1. A typical atom. All things, living and non-living, are composed of atoms, which are the smallest whole units of elements having the same chemical properties of the element. Atoms themselves are made up of protons, neutrons, and electrons.

C. The Building Blocks of Our Physical Life

Our physical bodies are made up of many tiny individual building blocks, or units, called cells. In turn, each cell is composed of **molecules** and **macromolecules**. Macromolecule is another term for large, complex molecules. These macromolecules play a dual role within the cell. They are part of the cell, but they are also constantly undergoing complex chemical reactions. In this chapter, we will learn about (1) the molecular basis of cells, (2) what cells are, (3) how cells function, (4) how cells are organized into **tissues**, (5) how cells are organized into **organs**, and (6) how cells are organized into **systems**.

2.2 Cells: the Building Blocks of Physical Life

An organism is a form of life. If we interpret the word "form" using the Book of Genesis, we might understand it as "kind." There are many organisms, and scientists called taxonomists place these in five kingdoms: Monera, Protista, Fungi, Plantae, and Animalia. Some organisms are unicellular (single-celled), very small, and are called microorganisms. One such microorganism that lives in our digestive tracts is a bacterium known as *Escherichia coli*. It is in the Kingdom Monera.

> **Section Objectives**
>
> - Discuss the three parts of the cell theory.
> - Name five functions of the cell.

Figure 2.2.
E. coli bacteria

Obviously, other organisms with which we are familiar, especially plants and animals, are **multicellular**, or composed of many cells. Humans are multicellular, because the human body contains trillions of cells. The complexity, order and functions of these trillions of tiny cells are evidence of the existence of God, as well as God's creativity and intellect.

A. Cell Theory

In 1665, Robert Hooke, an English scientist, examined a slice of cork under a microscope. Hooke noticed that the cork was composed of many small compartments. Hooke named these compartments "cells" because they resembled the little rooms, or cells, of a monastery. As you may recall from your reading of the lives of the saints, many monks lived in large monasteries, but they slept or prayed in tiny rooms, or cells, often with only a bed and a table.

The cork cells seen by Robert Hooke were nonliving. Hooke did not pursue the study of the structure and function of living cells. However, almost 200 years later, in 1835, Felix Dujardin, a French biologist, determined that many living microorganisms were composed of a single cell.

He also observed that the internal substance of all living cells was similar.

In 1838, Matthias Schleiden, a German botanist, reported that all plants were composed of cells. Shortly thereafter, Theodor Schwann, a German zoologist, concluded from his studies that all animals were composed of cells. About two decades later, Rudolf Virchow, another German, observed

Figure 2-3.
Drawing of the structure of cork as it appeared under the microscope to Robert Hooke, from his work Micrographia (1665)

Biology For Life

cells that were in the process of dividing and concluded that cells can arise *only* from other cells.

The observations from the above scientists formed the basis of Cell Theory, which states:

1. The cell is the basic unit of structure for all living things.
2. The cell is the basic unit of function for living things.
3. All cells come from other cells by the process of cell division.

B. Cell Functions

Most cells are a little less than 10 micrometers in length. One meter is equal to 39.37 inches, and 1 micrometer is one millionth of a meter, or one millionth of 39 inches. About 2,000 cells can fit across the width of your fingernail. As tiny as these cells are, their functions are truly amazing to study. All cells require energy from food in order to perform their functions. The process by which organisms obtain and use food is called **nutrition**. Some cells are able to make their own food, while other cells must obtain food from their environment.

Cells obtain energy chiefly by processing glucose molecules. The cell converts glucose into a form of energy usable by the cell. This process of energy conversion is known as **cellular respiration**. As cells live in an aqueous, or fluid, environment, they are able to absorb water, minerals, and other materials that are essential for life. This process is called **absorption**.

Cellular processes result in the accumulation of toxic wastes. These must be removed for the cell to function properly. Cells eliminate, or excrete, waste products through a process called **excretion**.

One of the most amazing functions of cells is that they are able to build complex chemicals from simple molecules in a process known as **biosynthesis**. For example, when you eat a hamburger, complex proteins and fats are broken down within your digestive tract into their precursors; that is, amino acids and fatty acids, respectively. These amino and fatty acids are transported by blood to other areas of your body, where they are used to build proteins and fats.

Cells quickly respond or adapt to changes in their environment. As conditions change so do their responses and functions. Cells even replace themselves. Multicellular organisms must continually produce new cells to replace old worn-out cells. In addition, cell reproduction can cause an increase in the number of cells in an organism, which results in its growth. Adults are larger than infants because they have grown and accumulated many more cells.

Figure 2-4.

Early Scientists Behind Cell Theory

Felix Dujardin
French Biologist

Matthias Schleiden
German Botanist

Theodor Schwann
German Zoologist

Rudolf Virchow
German Pathologist

In summary, cells convert energy from glucose, absorb materials necessary for life, excrete waste products, make complex chemicals from simple chemicals, and reproduce. The functions of the tiny, living cell are evidence for the amazing intelligence of our almighty and loving God!

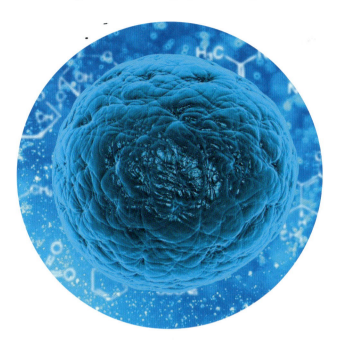

> **Section Review — 2.2**
>
> 1. What should we remember about Robert Hooke?
> 2. What two things did Felix Dujardin observe about living cells?
> 3. What did Matthias Schleiden report?
> 4. What did Theodor Schwann conclude?
> 5. What did Rudolf Virchow conclude?
> 6. What are the three parts of the Cell Theory?
> 7. What is the definition of nutrition?
> 8. What is cellular respiration?
> 9. What do cells absorb?
> 10. What do we call the process by which cells eliminate waste products?
> 11. Define the function of cells called biosynthesis.

2.3 The Parts of a Cell

Section Objectives

- Describe the molecular structure of the cell membrane and cell wall.
- Describe the function of the nucleus.
- Describe the structure and function of the following cellular organelles: mitochondria, ribosomes, endoplasmic reticulum, plastids, Golgi bodies, vacuoles, lysosomes, cytoskeleton, and centrioles.
- Compare and contrast prokaryotes and eukaryotes.

Each cell contains small organ-like structures called **organelles** that have specific jobs to carry out during the life of the cell. These cells, then, can be likened to factories. The functions of the cell depend upon the number and kinds of tiny organelles that are present inside the cell. Look at Figure 2-5 to see structures within a cell. A typical cell has a cell membrane, cytoplasm, a nucleus, and various organelles. Each of these components plays an important role in its activity.

A. The Cell

Cell walls are found in plants, fungi, and some single-celled organisms. These are the boundaries of cells and they have two functions: to protect the cells from their environment and to give the cells support. The cell wall of land plants is composed of cellulose, a complex **polysaccharide** [poly-sa-ka-rīd] that is made from glucose.

When a plant cell begins to form, it produces a primary cell wall, which expands as the cell grows. When the cell reaches its full size, it then produces a secondary cell wall inside of the primary cell wall. The secondary cell wall adds strength to the cell. The primary cell walls are not in direct contact with one another. They are separated by the **middle lamella** [lă-mĕ-la], which is a gel-like layer of polysaccharide called **pectin**.

B. Membranes

The **plasma membrane** serves as a barrier between the inside of the cell and the cell wall of plants. Within human and animal cells, the plasma membrane or cell membrane is located on the outside of the cell. It is a boundary for the cell and separates it from other cells. The plasma membrane also plays an important role in determining which substances may enter or leave the cell.

When we hear the term membrane, we might think of the plastic that is used to wrap a sandwich. We might think of the thin membrane that we find after cracking open and peeling a hard-boiled egg. The plasma membrane does not resemble either of these. It is membrane-like insofar as it encloses the contents of the cell. However, it is not a fixed structure but is fluid-like, and is composed of two layers of molecules called **phospholipids** [fos-fō-li-ped]. Each phospholipid molecule is made up of one head and one tail. The head is composed of a phosphate [fos-fāt] group, and the tail consists of a **lipid** or fatty acid. The phosphate group is attracted to water and faces away from the center of the membrane. The fatty acid part of the molecule is repelled by water and faces the interior of the membrane. Because the plasma membrane is double-layered, the tail ends of the phospholipids will position themselves toward the center of the membrane. Conversely, the head ends will orient

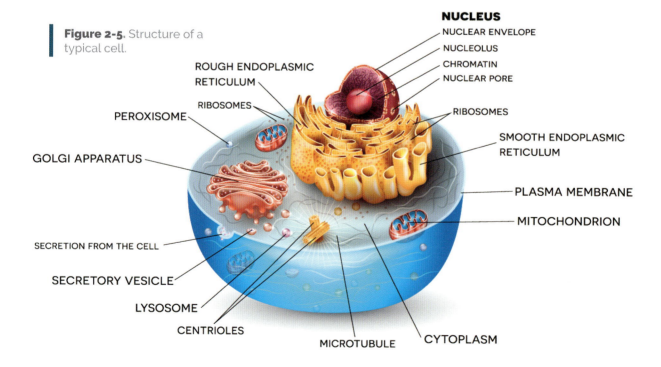

Figure 2-5. Structure of a typical cell.

themselves outwards. Restated, one end is directed towards the interior of the cell and one end towards the exterior. The plasma membrane is more commonly referred to as the **lipid bilayer**.

Water and some dissolved substances easily pass through the membrane in a process called **passive transport**. This process is driven by concentration differences of molecules on each side of the membrane. Some substances do not easily pass through the membrane and require assistance. This help comes from proteins that move about freely within the plasma membrane. These proteins, called transmembrane proteins, require energy to help shuttle complex molecules in and out of the cell by a process called **active transport**.

The phospholipids and proteins which make up the plasma membrane are constantly moving within the membrane. Phospholipids move sideways and proteins move in and out or stay put. If the plasma membrane becomes disrupted or broken, the phospholipids can quickly self-assemble to repair the damage. Because of the fluid-like and flexible nature of the plasma membrane, scientists have described it as the "fluid mosaic model."

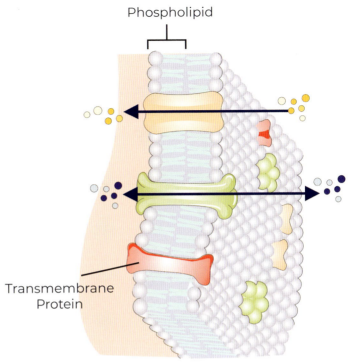

Figure 2-6. The plasma membrane. It is composed of phospholipids and embedded protein molecules. The transmembrane proteins are the sites of active transport, where substances are pulled across them in order to move in or out of the cell.

C. The Nucleus

The **nucleus** [nü-klē-us] is a spherical structure which is located near the center of the cell. Its job is similar to a factory that makes proteins for the cell. Proteins are among the most important kinds of molecules in the chemistry of life. They regulate all chemical reactions within the cell. The nucleus is enclosed within a **nuclear membrane**, a lipid bilayer that separates it from the rest of the cell. It contains pores that allow materials to pass in and out of the nucleus.

A material called **chromatin** [krō-ma-tin] is visible within the nucleus. It contains the hereditary information or DNA of the cell and becomes visible when the cell replicates or divides. Before the cell divides, the chromatin forms pairs of long strands called chromosomes. As you read in the previous chapter, genes are specific portions of the DNA that contain the information codes or blueprint for making everything the body needs. If all forty-six chromosomes were laid out end-to-end, they would measure a little over six feet in length. Keep in mind that this is in one tiny cell!

After cell division, an exact copy of the chromosome is present within each new cell. One region of the nucleus contains the **nucleolus** [nü-klē-ō-lus]. The nucleolus is involved in the production of **ribosomes** [rī-bo-sōm], which are composed of **nucleic acids** and proteins. Ribosomes are transported from the nucleolus and into the cytoplasm to make proteins.

Biology For Life

CHAPTER 2: CELLS

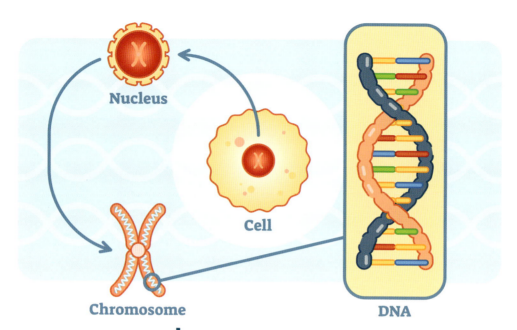

Figure 2-7. Genes of DNA are found in chromosomes that are formed by the chromatin in the nucleus of a cell.

Section Review — 2.3 A, B, & C.

1. Draw your own picture of the structure of a typical cell.
2. What are the two functions of the cell wall?
3. What is the composition of the cell wall in land plants?
4. What is the important role of the plasma membrane?
5. Each phospholipid molecule is composed of a head and a what?
6. The phosphate group head is attracted to what?
7. The fatty acid tail of the phospholipids is repelled by water and faces what part of the membrane?
8. What is meant by passive transport?
9. What is meant by active transport?
10. Phospholipids move sideways; how do proteins move in the plasma membrane?
11. What is a nucleus?
12. What is its function?
13. Why is a nucleus an important structure in the cell?
14. What is contained in the chromatin which is so important?
15. The nucleolus is involved in the production of what?
16. Why are ribosomes transported into the cytoplasm.

Biology For Life

D. Cytoplasm and Organelles

The **cytoplasm** [sī-tō-pla-zm] is a gel-like substance inside the cell that consists of many types of proteins and other macromolecules. It is a vehicle or medium which makes possible the activities of the cell. Many types of organelles (little organs) are suspended within the cell's cytoplasm. The **mitochondria** [mī-tō-kän-drē-a] are bean-shaped organelles that release the energy stored in food during the process known as **cellular respiration**. For this reason, they are often referred to as the "powerhouses of the cell." Mitochondria are surrounded by two membranes. The outer membrane separates the mitochondrion from the cytoplasm. The inner membrane is folded into ridges that stretch across it.

Golgi [gol-jē] **bodies** are large organelles. They look like wide rubber bands stacked together like pancakes. They are found in both animal and plant cells. Macromolecules are assembled in Golgi bodies and may also be stored in them. The "folded" ends of the rubber-band-like sacs, called **cisternae** [sis-ter-nā; singular - cisterna], can fill up with macromolecules and then break off and transport the "cargo," mostly proteins, to other parts of the cell or even to the outside. **Secretion** is the term used to describe the transport of these macromolecules. The ends that break off from the cisternae are tiny vacuoles called **vesicles** [ve-si-kal]. Because of their importance in transporting materials throughout the cell, Golgi bodies can be considered the "shipping department" of the cell.

Due to the relatively large size of Golgi bodies, they were among the first organelles ever discovered within cells. A Catholic Italian physician named Camillo Golgi was the discoverer.

Ribosomes are scattered throughout the cell's cytoplasm. They are often attached to folded sections of membrane known as the **endoplasmic reticulum** [en-dō-plaz-mik] [rǐ-tǐ-kū-lum]. The endoplasmic reticulum is a network of canals that connect the nuclear membrane with the plasma membrane. The job of the endoplasmic reticulum is to prepare proteins for transport through a process called **synthesis**.

There are two types of endoplasmic reticulum, rough and smooth. Rough endoplasmic reticulum contains ribosomes. It makes proteins that will be secreted, or released, from the cell through the plasma membrane. Some ribosomes are not attached to the endoplasmic reticulum and are used for making proteins within the cell. Smooth endoplasmic reticulum does not contain ribosomes and is used for the production of lipids that become part of cell membranes. In addition, smooth endoplasmic reticulum breaks down some fatty acids and degrades a variety of substances that could be injurious to the cell or organism.

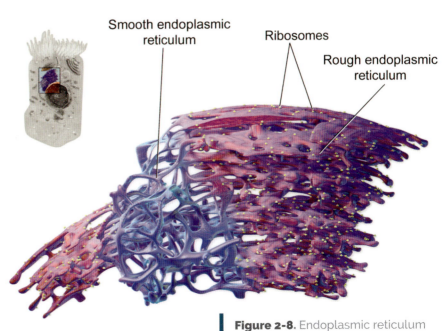

Figure 2-8. Endoplasmic reticulum

Biology For Life

Small sections of endoplasmic reticulum membrane bulge and break away as sphere-like structures called **vesicles**. These contain proteins or lipids. The vesicles then fuse with the Golgi body, which chemically modifies the proteins or lipids. These pinch off from the Golgi body as vesicles and move towards the plasma membrane. Each vesicle fuses with the plasma membrane and releases its contents to the exterior of the cell.

Lysosomes [lī-so-sōm] are a type of vesicle formed by the Golgi body. These organelles contain enzymes that are used by the cell for digesting almost all biological molecules. For example, certain proteins stay within the cell membrane for only a short time. These proteins may then be broken down into amino acids, which the cell reuses to make new proteins.

The **cytoskeleton** [sī-tō-ske-la-ton], which is made up of **microtubules** [mī-krō-tü-byül] and other small proteins, is a miniature internal support system for the cell. It gives the cell its shape, much like wooden beams give shape to a church or your skeleton gives shape to your body. The cytoskeleton divides the cell into compartments, like you might see in a beehive. In addition, it keeps organelles in place or helps move them to new locations. The microtubules are formed and organized by a cylinder-shaped organelle called a **centriole** [sen-trē-ōl]. Once the centriole has finished its job, it remains below the microtubules as a **basal body**.

Vacuoles [va-kyu-ōl] are organelles that are filled with fluid (mostly water) and surrounded by membranes. They are formed by the fusion of multiple vesicles. Vacuoles do not have uniform shapes or sizes, and their structures change with the needs of the cell. Vacuoles are filled with fluids and may contain small molecules, salts, and waste products. They have several jobs that include isolating and exporting of materials that may be harmful to the cell. In addition, they help maintain both the internal pressures of the cell and its pH requirements. You may recall from your earlier science classes that pH is a measure of the acidity or alkalinity of a solution. The normal pH of human cells is 7.2, which is slightly alkaline or basic. A pH of 7.0 is neutral. A ph that is either too high or too low can affect cellular functions and be dangerous to the cell. Therefore the regulation of pH by vacuoles is very important.

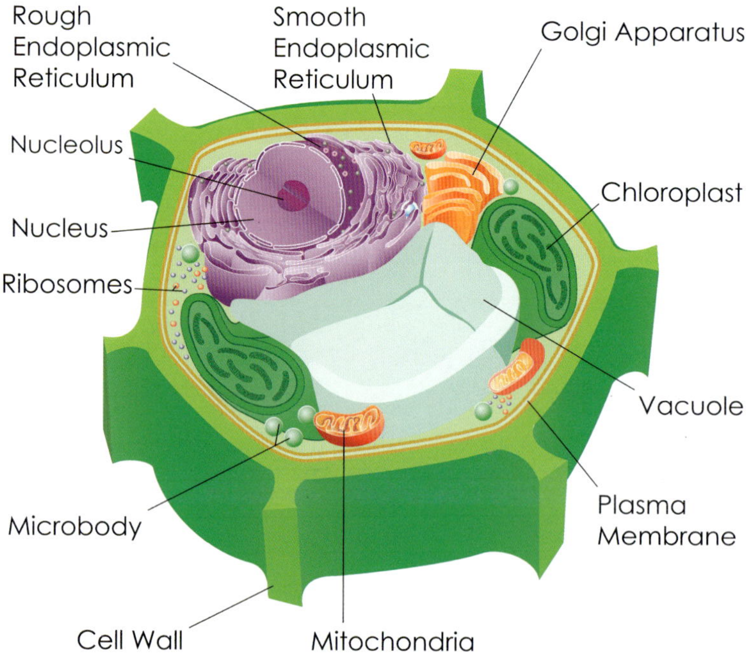

E. Plastids

Plants contain organelles called **plastids**. These are used by the cell for storing food. In addition, plastids contain pigments that give plants color and are used for photosynthesis. **Chloroplasts** [klȯr-ō-plast] are one type of plastid that contain the green pigment, **chlorophyll** [klore-ōh-fill]. It is during the

Figure 2-9. A typical plant cell

process of photosynthesis that chlorophyll captures the energy from sunlight to manufacture glucose. **Chromoplasts** [krō-mō-plast] contain red, orange, or yellow pigments that give many flowers and fruits their distinctive colors. They are also responsible for the brilliant colors of leaves in autumn. Like chloroplasts, these pigments produce glucose during photosynthesis. Finally, the last group of plastids is called **amyloplasts** [a-mill-oh-plast]. Their principal job is to convert starch from glucose and then store it. Some ornamental plants that you may find in your house contain amyloplasts. These are visible as "white specks" of starch in the leaves. A more common example comes from the potatoes that we eat. Potatoes rank among the most nutritious of foods and have a very high starch content. As you might guess, potatoes have a large number of amyloplasts that give the potato its white color.

F. The Living Cell

It is almost impossible to describe a living cell by illustrations in a book. The average cell carries out many chemical reactions every second and can reproduce itself every twenty minutes or so. The contents of the cell are moving, shifting, and always changing. This is what we mean by life. A cell takes non-living chemicals and through **information**, arranges them in such a manner to grow, reproduce, and carry out the functions of the cell, not only for itself but for the organism too.

G. Differences in Cells

The cells of all humans, animals, and plants contain chromatin in a well-defined nucleus. (In mammals, like humans, active red blood cells do not have a nucleus). Cells that contain a nucleus are called **eukaryotic cells** and the organisms that have these are referred to as **eukaryotes**

TYPES OF CELL STRUCTURE

Figure 2-10.

A **prokaryotic cell** is a single-celled organism that does not contain a nucleus.

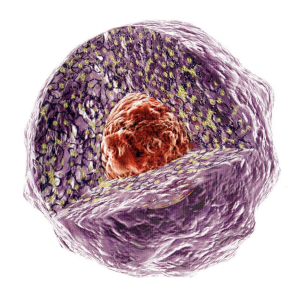

An **eukaryotic cell** contains a nucleus where the hereditary information is stored.

[yü-ka-rē-ōt]. Single-celled organisms such as bacteria do not contain a nucleus and are called **prokaryotes** [prō-ka-rē-ōt]. In these organisms, the chromatin is located in an area called the nucleoid region. In addition, the organelles of prokaryotes are not covered with membranes as are the eukaryotes.

Section Review — 2.3 D, E, F, & G.

1. The cytoplasm is a _____ substance. It makes possible the _____ of the cell.
2. _____ are "little organs" found throughout the cytoplasm of a cell that perform specific functions.
3. The mitochondria release _____ stored in glucose and are often called the _____ of the cell.
4. Golgi bodies look like stacks of _____.
5. Golgi bodies store _____.
6. The cisternae _____ macromolecules.
7. When macromolecules are carried outside the cell, it is called _____.
8. What are vesicles?
9. What are scattered throughout the cell's cytoplasm?
10. The endoplasmic reticulum is a network of _____ that connect the nuclear membrane with the _____ membrane.
11. The job of the endoplasmic reticulum is to prepare _____ to be transported through a process called _____.
12. Rough endoplasmic reticulum contains ribosomes which make proteins which will be _____ from the cell.
13. Smooth endoplasmic reticulum is used for the production of ____ that become part of the cell membrane.
14. _____ are sections of endoplasmic membrane that break away from the cell.
15. _____ are important because they fuse with the Golgi body and release its content of large molecules to the exterior of the cell.
16. _____ are a type of vesicle which contain enzymes used by the cell for digesting.
17. The _____ gives the cell its shape; it is a miniature internal support system.
18. Vacuoles are formed by vesicles; they isolate and export materials that may be _____ to the cell.
19. The regulation of pH, that is the _____ or alkalinity of a cell, is an important job of the vacuoles.
20. _____ are organelles in plants.
21. There are several kinds of _____ for plants, such as chloroplasts, chromoplasts, and amyloplasts.
22. During photosynthesis, _____ in plants captures energy from sunlight to manufacture glucose.
23. Cells which contain a nucleus are called _____.
24. Some single-celled organisms do not contain a nucleus and are called _____.

2.4 The Arrangement of Living Matter

A. The Variety of Unicellular Organisms

Organisms that consist of a single cell are called **unicellular organisms**. These microscopic organisms are capable of carrying out all necessary functions of life. That is, unicellular organisms can obtain and digest food, make proteins, excrete wastes, respond to environmental changes, and reproduce. In addition, unicellular organisms have a variety of specialized structures. For example, some have long, threadlike flagella or cilia, which can propel them through water. The flagella are more complex than cilia and have been compared to an outboard motor of a boat. Other unicellular organisms have specialized cell walls for protection or eyespots that are light sensitive. Such tiny organisms cannot be seen with the unaided eye and are too complex to comprehend. These are evidence for an Omniscient and Omnipotent Creator God!

B. Tissues to Systems

Within multi-cellular organisms, there are many types of similar cells, which work together to perform a certain function. These groups of similar cells are called **tissues**. For example, nervous tissue is composed of many nerve cells; muscle tissue is composed of muscle cells; and bone tissue is composed of bone cells. Blood is made of blood cells and is considered a specialized type of connective tissue. Plants also contain tissues. One example is the wood of a tree, which provides the tree with support. A network of tubes called xylem and phloem run throughout the woody tissues of a plant or tree. The xylem transports water and minerals, and phloem transports complex compounds, including sugars and proteins.

Figure 2.11. Bacteria

> **Section Objectives**
> - Explain the difference between the function of a unicellular organism and a multicellular organism.
> - List the levels of organization of living matter from the simplest to the most complex.

Several different types of tissues sometimes work together as a unit to perform a certain function. We know these units as **organs**. The stomach, heart, liver, and kidneys are examples of organs. Plants also have organs, including roots, stems, flowers, and leaves. A group of organs working together to perform a specific function is called an **organ system**.

The organ systems of the human body include the following:

The **skeletal system** supports and protects the body. Blood cells are manufactured inside the bones, and calcium and phosphorus are stored in bone tissue.

The **muscular system** works with the skeletal system to make the body move. Muscles also protect some of the body's organs.

The **digestive system** includes the canal that extends from the mouth and through the abdomen. In this system, food is broken down into essential nutrients and becomes available for distribution to the body.

The **circulatory system** transports nutrients, gases, and other dissolved compounds to all parts of the body. It also collects waste products from cells. Blood is circulated through blood vessels by the pumping action of the heart.

The **lymphatic system** is part of the circulatory system. It collects fluid from tissue and returns it to the blood. Both systems work together to fight infections and diseases.

Biology For Life

CHAPTER 2: CELLS

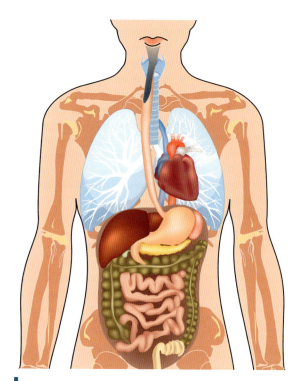

Figure 2-12. An illustration of human anatomy showing the various organ systems.

The **respiratory system** takes oxygen from the lungs and transports it to all of the organs and tissues of the body. It also eliminates carbon dioxide and water.

The **excretory system** removes both liquid and solid wastes from the body that are primarily the products of digestion but include other harmful substances. The excretory system also helps maintain the body's fluid and chemical balance.

The **nervous system** monitors the internal and external environments of the body and coordinates its activities.

The **integumentary (in-te-gyu-men-tary) system** is the outer, protective layer of the body. It consists of the skin, hair, and nails.

The **endocrine (en-dō-krin) system** helps control body functions through chemicals called hormones. Hormones regulate all processes within the human body; for example, growth and maturation.

The **reproductive system** provides a means of producing offspring in order to maintain the species.

It is difficult to examine any of the above systems as isolated, separate units. For example, as we examine the skeletal system, we will note the muscles that are attached to it. In addition, we might observe its production of blood cells. All organ systems of the body are interrelated and operate in unison. In summary, there are five levels of organization in a living organism. The simplest are the cells that are used to construct tissues, tissues form organs, and organs connect to produce organ systems, and finally the whole organism. These remind us of the complexity of the world which God created and how the body is, according to St. Paul, similar to the Church: One Body with Many Members (1 Cor 12: 12-27).

Section Review — 2.4 A & B

1. Unicellular organisms are _____.

2. The necessary functions of life performed by cells are the following five:
   _____,
   _____,
   _____,
   _____,
   _____.

3. What are the two kinds of organs that propel unicellular cells through water?

4. Groups of similar cells are called _____.

5. When different types of tissues work together to perform a certain function, we call these _____.

6. A group of organs working together is called a _____.

7. Name the eleven kinds of systems of the human body.

Biology For Life

CHAPTER 2: CELLS

2.5 The Chemistry of Carbon

Section Objectives

- Describe the types of bonding which can occur with the element carbon.
- Identify the following structural units of organic molecules: alcohol, organic acids and bases, and amino groups.

Figure 2.13. Carbon atom

Many of the macromolecules that make up the chemistry of living things are based on the element **carbon**. It is carbon that gives these macromolecules their special properties. The study of the chemistry of carbon compounds is referred to as **organic chemistry**.

A. The Properties of Carbon

In previous science classes, you should have learned about the periodic table. The periodic table lists all the elements that scientists have discovered and produced synthetically in the laboratory. They are listed so that elements with similar properties are listed in the same column. Each element has a periodic table number known as the atomic number.

Carbon is the sixth element on the periodic table; its symbol is C. The periodic table lists all known elements and ranks the elements based on the number of protons in each atom. A carbon atom has six **protons** and six **electrons**. Two of these electrons are located within the inner shell of the electron cloud around the nucleus of the atom, and four electrons are found in its outer shell.

All atoms, except hydrogen (symbol H) and helium (symbol He), require eight electrons in their outer shell to produce a stable arrangement. Since carbon needs four more electrons for its outer shell, it achieves this arrangement by sharing four electrons with one or more atoms. The arrangement made by sharing electrons is called a **covalent [kō-vā-lent] bond**. Carbon can form four covalent bonds with up to four atoms. These can include other carbon atoms or the atoms of other elements. For example, if carbon is bound to four hydrogen atoms, a compound called methane is formed. Methane, also known as natural gas, is used for cooking on gas stoves and for heating our homes.

Figure 2.15. The structural formula of ethane

The bonds of atoms are either described as **molecular formulae** or depicted as **structural formulae**. The molecular formula tells the number of each kind of atom in a molecule of the compound. Methane has a molecular formula of CH_4. This symbolic formula means that one molecule of methane, CH_4, contains 1 atom of carbon and 4 atoms of hydrogen. The structural formula shows the bonds connecting the atoms and their arrangement within each molecule. Methane has the structural formula shown in Figure 2-14.

Methane is the simplest compound in a group of substances known as the hydrocarbons. **Hydrocarbons** are molecules made from carbon

Figure 2.14. The structural formula of the simple hydrocarbon known as methane. The carbon has formed four separate covalent bonds with hydrogen atoms.

Biology For Life 51

CHAPTER 2: CELLS

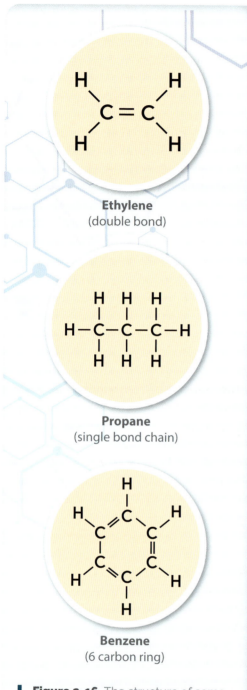

Ethylene
(double bond)

Propane
(single bond chain)

Benzene
(6 carbon ring)

Figure 2-16. The structure of some simple hydrocarbons. Notice the carbon atom's ability to bond with other carbon atoms, not only by means of single bonds, but also with double bonds. (The symbol = represents double bonds; the symbol - represents single bonds.)

and hydrogen. Coal, oil, natural gas, kerosene, and gasoline are types of hydrocarbons. These contain varying numbers of hydrogen and carbon atoms and have the general formula CnH2n+2. In this general formula, 'n' stands for a variable number. If the number of carbon atoms is 2, then the number of hydrogen atoms will be 2n + 2 or 2 * 2 + 2 or 4 + 2 = 6. All hydrocarbons have this basic formula. Several hydrocarbons are shown in Figure 2-16.

Carbon is a versatile atom and can form more than just single bonds. It can form double bonds and triple bonds and is capable of forming chains and rings. Notice the double bonds (two lines between atoms) in ethylene and benzene. Double bonds are very common in organic compounds. If the molecule has multiple alternating double and single bonds, it often produces color.

Organic molecules that contain only single bonds between carbons are called **saturated hydrocarbons**. Methane is one example of a saturated hydrocarbon. Organic molecules with double bonds have fewer atoms bonded to the carbon atom and are called unsaturated. Ethylene [eth-uh-leen] and benzene [ben-zeen] are types of **unsaturated hydrocarbons**. Saturation is a word that likely sounds familiar. It is often used to describe food products that are considered unhealthy. Saturated oils or fats used in cooking have been linked to diseases of the heart and circulatory system.

B. Alcohols, Organic Acids, and Organic Bases

Carbon can form covalent bonds with elements such as nitrogen and oxygen. Look at the compounds methanol and ethanol in Figure 2-17.

Notice that these compounds are like methane and ethane, except that an -OH group or **hydroxyl [hī-dräk-sil] group** replaces one of the hydrogen atoms. Methanol and ethanol are examples of compounds known as **alcohols**. **An alcohol is a carbon compound with a hydroxyl group attached to one or more carbon atoms**. The hydroxyl group is an example of a special group of atoms attached to carbon atoms in organic molecules. These groups are known as **functional groups** and give the molecules distinctive properties. For example, methane and ethane are gases and cannot dissolve in water. The addition of hydroxyl groups to these molecules, however, allows them to become **water soluble**.

Biology For Life

Two other functional groups that are found in many organic compounds and living organisms are the organic **acid group** (-COOH) and the **amino group** (-NH2 – the symbol N is for nitrogen). An **acid** is any substance that will release hydrogen ions in water, and a **base** is any alkaline substance that accepts hydrogen ions in water. Ions are atoms that display an electric charge because they have either gained or lost an electron.

The -COOH group, or acid group, can release hydrogen ions, so the organic compounds that it forms are acids. The -NH2, or amino, group can accept hydrogen ions, so the organic compounds that it forms are bases.

Some of these chemistry concepts may seem difficult to understand or to remember and will be studied in more detail in a chemistry course. However, keep in mind that carbon, hydrogen, oxygen, and nitrogen are the "common denominator" or common elements of organic compounds.

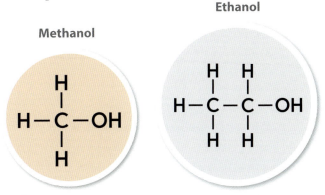

Figure 2-17. The structure of the alcohols methanol and ethanol. Notice the presence of the -OH group (O is the symbol of oxygen), which makes these compounds alcohols.

Section Review — 2.5 A & B.

1. What is organic chemistry?
2. What is the periodic table?
3. What is the symbol for carbon?
4. Explain the location of the electrons in a carbon atom.
5. The arrangement of sharing electrons in an atom is called what?
6. What is methane?
7. What is the difference between the molecular formulae and the structural formulae?
8. Define hydrocarbons. Name some types of hydrocarbons.
9. Carbon can form bonds with other kinds of atoms. It can form _____ and _____ bonds, as well as chains and rings.
10. What is the difference between saturated hydrocarbons and unsaturated hydrocarbons?
11. What are two examples of compounds known as alcohols?
12. What is an alcohol?
13. Why do methane and ethane not dissolve in water?
14. What is an acid?
15. What is a base?
16. What is an ion?
17. The acid group can release hydrogen ions, so that the compounds that it forms are _____.
18. The amino group can accept hydrogen ions, so that the compounds that it form are _____.
19. What are the four common elements of organic compounds? _____ _____.

2.6 Macromolecules

Section Objectives

- List the characteristics of carbohydrates and know the major classes of carbohydrates.
- Explain how polysaccharides are synthesized and how they are broken down.
- Describe the structural features of lipids, including fatty acids, triglycerides, and waxes.

All living things synthesize, or put together from smaller molecules, very large molecules called macromolecules. The properties of the larger macromolecules are determined by the types of molecules of which they are made.

A. Carbohydrates

Sugars and starches are examples of carbohydrates and are made from carbon, hydrogen, and oxygen. These carbohydrate molecules contain two hydrogen atoms for every oxygen atom (2:1). In living things, the main function of carbohydrates is to provide a source of energy. Perhaps the quickest source of energy can be obtained from fruits. Fruit has a high concentration of fruit sugar, or fructose, which is absorbed directly into your body.

The simplest carbohydrates are sugars called **monosaccharides** [mä-nō-sa-ka-rīd]. These monosaccharides are typically ring-shaped structures. The "ring" contains five or six carbon atoms and one oxygen atom. Monosaccharides also have many hydroxyl groups (-OH) attached to the carbon atoms. Glucose is a monosaccharide that is used for fuel in nearly all living things. The cell extracts energy from glucose during a process called cellular respiration. Figure 2-18 shows the structural formulae of two common monosaccharides, **glucose** and **fructose**.

Fructose is found in honey, tree fruits, berries, and melons. Glucose is the main product resulting from photosynthesis and is also found in fruits and in honey. Both fructose and glucose have the same molecular formula ($C_6H_{12}O_6$) but each has different structures. These differences are what give the molecules their unique properties and roles within living systems.

> **Figure 2-18.** The structure of the Common Carbohydrates Glucose and Fructose

GLUCOSE	FRUCTOSE
H−C=O H−C−OH HO−C−H H−C−OH H−C−OH H−C−OH H	H H−C−OH C=O HO−C−H H−C−OH H−C−OH H−C−OH H

Figure 2-19. Sugar Cubes

in liver and muscle tissues. Glycogen is made from many glucose molecules and can be broken down into glucose when necessary. Glycogen has a similar structure to starch but has many branches. The structure of glycogen looks like a branched tree, while the structure of a starch is like a tree that has had its branches pruned.

Polysaccharides and disaccharides can be broken down into monosaccharides by a process known as **hydrolysis** [hī-drä-le-sis]. In hydrolysis, a large molecule is split into two smaller molecules by a chemical reaction that involves the addition of water.

A molecule containing two monosaccharides bonded together is called a **disaccharide** [dī-sa-ka-rīd]. The disaccharide **sucrose** [sü-krōs], known as table sugar, is made from glucose bonded with fructose. Two glucose sugars can combine to form the disaccharide called maltose **[mȯl-tōs]**, or malt sugar. The disaccharide lactose, or milk sugar, is made from glucose and **galactose [ga-lak-tōs]**.

Polysaccharides [poly-sa-ka-rīd] are complex carbohydrates made by joining more than two monosaccharides together. Starch is a polysaccharide made from hundreds or even thousands of glucose molecules that are linked together. Plants make starch as a storage form of sugar molecules. When the plant needs fuel, its starch can be broken down into glucose molecules by a chemical reaction with water. Seeds such as rice, corn, and wheat contain starch that is used as a food source by the young germinating plant.

Cellulose [sel-yu-lōs] is a complex polysaccharide found in plants. It is made from thousands of glucose molecules that are linked together. Starch and cellulose differ in the way that the glucose units are linked together. Cellulose is a long, straight, rigid molecule and does not dissolve in water. It provides structural support for the plant cells and is the principal component of wood or woody tissues.

Glycogen [glī-kō-jen] or animal starch is the storage form of glucose for animals and is found

Section Review — 2.6 A

1. The main function of carbohydrates is to provide _____.
2. The quickest source of energy can be found in _____, which has a high concentration of fructose.
3. What are monosaccharides?
4. The cell extracts energy from glucose in a process called _____ _____.
5. What is a disaccharide?
6. Name three disaccharides given in the textbook.
7. Polysaccharides are made by joining more than _____ monosaccharides.
8. One example in our text of a polysaccharide is _____.
9. What is a complex polysaccharide found in plants?
10. What is the name of animal starch, which is the storage form of glucose found in animals?
11. What happens in a process known as hydrolysis?

Biology For Life

B. Lipids

Bacon fat, vegetable oil, and butter are examples of lipids. **Lipids** are organic molecules that will not dissolve in water but will dissolve in **non-polar** substances such as alcohols. What are non-polar substances? These are types of hydrocarbons that do not dissolve in water. You can examine this by using a couple of items that you can find in your kitchen. Take equal amounts of vegetable oil and vinegar and pour these into a small bottle with a lid or screw-type cap. Shake the bottle vigorously for a few seconds and let it rest on a countertop. Within a few seconds, the vegetable oil will form globules and float to the top. In a few more seconds, the vegetable oil will form a distinct, separate layer above the vinegar. What you just witnessed was the interaction of polar and non-polar substances. Vegetable oil (non-polar) and vinegar (polar) do not mix.

What is meant by a **polar** substance? Polar substances are compounds that have slight negative charges in one part of the compound and slight positive charges in the other part due to ionic action. Water is a polar compound; it has the molecular formula H_2O. The oxygen atom exerts a stronger pull on the electrons than does the hydrogen atom, which results in a partial negative charge on the oxygen atom and a partial positive charge on the hydrogen atom. Compounds that dissolve in water are called **polar** substances. When these compounds dissolve in water, they separate into their ions. If a teaspoon of table salt (symbol NaCl: Na is sodium, Cl is chlorine) is placed in water, the salt dissolves into sodium and chlorine ions. It is still table salt but exists in its **ionic form**.

Only polar substances will dissolve in water. Because lipid molecules are nonpolar, they will not dissolve in water. You may recall from the last chapter that all plasma or cell membranes are made from phospholipids, or lipids, with a phosphate tail. But lipids also play two other important roles within cells: energy storage and insulation. Look at Figure 2-21 to see the structure of an important kind of lipid, a fatty acid.

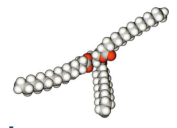

Figure 2-20. A structure of a triglyceride.

Fatty acids have an organic acid group (-COOH) at the end of a long chain of carbon atoms. The carbon chain may be saturated or unsaturated. Fatty acids are used to produce a lipid called triglyceride. **Triglycerides [trī-gli-sə-rīd]** are rich in energy and are stored in specific tissues within our bodies. During fasting or starvation, the body calls upon these energy stores to release their energy. Triglycerides also provide insulation by blocking the release of heat from the body. For a moment, stop and consider all of the living organisms that live in extremely cold climates by making use of a great deal of fatty tissue.

Triglycerides are produced by the addition of three fatty acid molecules with a three-carbon alcohol called **glycerol [gli-sə-ròl]**. When triglycerides are broken down by the chemical addition of water (hydrolysis), they produce three fatty acids and glycerol. The three fatty acids in a triglyceride can be identical, or each one can be different. The fatty acids can vary in length and can have varying numbers of double bonds. Fatty acids that have more than two double bonds are called **polyunsaturated [pä-lē-un-sa-chə-rā-təd] fats.**

Triglycerides such as butter and lard are made from saturated fatty acids and are solid at room temperature. A triglyceride that is solid at room temperature is called a **fat**. A triglyceride that is liquid at room temperature is called an **oil**. Oils usually have unsaturated

fatty acids and are more commonly made by plants than by animals. Some examples are corn oil and peanut oil.

Waxes are fatty acids combined with an alcohol which has a single -OH group. Beeswax, produced by honey bees, is an example of a naturally produced wax.

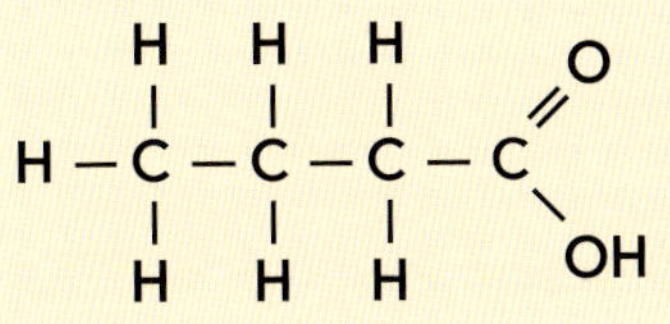

Figure 2-21. The structure of the lipid, butyric acid, a short chain fatty acid found in butter.

Section Review — 2.6 B

1. Lipids are organic molecules that will not dissolve in _____.
2. What are non-polar substances?
3. Non-polar and polar oils will not _____.
4. Polar substances have slight _____ charges in one part of the compound and have slight positive charges in the other part due to ionic action.
5. Compounds that dissolve in water are called _____ substances.
6. When polar compounds dissolve in water, they separate into ____.
7. Lipids store _____ in the cell.
8. Fatty acids are used to produce a lipid called _____.
9. The lipid triglycerides are rich and energy and are stored in specific _____ in the body.
10. A triglyceride that is solid at room temperature is called a _____.
11. A triglyceride that is liquid at room temperature is called an _____.
12. Waxes are _____ combined with an alcohol.
13. Beeswax is an example of a naturally produced _____.

2.7 Proteins and Nucleic Acids

Section Objectives

- Compare and contrast how the structure of proteins differs from that of other large molecules.
- Discuss how the shape of a protein molecule is determined.
- Describe the function of enzymes.
- Compare and contrast the structure and function of DNA and RNA.
- Discuss the function of ATP in the cell.

Proteins and nucleic acids are the most complex molecules in a living organism. Their structures have come to be understood only in the second half of the 20th century.

A. Amino Acids and Proteins

Proteins are one of the major molecules found in living things. Proteins are important components of every living cell and have many roles. For example, egg yolk is a complex protein that provides energy for the embryo growing inside the egg. Hair and fingernails are composed mostly of a protein known as keratin.

CHAPTER 2: CELLS

Proteins are made from smaller molecules called **amino acids**. These amino acids are linked together during a process called **protein synthesis**. In all amino acids, there is a central carbon atom simultaneously bonded to a hydrogen atom, an amino group (NH2), and an organic acid group (COOH). The fourth atom bonded to the carbon contains another organic group; the letter R is used to designate this fourth group. Each amino acid has its own unique **R group** which distinguishes it from every other amino acid. Figure 2-22 shows this basic structure of amino acids.

There are twenty different amino acids; therefore, there are twenty unique R groups that can be attached to the fourth position on the carbon atom. Amino acids are joined together by **peptide [pep-tīd] bonds**. These are formed by linking the acid group of one amino acid with the amino group of another. During peptide bond formation, one molecule of water is removed from the two amino acids. The R group and the single hydrogen atom bonded to the carbon are never involved in forming the peptide bond.

When two amino acids bond together, the result is the formation of a **dipeptide [dī-pep-tīd]**. If more than two amino acids join together, a **polypeptide** is formed.

Proteins are long polypeptides. An average protein is about 200 amino acids long. Proteins are distinguished by the particular order in which the amino acids are bonded together. Cell biologists refer to this order as the **primary sequence**. Different types of proteins are formed by particular sequences of the twenty amino acids.

Figure 2-22. The basic structure of an amino acid

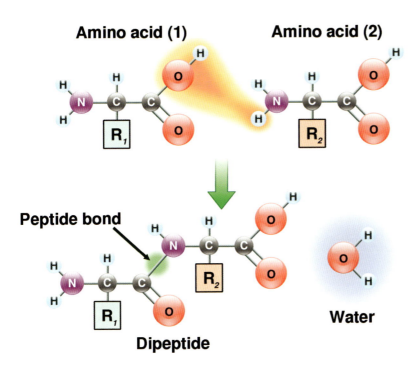

Figure 2-23. A peptide bond is a covalent bond that is formed between amino acids. The resulting CO-NH bond is called a peptide bond.

B. Protein Shape

The three-dimensional shape of a protein is determined by how the chain of amino acids folds. The R groups of the amino acids interact with each other to bend the polypeptide chain. Positively charged R groups are attracted to negatively charged R groups of other amino acids in the protein. The R groups that have like charges are repelled by each other. Conversely, R groups that are water-soluble move closer to water molecules that surround the proteins. The R groups that are insoluble in water (non-polar) move closer to each other and away from the surrounding water molecules. Forces of attraction and of repulsion between the different R groups cause the protein to bend and twist until all of the R groups are positioned in the most stable manner.

The bending of the protein to assume its best three-dimensional shape is very important. Shape determines the protein's function for the cell, and if even one amino acid is out of place in the sequence, the entire protein will fold differently. This change in protein folding can totally alter the function of the protein or even render the protein nonfunctional. The natural shape of a protein is referred to as its **conformation**.

C. Enzymes

Enzymes are proteins and the biological catalysts for living organisms. A catalyst is a substance that, without itself changing, allows chemical reactions to occur faster with lower energy. This energy is referred to as **activation energy**. For example, if you were to take a teaspoon of glucose and warm it in your hands at 98.6F, it would remain unchanged. You would have to place it over an open flame to release its

Section Review — 2.7 A

1. Proteins are made from smaller molecules called _____.
2. Amino acids are linked together during a process called _____.
3. In all amino acids, there is a central _____ atom.
4. There are _____ different amino acids.
5. Amino acids are joined together by _____ bonds.
6. When two amino acids bond together, a _____ is formed.
7. When more than two amino acids are joined together, a _____ is formed.

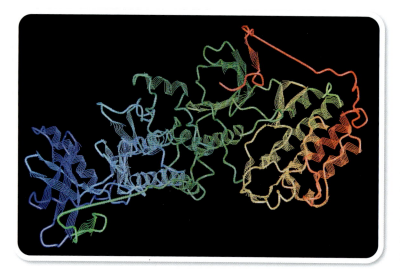

Figure 2-24. Proteins unfolding and refolding

energy in the form of heat. However, because of enzymes that regulate reactions, your body is able to release and capture the energy of glucose within its cells. All reactions that take place in living cells require specific enzymes. Enzymes serve many other functions within living cells, including the clotting of blood and the formation of proteins.

To understand how enzymes work, let us examine the enzyme **maltase**. Maltase is the enzyme that hydrolyzes, or breaks down, maltose, which is a disaccharide composed of two molecules of glucose. The molecule on which an enzyme acts is called a **substrate**. Maltose is the substrate of maltase.

There is a small area on the surface of maltase where the chemical reaction occurs. This location is called the **active site**. The shape of the active site is determined by the structure of the enzyme. The active site of maltase matches the shape of the maltose molecule, so that the maltose can fit into it.

However, maltose does not fit perfectly into the active site. Upon entering the active site, the maltose molecule becomes slightly distorted. This distortion makes the bond between the glucose portions of the molecule easier to break by one water molecule, and, the maltose is reduced into two glucose units. As the maltose molecule is broken down and the two glucose molecules are released, the active site of maltase is ready to accept another maltose molecule. In summary, the enzyme maltase lowers the activation energy of the reaction and therefore, is a catalyst.

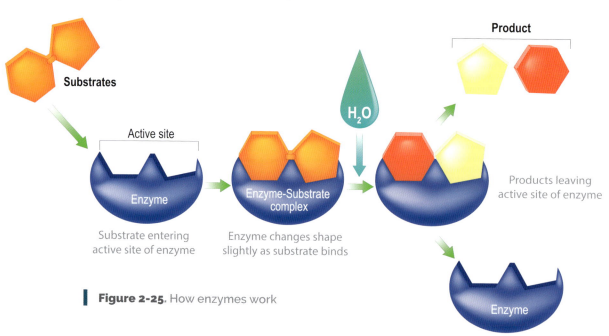

Figure 2-25. How enzymes work

Each type of enzyme has an active site that matches only a specific molecule. The active site of a particular enzyme can combine with only one set of substrate molecules; therefore, there is a different enzyme for every type of reaction in a cell.

Not all enzymes are made entirely of protein. Some enzymes contain a non-protein molecule attached to the amino acid chain. This non-protein molecule is known as a **coenzyme**. Coenzymes are attached near the active site. Certain proteins will not catalyze a reaction unless they are combined with a coenzyme.

Thousands of different chemical reactions occur in every cell, and each cell needs thousands of different enzymes to control these reactions. Anyone who recognizes this should admit to the greatness and the intellect of God!

Section Review — 2.7 B & C

1. _____ and _____ between the different R groups of a protein cause the protein to bend and twist.
2. The bending of a protein determines the protein's _____ for the cell.
3. The natural shape of a protein is referred to as its _____.
4. Enzymes are _____ and the biological _____ for living organisms.
5. A catalyst is a substance that allows _____ reactions to occur.
6. All reactions that take place in living cells require specific _____.
7. _____ is an enzyme which breaks down maltose.
8. The area on the surface of the enzyme where the chemical reaction occurs is called the _____ site.
9. The molecule on which an enzyme acts is called a _____.
10. Some enzymes include non-protein molecules known as _____.
11. _____ of different chemical reactions occur in every cell.
12. Each cell needs _____ of different enzymes to control these reactions.

D. Nucleic Acids

In 1869, Johann Friedrich Miescher, a Swiss chemist, discovered a new class of acid-like molecules in cells. These were found in the center, or nucleus, of the cell and given the name "**nucleic acids.**" They contain the necessary information which controls the activities of a cell.

The basic unit of a nucleic acid is called a **nucleotide** [nü-klē-ō-tīd]. Each nucleotide consists of three parts: a five-carbon sugar, a phosphate group, and a nitrogen base. There are five types of nitrogen bases: adenine, cytosine, guanine, thymine, and uracil.

Nucleotides are linked together to form long-chained nucleic acids. The sugar of one nucleotide is bonded to the phosphate group of the next nucleotide. The nitrogen bases are left exposed with this type of bonding arrangement. This means that one end of the nitrogen base is attached to the sugar-phosphate group, but the other end is not attached

Figure 2-26. Johann Friedrich Miescher discovered nucleic acids.

CHAPTER 2: **CELLS**

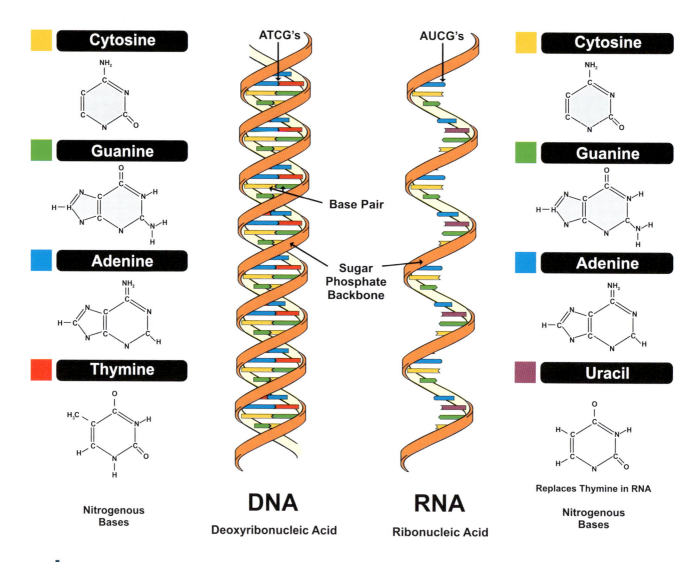

Figure 2-27. Nucleic acids are made up of smaller molecules linked together. They contain the master plan which controls all the activities of the cell. The above diagram shows the structure of DNA, consisting of two long nucleotides, with backbones made of sugars and phosphate groups joined by specific chemical bonds known as esters. These two strands run in opposite directions to each other. One of the four types of molecules called bases is attached to each sugar. The RNA molecule has only one long strand of nucleotides; it looks like one half of a DNA molecule.

Biology For Life

to anything. In DNA, the other nitrogen base attaches to the free end of a second nitrogen base. In RNA there is no attachment.

One of the most important nucleic acids for the cell is **DNA**. It is much larger than any other molecule in the cell. The five-carbon sugar in DNA is known as **deoxyribose**. As you learned in the previous section, it contains all of the heritable information in the genome; that is, the genetic code, or the set of instructions, that will be used for making a specific product.

Ribonucleic acid, or **RNA**, is also very important. RNA is similar to DNA but much smaller. DNA is double stranded but RNA is single stranded. The main sugar in DNA is deoxyribose, while the sugar in RNA is ribose. Three of the nitrogen bases in DNA and RNA are identical, but RNA has uracil as the fourth, not the thymine of DNA.

Protein synthesis is made possible through RNA. There are several types of RNA: **messenger RNA** (or mRNA), **transfer RNA** (or tRNA), **ribosomal RNA** (or rRNA), and others. These various types of RNA perform different functions in protein synthesis and DNA duplication.

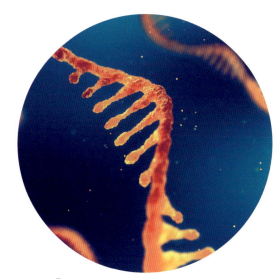

Figure 2-29. A strand of RNA

E. ATP and ADP

ATP, or **adenosine triphosphate** [a-de-nō-sēn] [trī-fos-fāt], is a nucleotide with two additional phosphate groups attached to it. It is formed in the mitochondria using energy produced during **cellular respiration** in which the cells "burn" glucose. This energy is added to **adenosine diphosphate**, or ADP, , and forms a bond with an inorganic phosphate group (Pi). During ATP

Figure 2-28. The structure of ATP, the energy storage and transfer molecule for the cell.

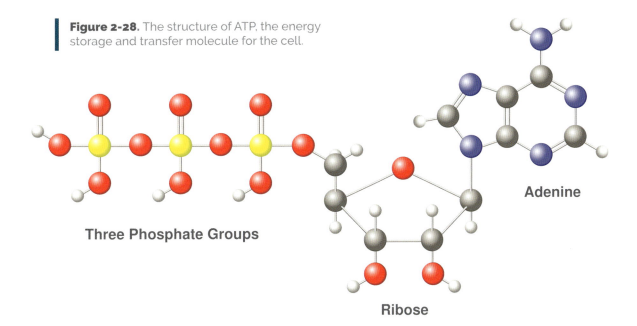

formation, energy becomes stored in the bonds between the phosphate groups. (Phosphate is made up of phosphorus, symbol P, and oxygen, in the form of PO4.)

ADP is a nucleotide that contains less energy than ATP and is both a building block and a product of ATP construction and use. When glucose is "burned," energy is added to ADP to form ATP. When ATP is used as an energy source within a cell, ADP is formed as a byproduct to be reused again.

In terms of energy, ATP is known as the "universal currency." That is, it is involved in nearly every cellular process that requires energy. Cells "spend" this energy on biochemical reactions and invest it in processes that keep them alive. During cellular respiration, ATP breaks down into ADP, and water and energy is released. ADP can be changed into ATP by the addition of water and energy. This whole process takes place in the mitochondria.

The sugar, glucose ($C_6H_{12}O_6$), is the main source of energy for making ATP. The mitochondrion can convert one glucose molecule into 36 ATP molecules through both anaerobic and

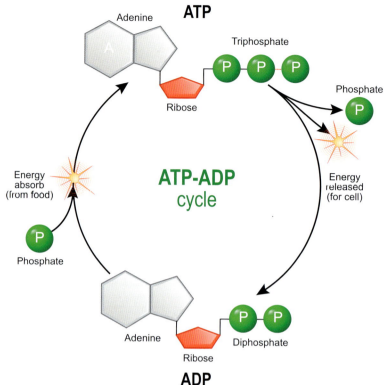

Figure 2-30. The ATP - ADP Cycle

aerobic processes. The glucose is first broken down into 2 ATP and 2 pyruvate [pī-rü-vāt] molecules anaerobically, that is, without oxygen. Then the pyruvate is further broken down into 32 more ATP through the Krebs Cycle and the Electron Transport System. The general formula is: $C_6H_{12}O_6 + 6O_2 > 6CO_2 + 6H_2O + 36ATP$ (for cellular energy).

Section Review — 2.7 D & E.

1. In 1869, a new class of acid-like molecules were discovered in the nucleus of the cell. They were given the name _____ acids.

2. The basic unit of a nucleic acid is called a _____.

3. One of the most important nucleic acids for the cell is deoxyribonucleic acid, or _____.

4. Ribonucleic acid, or _____, is a major factor in protein synthesis.

5. Adenosine triphosphate, or _____, is a nucleotide.

6. Adenosine diphosphate, or _____, can be changed into ATP by the addition of water and energy.

Biology For Life

Chapter 2 Supplemental Questions

Answer the following questions.

1. Define what it means for a thing to be alive.
2. Name some differences between living things and machines.
3. Discuss the three parts of the Cell Theory.
4. Name five functions of the cell.
5. Describe the molecular structure of the plasma membrane and the cell wall.
6. Describe the function of the nucleus.
7. Describe the structure and function of the following cellular organelles: mitochondria, ribosomes, endoplasmic reticulum, plastids, Golgi bodies, vacuoles, lysosomes, cytoskeleton, and centrioles.
8. Compare and contrast prokaryotes and eukaryotes.
9. Explain the difference in function between a unicellular organism and a multicellular organism.
10. List the levels of organization of living matter from the simplest to the most complex.
11. Describe the types of bonding which can occur with the element carbon.
12. Identify the following structural units of organic molecules: alcohol, organic acids and bases, and amino groups.
13. List the characteristics of carbohydrates and know the major classes of carbohydrates.
14. Describe the structural features of lipids, including fatty acids, triglycerides, and waxes.
15. Discuss how the shape of a protein molecule is determined.
16. Describe the function of enzymes.
17. What is chromatin? Where is it found?
18. Compare and contrast the structure and function of DNA and RNA.
19. Discuss the function of ATP in the cell.
20. Study all diagrams in this chapter. Be able to draw and label all diagrams.

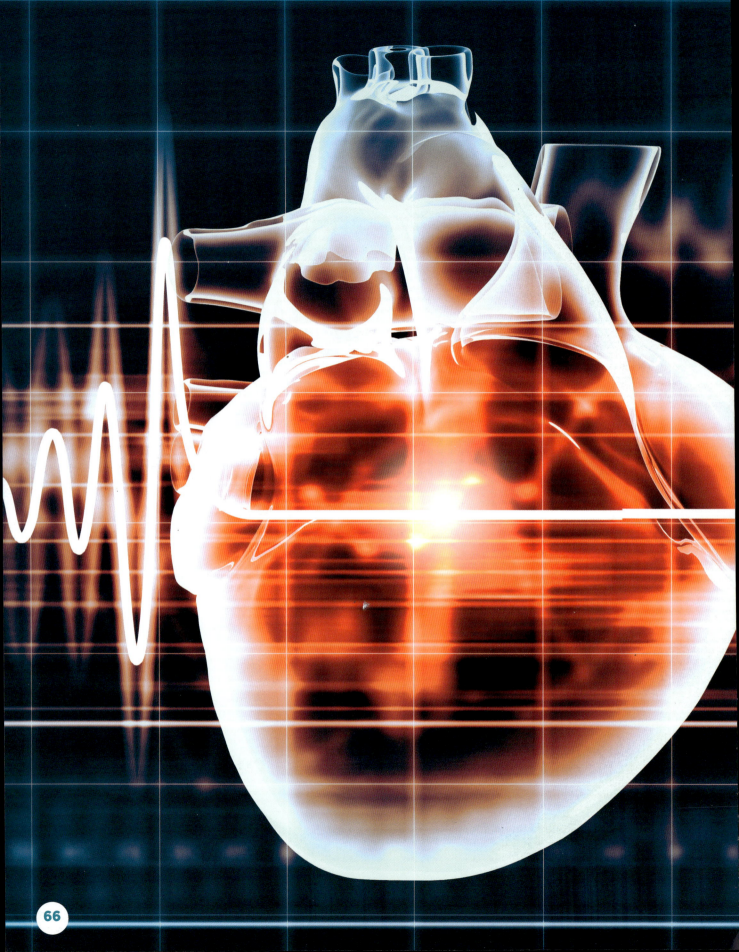

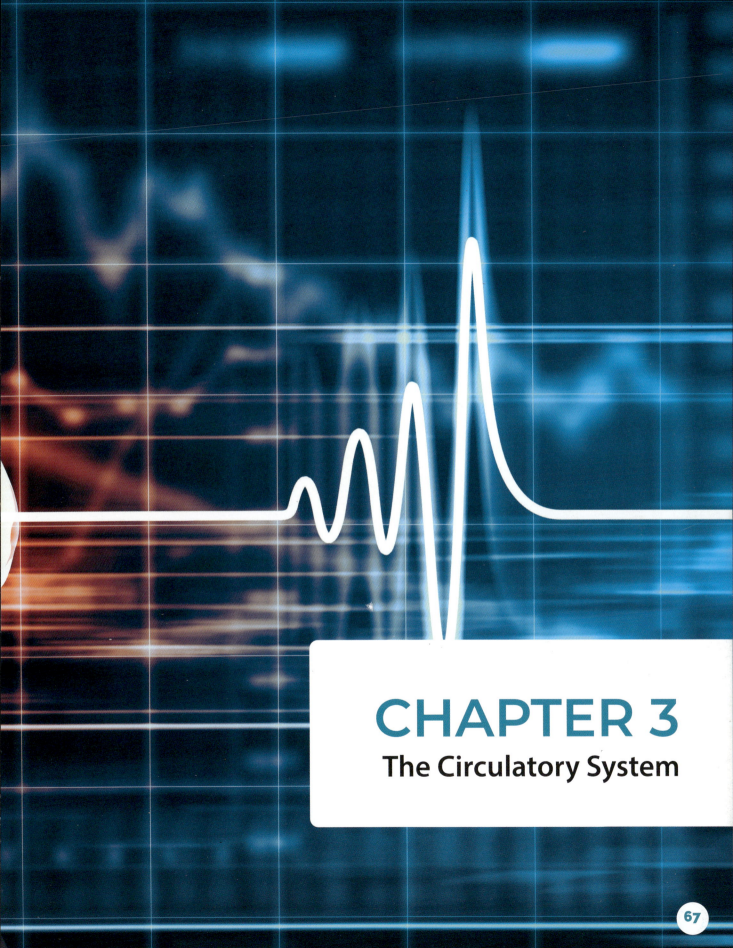

CHAPTER 3
The Circulatory System

Chapter 3
The Circulatory System

Chapter Outline

3.1 Introduction

3.2 The Heart and Blood Vessels
- A. The Heart
- B. The Path of Flow in the Heart
- C. The Path of Circulation
- D. Blood Vessels
- E. Blood Pressure
- F. Heartbeat Rate

3.3 Blood
- A. Blood Plasma and Clotting
- B. Red Blood Cells and Blood Types
- C. White Blood Cells

3.4 The Lymphatic System
- A. The Immune System
- B. Non-specific Defenses
- C. Specific Defenses
- D. Immunity
- E. Problems

3.5 Special Concerns Regarding the Circulatory Systems of the Young and Elderly
- A. Defects and Diseases of Blood Circulation
- B. Changes in Arteries
- C. Strokes and Heart Attacks
- D. Hypertension
- E. The Circulatory System of the Newborn
- F. Conclusion

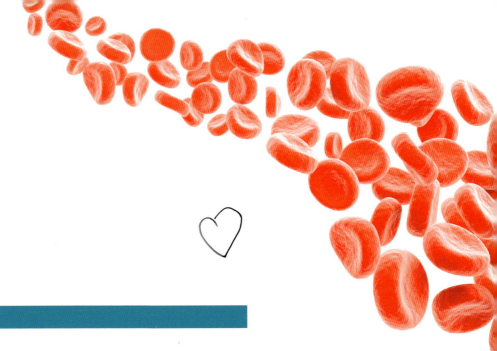

3.1 Introduction

The heart has long been regarded as a symbol of that which is most personal in man – representing man in many of his qualities. For example, someone with a "big heart" is generous; someone with "no heart" is a coward; someone with a "cold heart" is loveless, etc. Our Lord used this understanding of the heart too. In St. Matthew's Gospel, Our Lord used the heart to represent the Christian person. He advised us to put our treasures in Heaven, because where our treasure is, there is our heart. He wants the whole person in Heaven, not just the heart, but where the heart is, the person must be.

"Lay not up to yourselves treasures on earth: where the rust and moth consume, and where thieves break through and steal. But lay up to yourselves treasures in heaven: where neither the rust nor moth doth consume, and where thieves do not break through, nor steal. For where thy treasure is, there is thy heart also" (Mt. 6:19-22).

Our Lord revealed Himself to us over the centuries through the special means of His Sacred Heart. For this reason, the Sacred Heart of Our Lord is represented in many images and statues. Our Lord is shown with His Heart exposed to illustrate His deep and abiding love for us. When we adore and honor the Sacred Heart of Jesus, we recognize the central part of His human nature and His great love for us. His love was so great that he suffered and then died on the cross for our sins.

The love of Christ is also represented as a flaming heart surrounded by a crown of thorns. It reminds us of how much He suffered because love was not returned to Him. Likewise, Our Lady's Immaculate Heart is shown as being pierced by a sword and is sometimes encircled by roses.

Some of the saints have a heart as their emblem. Saint Augustine's heart symbolizes his great love for God. Saint Theresa of Avila is often pictured with a pierced heart, recalling the wound of the seraph that she received in ecstasy. Saint Margaret Mary is pictured with a heart because of her role in extending devotion to the Sacred Heart in the modern world.

Biology For Life

CHAPTER 3: THE CIRCULATORY SYSTEM

3.2 The Heart and Blood Vessels

Section Objectives

- Identify the parts of a human heart and label them on a diagram.
- Know which parts of the heart, as well as which blood vessels, carry oxygenated and deoxygenated blood.
- Describe the differences in structure and function of arteries, capillaries, and veins.
- State the point in the circulatory system where blood pressure is highest and the point where it is lowest.
- Describe how heart rate is controlled.

God designed our heart to be a powerful pumping machine which circulates life-nourishing blood throughout our bodies for an entire lifetime. Each day, a healthy human heart pumps blood by beating about 70 times per minute. Each beat forces the blood to move throughout the entire body. During the average lifetime of a person, the heart will beat more than 2.5 billion times. The heart pumps enough blood in a lifetime to fill about 2,000 swimming pools. In this chapter, you will learn about this vital organ, the heart, and you will also learn about the blood and its life-sustaining path of circulation.

A. The Heart

The **heart** is a very complex organ located between the lungs and above the diaphragm. It is surrounded by the **pericardium [per-ē-kär-dē-um]**, which is a protective membrane. The human heart consists of four chambers: the left and right **atria** (plural of **atrium [ā-trē-əm]**) and the left and right **ventricles** [ven-tri-kəl]. The atria can also be called auricles [ȯr-i-kəl]. The atria have thin walls and are located in the upper portion of the heart. These collect blood that enters the heart and continuously pump it to the ventricles that are located below them. The blood that enters the right atrium is lacking in oxygen and is called **deoxygenated blood**. Conversely, the blood that enters the left atrium is rich in oxygen and is known as **oxygenated blood**.

The ventricles are responsible for pumping blood throughout our bodies through vessels known as **arteries**. The left ventricle is larger than the right ventricle, because it must pump blood to all parts of the body except the lungs. The right ventricle pumps blood exclusively to the lungs. When deoxygenated blood travels to the lungs it releases carbon dioxide and acquires oxygen.

There are four valves in the heart that prevent blood from going backwards. The bicuspid and tricuspid valves separate the atria from the ventricles. The **tricuspid valve** separates the right atrium from the right ventricle on the "lung" side of the heart, and the **bicuspid valve** separates the left atrium from the left ventricle on the "body" side of the heart. When the atria beat, or contract, these valves open, and blood flows from the atria to the ventricles. When the ventricles contract, two other valves, the pulmonary semilunar valve and the aortic semilunar valve, open, and blood flows out of the ventricles into the pulmonary artery and the aorta respectively.

The other major parts of the heart are the "plumbing pipes," the **arteries** and **veins**. Blood flows into the heart through two large veins known as the superior and inferior vena cava. The **superior vena cava** drains the upper part of the body into the right atrium, while the **inferior vena cava** returns deoxygenated blood from the lower part of the body. The **pulmonary artery** takes deoxygenated blood from the right ventricle and brings it to the lungs so that it can be oxygenated. The newly oxygenated blood from the lungs returns to the left atrium through the pulmonary veins, and finally the left ventricle pumps oxygenated blood through the aorta, the largest artery in the body, where the blood enters

CHAPTER 3: **THE CIRCULATORY SYSTEM**

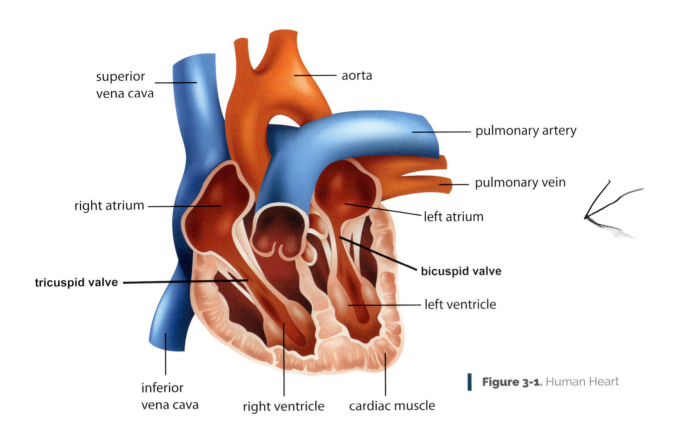

Figure 3-1. Human Heart

into smaller arteries and is transported to the whole body.

NOTE: The view of the heart in Fig. 3-1 is as you would see it in a person, so the left side looks like it is on the right.

B. The Path of Blood Flow in the Heart

The path of blood flow is as follows:

1. As blood is transported to all cells, tissues, and organs of the body, it gives up its oxygen and becomes deoxygenated.
2. Deoxygenated blood returns to the heart from all parts of the body through the vena cava and enters the right atrium.
3. Blood in the right atrium flows past the tricuspid valve and into the right ventricle.
4. The right ventricle contracts, forcing blood up and out of the ventricle.
5. The force of blood causes the tricuspid valve to close and the **pulmonary semilunar valve** to open.
6. Next, the blood travels from the right ventricle, through the pulmonary arteries, to the lungs, where it picks up oxygen.
7. Oxygenated blood returns to the heart from the lungs through the pulmonary veins and enters the left atrium.
8. Blood moves past the **bicuspid** (also called **mitral**) **valve** and enters the left ventricle.
9. When the left ventricle contracts, blood opens the aortic semilunar valve. The bicuspid valve closes, and the oxygenated blood travels to the body through the aorta.

Biology For Life

10. After depleting its oxygen, the deoxygenated blood returns to the right atrium of the heart, and the entire path of flow through the heart repeats itself.

The muscular walls of the heart are too thick to receive oxygen and nutrients directly from the blood in the heart chambers. For this reason, the heart muscle is supplied with blood by coronary arteries. These are located above the aortic semilunar valve.

> **Section Review — 3.2 A & B.**
>
> 1. Draw a picture of the human heart and label the parts.
> 2. What is the purpose of the atria?
> 3. What is the purpose of the ventricles?
> 4. Name the four valves of the heart and describe their purpose.
> 5. Describe the path of blood as it circulates through the heart.

C. The Path of Circulation

The blood circulates through the blood vessels of the body in the **circulatory system**, which has several subsystems within it that function as a closed loop. The pathway of blood from the heart to the lungs and back is called the pulmonary circuit. In **pulmonary circulation**, blood travels to the lungs from the right side of the heart and eventually returns to the left side of the heart. The pathway of blood from the heart to all parts of the body, its organs, and back to the heart is called the systemic circuit. In **systemic circulation**, blood travels from the left side of the heart to all tissues of the body and then returns to the right side of the heart. In each circulatory loop, blood passes through the heart. It pumps blood in one direction through the blood vessels of the body. The total volume of blood in the circulatory system is about five liters, or a little over one gallon.

After blood leaves the heart, it enters an **artery**. An artery is a blood vessel which carries blood away from the heart and to the body tissues. Conversely, veins are the vessels that transport blood back to the

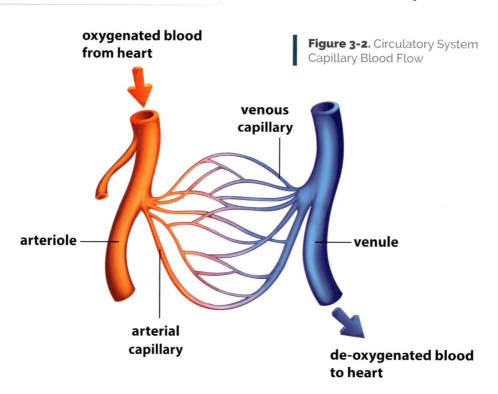

Figure 3-2. Circulatory System Capillary Blood Flow

heart. A good way to remember this is that "artery" and "away" begin with the letter "a." Arteries branch into smaller blood vessels known as **arterioles**. The blood then travels from arterioles into even smaller vessels called **capillaries**. These are the smallest vessels in the body and allow the exchange of oxygen and other materials between blood and tissue cells. When you injure yourself with a minor scratch, the blood that issues forth is mainly from the capillaries. Capillaries also provide the connection between large arteries and large veins. Blood enters vessels called **venules** by way of the capillaries. These tiny venules join at various intervals with the larger vessels called **veins**.

Deoxygenated blood is carried to the right atrium by the largest veins, the superior vena cava and inferior vena cava. Next, the right ventricle pumps deoxygenated blood through the pulmonary artery to the lungs. Oxygenated blood from the lungs returns to the heart via the pulmonary veins. The blood is then pumped out of the left ventricle and through the aorta, which is the largest artery in the body.

All of our organ systems depend on oxygen and materials transported by the blood in our circulatory system. Sometimes it is easy to forget about our heart and its work. We count on it working and do not think much about it. However, we should not forget that God, because He loves us and is thinking about us all the time, is keeping our heart beating each and every second of every day! Let us not forget to thank Him for giving us each day of our life.

D. Blood Vessels

Arteries and veins are made up of the same number of layers of tissue. However, most of the tissue layers are thicker in arteries. Since blood in the arteries comes from the heart, the blood is under greater pressure in these vessels. A person can bleed to death in a few minutes from a cut artery. However, an artery's thick walls and elastic layer make it resistant to injury. The location of arteries also protects them, as the arteries are found buried deep within the body. This is a lovely example of God's omniscience in designing our bodies.

There is only a single layer of cells making up the wall of a capillary. They are so small that blood cells

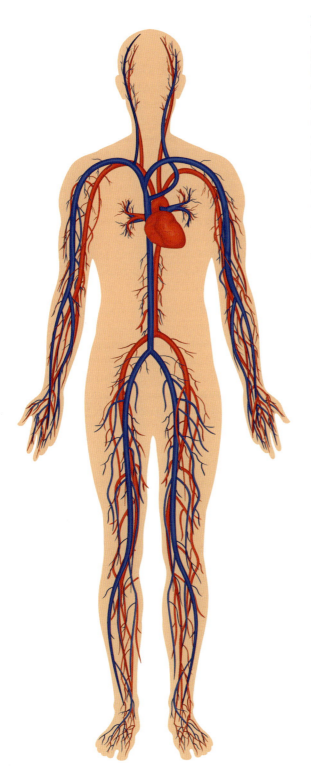

Figure 3-3. The human circulatory system

can move through them only in single file. Small molecules can easily diffuse (pass through) this thin wall. The exchange of materials between the blood and tissue fluid takes place in the capillaries.

The blood in the veins is under very little pressure and would flow backwards if it were not for valves. Each vein has **valves**, and these are spaced about two centimeters apart (about 3/4 in.). Valves open when blood flows in the direction of the heart. If the blood begins to flow backward, the valves close.

It is vital for physical life that the blood flows through the circulatory system just as God planned it. In the same way, it is vital for the spiritual life that graces flow from the sacraments, not only for ourselves but also for the benefit of others with whom we come in contact. "See how they love one another!" was the cry of the pagans when they saw the early Christians living the Christian life.

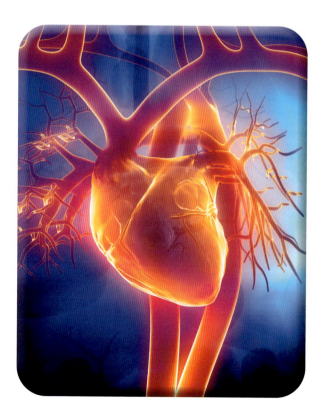

E. Blood Pressure

You have probably heard about people who have high blood pressure. As a remedy, people with high blood pressure take medicine to keep it at safe levels. You may have a grandfather who is taking medicine for high blood pressure. You may know a mother who gets high blood pressure when she is expecting a baby. What is blood pressure? It is the result of force applied to the blood by the beating heart. Blood pressure is highest in the arteries when the ventricles contract; pressure drops in the arteries when the ventricles relax. You can simulate this contraction by squeezing a rubber ball with your hand. Release the force on the ball, and then squeeze it again. The movement of your hand is similar to the contraction of the heart.

The force exerted by the contraction of the ventricles is known as **systole** [sĭs-tō-lē]. When the ventricles relax and are not pumping blood, this phase is known as **diastole** [dī-ăs-tō-lē]. During diastole, the ventricles fill with blood to prepare for the next contraction. As blood is forced out of the ventricles, the artery walls stretch because of blood pressure. When you feel the pulse in your wrist, you are feeling the stretching of an artery in the arm.

The measurement of blood pressure is taken in the upper arm. The result is stated as a sequence of two numbers usually, "120 over 80," expressed as a fraction like this: 120/80. The top number (120) represents the pressure that results from the pumping of the heart during systole. This is known as **systolic pressure**. The bottom number (80) represents the pressure that results from the heart relaxing during diastole. It is called **diastolic pressure**. Blood pressure that differs significantly from 120/80 is considered to be abnormal. There are a variety of reasons for abnormally high or low blood pressure. These typically include inherited traits from parents, stresses of everyday life, and diet. A person cannot change the traits they were born with and may not be able to change the stresses of everyday life. However, diet is something that can be managed by each person. If you know someone with high blood pressure, it cannot hurt to remind him or her to eat healthy foods, especially those low in salt and saturated fats.

F. Heartbeat Rate

A group of specialized muscle cells is buried within the wall of the right atrium. These are called the **sinoatrial** [sī-nō-ā-trē-əl] **node**. They generate synchronized electrical impulses that cause the atria and ventricles to contract at regular intervals. This group of cells is often referred to as the **pacemaker** of the heart.

The rate of the heartbeat is dependent upon how often the pacemaker generates its signals. There are two nerves from the brain that connect to this pacemaker. One nerve increases the signal rate from it, and the other nerve slows it down. When you are at rest, your heart rate slows down. When you exercise, your heart rate increases. Without the pacemaker, the heart rate would not be able to respond to the activity of the body. In a condition called **fibrillation**, the contractions become irregular and rapid. Fibrillation of the atria can affect the ventricles and overall blood flow to the body. This can be a chronic condition and can cause fatigue. Fibrillation of the ventricles can cause heart attack, cardiac arrest, and often death.

This whole system of blood vessels, arteries, veins, and capillaries reminds us of our role in conveying the love of Jesus. Just as the blood from the heart is conveyed throughout our bodies to keep us physically healthy, so each one of us is a vessel to carry the love from the Sacred Heart of Jesus to others.

Section Review — 3.2 C, D, E & F

1. What are the differences between arteries, veins, and capillaries?
2. Arteries branch into arterioles, and arterioles branch into _____.
3. What is the difference between pulmonary circulation and systemic circulation?
4. What are the names of the veins which carry deoxygenated blood to the right atrium?
5. What is the largest artery in the body?
6. All of our organs depend on _____ and other materials transported by the blood.
7. Draw a picture of the human body with the circulation system in red and blue.
8. Why are the tissue layers in arteries thicker than in veins?
9. Which are closer to the surface, arteries or veins?
10. Where does the exchange of materials between blood and tissue fluid take place?
11. What do veins have to keep the blood in them from flowing backwards?
12. What is blood pressure? Where is blood pressure highest?
13. At what point in the circulatory system is blood pressure highest and when is it the lowest?
14. What is the difference between a systole and a diastole?
15. In measuring blood pressure, what does the top number represent? What does the bottom number represent?
16. A group of muscle cells in the heart are called the _____ node. They are often referred to as the _____ of the heart.
17. How is heart rate controlled?
18. What happens in the condition called fibrillation?

3.3 Blood

All of our body cells depend on blood to deliver needed nutrients and to carry away waste products. These functions keep us healthy. You may have a grandparent who says he has a "circulation problem." This may mean that his circulation is slow. If so, cuts and wounds that he may acquire tend to heal slowly. This is because his blood is not able to quickly deliver the nutrients he needs for wounds to heal in a normal, healthy manner.

A. Blood Plasma and Clotting

Approximately 55% of the blood's volume consists of a clear yellow fluid called **plasma**. The remaining 45% consists of blood cells. Some of the plasma and blood cells pass out of the capillaries. These materials are returned to the circulation via the **lymphatic system**.

Plasma carries molecules in solution. The word solution refers to a mixture of dissolved substances in a liquid. Some of the molecules in plasma are glucose, amino acids, carbonic acid, and urea. **Urea** is a waste product, which will end up in the urinary bladder. The other substances listed are needed for life processes in the body.

About six to eight percent of the plasma is protein. Some of this protein consists of **antibodies**. These are proteins that attack and neutralize foreign substances that may enter the body, such as bacteria and viruses. The plasma also contains proteins that are involved in the clotting of blood. When a blood vessel gets damaged, a blood clot forms at the site of the injury to prevent the vessel from losing too much blood.

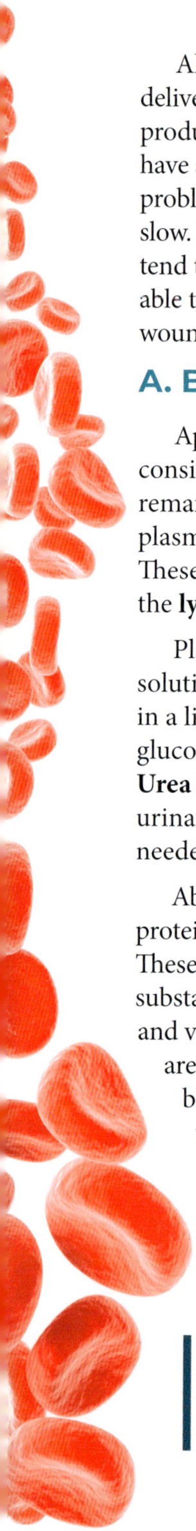

> **Section Objectives**
> - List the major components of blood plasma.
> - Describe the structure and function of red blood cells and white blood cells.
> - List the five types of white blood cells.
> - State the function of blood clots and describe in detail the blood clotting mechanism.
> - Describe the differences among the four major blood types.

Blood clotting is an amazing phenomenon that involves many steps. It begins with substances in the plasma known as **platelets**. When blood vessels are damaged, platelets break apart and release a chemical into the blood plasma. The chemical converts a plasma protein called **prothrombin** into a substance called **thrombin**. Calcium ions must also be present in order for prothrombin to be converted to thrombin. In the presence of thrombin, many molecules of another blood protein, known as **fibrinogen**, join together. The fibrinogen forms long, strand-like molecules called **fibrin**. Many strands of fibrin gather at the end of a cut blood vessel. The fibrin strands trap blood cells in a tangled mesh, which is called a clot. It seals the cut and prevents the escape of blood from the vessel. The clotting of blood provides another example of the omniscience of our creator God.

Figure 3-4. Red Blood Cells. Their red color is due to the color of hemoglobin, and they have a biconcave disk shape.

Figure 3-5. Red and white blood cells traveling through a vein.

Biology For Life

CHAPTER 3: **THE CIRCULATORY SYSTEM**

Blood Type	Antigens present in Red Blood Cells	Antibodies present in Blood Plasma
A	A	Anti - B
B	B	Anti - A
AB	A and B	None
O	None	Anti - A & Anti - B

B. Red Blood Cells and Blood Types

Red blood cells are known as **erythrocytes**. A mature red blood cell has no central nucleus. Rather, the central area of a red blood cell contains molecules of hemoglobin. It is the protein in red blood cells which carries oxygen from the lungs to all cells of the body. Each red blood cell contains more than 200 million molecules of hemoglobin Because there are about 4.5 to 5.5 million red blood cells per cubic millimeter of blood, the oxygen-carrying capacity of the blood is enormous! New red blood cells are continually being made in the **bone marrow**. They are produced at the rate of 10 billion per hour and have a life span of 120 days. The dead cells are dismantled, and the hemoglobin is incorporated into new red blood cells.

In 1903, Karl Landsteiner, an Austrian physician, discovered the different blood types. He found that whenever different types of blood were mixed, the red blood cells clumped together. This clumping is known as agglutination. Agglutination occurs because of special proteins called **antigens** [AN-ti-jens] on the red blood cell membranes. The presence or absence of these antigens

Figure 3-6. Karl Landsteiner discovered the different blood types.

determines the four different types of red blood cells: A, B, O, and AB. When identical blood types are mixed, no agglutination occurs. However, when an individual blood type is exposed to a different blood type, the antigen is not recognized as "self" and causes the synthesis of an antibody.

The table above shows the various blood types and the antigens on the red cell, as well as antibodies present in the plasma.

The anti-A antibody will react with A antigen; anti-B antibody will react with B antigen. These reactions are especially important for people who require blood following surgery or after experiencing severe injury or burns. The transfer of blood to a person is known as a **blood transfusion**. If a person with type B blood receives a blood transfusion from a person with type A blood, the anti-A antibody in the person's plasma will react with the A antigen from the donated blood. As a result, **agglutination**, or clumping of the blood, would occur. Blood that is clumped cannot flow throughout the circulatory system or carry oxygen to the body's cells. This is why blood must be carefully typed and matched with the person who is to receive the donated blood.

People with type O blood are known as **universal donors**. Universal donors have no antigens on their blood cells. Therefore, antibodies of the other blood types have no antigens with which to react, and agglutination cannot take place. Moderate amounts of type O blood can be transfused into patients with any blood type.

Persons with type AB blood are known as **universal recipients**. Their blood cells do not contain antibodies and, consequently, agglutination cannot occur. They can receive moderate amounts of any type of blood.

Biology For Life

Other antigens that may be present on red blood cells are called **Rh factors**. People who have these antigens are said to be Rh positive (Rh+), and those who don't are considered Rh negative (Rh-). The Rh factor is also known as the D antigen.

Rh factors are particularly important when a woman is expecting a baby, because it can cause a problem to children of an Rh- mother. If the father is Rh+, the child could have Rh+ blood. If some of the Rh+ blood antigens from the unborn child enter the mother's bloodstream, her body produces anti-Rh antibodies. During any subsequent pregnancy, the mother's anti-Rh antibodies may pass into the unborn child's bloodstream. If the unborn child is Rh+, the could antibodies cause clumping and destruction of the child's red blood cells. The Rh problem can even be a critical one, since the results can be anemia, brain damage, or even death to the baby.

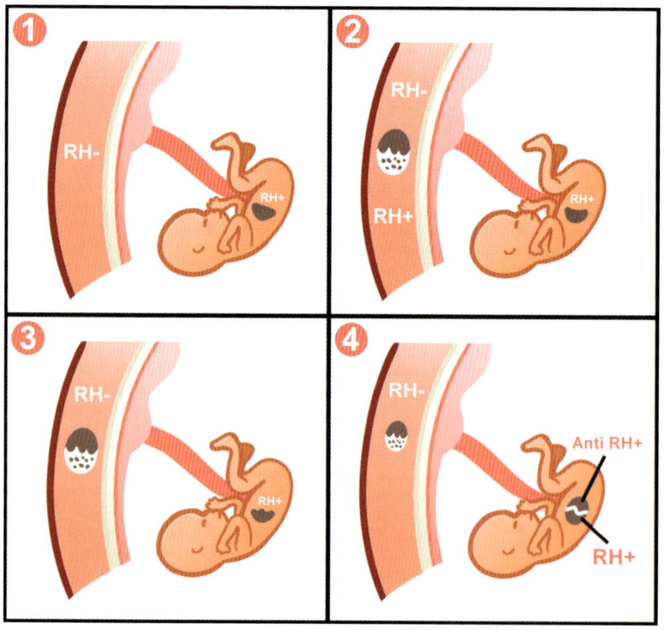

Blood Type	Rh	How Many Have It	Frequency
O	Rh+	1 Person in 3	37.4%
O	Rh-	1 Person in 15	6.6%
A	Rh+	1 Person in 3	35.7%
A	Rh-	1 Person in 16	6.3%
B	Rh+	1 Person in 12	8.5%
B	Rh-	1 Person in 67	1.5%
AB	Rh+	1 Person in 29	3.4%
AB	Rh-	1 Person in 167	0.6%

Figure 3-7. Rh Factors during Pregnancy. The mother's anti-Rh antibodies can pass into the unborn child's bloodstream and cause problems.

C. White Blood Cells

The white blood cells, or **leukocytes**, are fewer in number than the red blood cells. There are about 5,000 to 9,000 white blood cells per cubic millimeter of blood. There is approximately 1 white blood cell for every 750 red blood cells. Many white blood cells are made in the bone marrow like red blood cells. The white blood cells protect the body from infection by foreign microorganisms. They travel throughout the body via the bloodstream and move through the walls of capillaries to destroy foreign material.

White blood cells can be divided into two major groups: **phagocytes** [fa-go-sīt], and **lymphocytes**. When the white blood cells in the phagocyte group come in contact with foreign material, such as disease-causing microorganisms, the blood cells phagocytize (digest) the material. In this manner, foreign bodies are eliminated from the body.

The Eucharistic Miracle of Lanciano

Over the centuries of the Church, there have been many Eucharistic miracles. One of the most popular is known as the miracle of Lanciano.

"In the eighth century, there was a Basilian monk who had been having doubts about the real presence of Jesus Christ in the Holy Eucharist. One day, while celebrating Holy Mass, the troubled priest elevated the unleavened host, as he had done perhaps hundreds of times before. Immediately after the double consecration, in an instant, the Host changed from its bread-like appearance into flesh, and wine was changed into blood, which then coagulated and split up into five irregular globules that also differed in form and size. At first, the frightened monk tried to hide what had happened, but later, unable to conceal his emotion, he revealed it to the faithful, who as direct witnesses, spread the miraculous news throughout the city. Throughout the centuries, the holy relic, preserved in a monstrance, has remained physically intact." (*The Eucharistic Miracle of Lanciano*, Bruno Sammaciccia, 1976)

Scientific investigations of this miracle have taken place since 1574.

These are some of the conclusions from the scientific investigation that occurred in 1970:

1. The blood of the Eucharistic miracle is real blood, and the flesh is real flesh.
2. The flesh consists of the muscular tissue of the heart.
3. The blood and the flesh belong to the human species.
4. The blood and the flesh of the Eucharistic miracle belong to the same blood type: AB.

So based on this Eucharistic Miracle, we may say that Our Divine Lord had an AB type blood – a universal recipient. As the author of the report of the 1970 investigation notes:

"For the first time in the course of history, at least in as much as the Eucharistic miracle of Lanciano is concerned, science, equipped with exceptional and precise means, offers us categorical data which confirm the validity and the certainty about the Eucharistic Miracle … now all that remains is to accept and meditate."

Figure 3-8. Eucharistic Miracle of Lanciano. Note the Host's flesh appearance and the globules of Blood.

Biology For Life

The lymphocytes manufacture antibodies (proteins found in the blood to detect and destroy foreign invaders, like bacteria). These lymphocytes help protect the body against disease, which could be caused by microorganisms. These antibodies will destroy the germ and will recognize it if it enters the body again. Upon re-infection by the same germ, the antibody that was formed previously is available to immediately combat the infection.

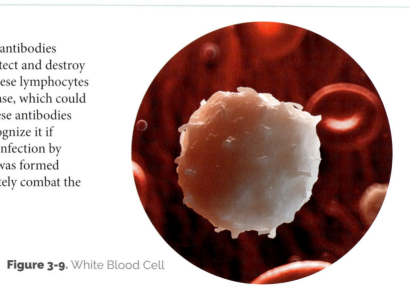

Figure 3-9. White Blood Cell

Section Review — 3.3 A, B, & C

1. What percent of the blood's volume contains plasma?
2. What are contained in the plasma molecules?
3. What percent of plasma is protein?
4. What are antibodies? What do they do?
5. Describe how blood clotting works. Red blood cells are also called _____.
6. A red blood cell has no nucleus but contains molecules of _____.
7. Each red blood cell carries more than _____ million molecules of hemoglobin.
8. New red blood cells are continually being made in the bone _____.
9. New red blood cells have a life span of _____ days.
10. The clumping of red blood cells is called _____.
11. The presence or absence of _____ determines the blood types.
12. What do we call the four different blood types?
13. The transfer of blood to another person is called a blood _____.
14. Which blood type is known as the universal donor?
15. Which blood type is known as the universal recipient?
16. A mother's anti-Rh antibodies can pass into the unborn child's _____ and cause problems.
17. From the scientific studies of the Miracle of Lanciano, what type of blood did Our Lord have?
18. White blood cells are also called _____.
19. There is one white blood cell for every _____ red blood cells.
20. What is the function of white blood cells?
21. How do the phagocytes fight infection?
22. How do lymphocytes fight infection?

CHAPTER 3: **THE CIRCULATORY SYSTEM**

3.4 The Lymphatic and Immune Systems

> **Section Objectives**
> - List three functions of the lymphatic system.
> - Describe the body's immune system.

Lymphatic System

The **lymphatic system** is another part of the body's circulatory system. Body fluids are carried in lymphatic system vessels and blood vessels. Together these vessels form the vascular system of the body.

Every cell in the body is surrounded by **tissue fluid**. This fluid consists of water and small molecules that have escaped from capillaries. It also contains plasma proteins and white blood cells from the circulatory system. Tissue fluid allows the passage of materials between the blood and body cells and makes possible cellular respiration and the elimination of wastes. Each day a little more fluid filters out of the capillaries than is re-absorbed by them.

Tissue fluid returns to circulating blood by vessels of the lymphatic system. Lymphatic vessels have thin walls, and tissue fluid can easily pass into these vessels. When the tissue fluid is inside the lymphatic vessels, it is called **lymph**.

The largest lymphatic vessels join with two large veins that are near the heart. Lymph drains back into the circulatory system at this location. Since the lymphatic system has no pump (the heart is the circulatory system's pump), the system depends upon one-way valves to move tissue fluid uphill from the legs and arms. As with the veins, the movement of the skeletal muscles helps propel the lymph fluid up to the top of the body so that the fluid can enter the circulation.

The two main functions of the lymphatic system are absorption of fats and combating disease. The lymphatic system combats disease

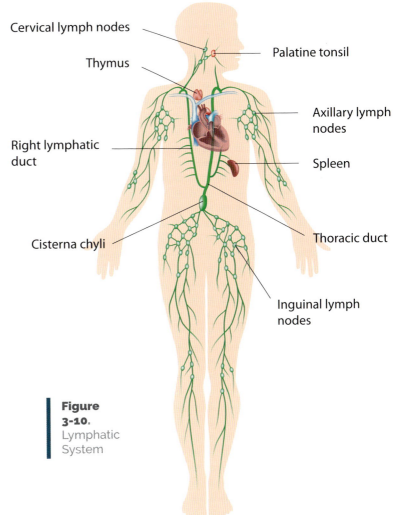

Figure 3-10. Lymphatic System

Biology For Life

because white blood cells in the **lymph nodes** (areas of swelling in the lymph vessels where white blood cells are found) are available to attack foreign microorganisms.

A. The Immune System

Blood and tissue fluid keep cells alive by carrying material to them. The blood also carries substances that defend the body against diseases.

B. Non-specific Defenses

The body has general, **non-specific defenses** that it employs against all disease-causing organisms (pathogens). The skin and mucus lining on many passageways within the body are part of this non-specific system. These provide a mechanical barrier against disease-causing agents such as viruses and some bacteria. These agents get caught in sticky mucus and are then eliminated from the body.

If pathogens do enter the body, they are engulfed by phagocytes and then broken down, or destroyed, by a process known as **phagocytosis** [fa-go-sī-tō-sis]. The dead bacteria and white blood cells may become **pus**, which often indicates an infection in the body. Virus-infected cells may also release a protein known as **interferon**, which neutralizes attacking viruses by preventing them from reproducing. Interferon is also effective against many types of bacteria.

Section Review — 3.4 A & B

1. The _____ system is part of the body's circulatory system.
2. Every cell in the body is surrounded by _____ fluid.
3. _____ allows the passage of water, molecules, protein, and white blood cells between the blood and body cells.
4. When tissue fluid is inside the lymphatic vessels, it is called _____.
5. The movement of the _____ muscles helps propel the lymph fluid up to the top of the body.
6. The two main functions of the lymphatic system are absorption of _____ and combating _____.
7. White blood cells are available in the _____ nodes to attack foreign microorganisms.
8. Non-specific defenses against diseases are the _____ and _____ linings.
9. Non-specific defenses provide a barrier against _____ and some bacteria.
10. Dead bacteria and white blood cells may become _____, which indicates an infection.
11. The protein _____ neutralizes attacking viruses; it is also effective against _____.

C. Specific Defenses

The body also has methods by which it defends itself against specific pathogens. **Lymphocyte** white blood cells constantly circulate in the bloodstream and tissue fluid, and track down harmful microorganisms and diseased cells. When these are located, the lymphocytes initiate a precisely targeted attack on them.

Each and every cell has molecules on its surface that are called **surface molecules**. These help identify the cell as belonging to the body of which it is a member. Foreign substances have surface molecules that tag them as not belonging to the body. If a surface molecule contacted by a lymphocyte is recognized as a "belonging" marker, nothing happens, and the lymphocyte moves on. If the molecule is identified as a "not-belonging" marker (antigen), then the lymphocyte does not move on. Immediately, an **immune response** is triggered, by which tissue fluid carries the antigen to a lymph node. It is there that the phagocytized foreign body is presented to a type of lymphocyte called a B-cell. The B-cell may become a type of cell that manufactures proteins that exactly fit the surface markers of the antigen. These proteins are called **antibodies** and fit one specific antigen.

Circulating antibodies are another type of antibody that moves through the fluids of the body. When the antibody encounters an antigen, it attaches to it and signals for assistance by **phagocytes**. The antigen becomes surrounded. In this manner, the antigen and pathogen are destroyed. These phagocytes are like an army within our body helping to battle an enemy.

A type of lymphocyte called a **T-cell** carries antibodies on its surface. These cells can recognize body cells that have been invaded by cancer cells and certain viruses. These cells and pathogens have markers on their surface that are best described as "not-quite-belonging." The T-cells either activate phagocytes or directly kill the cancerous or virus-infected cells.

D. Immunity

The body generally requires several days to form antibodies when it first detects an antigen of invading microorganisms. Future responses to an attack are much more rapid because of **memory cells**. These are either B-cells or T-cells that "remember" antigen patterns and produce antibodies immediately if the antigen is detected again. The antibodies produced are much stronger and longer lasting than the original ones.

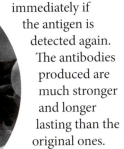

Figure 3-11. Blood Cells

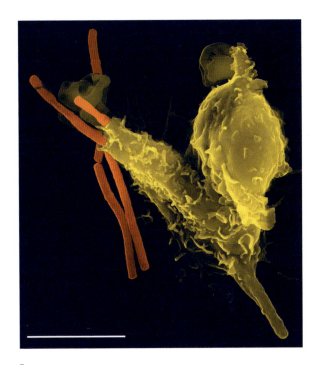

Figure 3-12. A scanning electron microscope image of a single white blood cell (yellow) engulfing anthrax bacteria (orange). Scale bar is 5 micrometers.

Memory cells persist for the life of the individual. New ones are produced during each response to a new attack. This process of warding off disease through **antibodies** is known as **immunity**. Immunity prevents a person from repeatedly getting certain diseases, such as measles or chicken pox.

E. Problems

Not all "not-belonging" markers are harmful. Sometimes the body cannot distinguish between harmful and harmless ones. For example, the body may fail to recognize certain pollens as harmless. This results in reactions called allergies. Generally these are harmless, but some can be life threatening. Bee stings and shellfish allergies occasionally cause violent immune responses.

The body occasionally fails to recognize some body cells as "belonging" and attacks them as antigens. Such misdirected attacks occur in **autoimmune diseases**. Rheumatoid arthritis, which affects the joint tissue, is such a disease.

Biology For Life

CHAPTER 3: THE CIRCULATORY SYSTEM

Section Review — 3.4 C, D & E.

1. White blood cells called _____ track down harmful microorganisms and diseased cells and attack them.
2. Every cell has surface _____, which help identify the cell as belonging to the body.
3. If a cell in the body is identified as not belonging to that body, an _____ response is triggered, and the tissue fluid carries the antigen to a lymph node.
4. When an _____ encounters an antigen or foreign body, it attaches to it and signals for assistance.
5. The _____ are like an army within our body to battle an enemy.
6. A _____ carries antibodies and can recognize cells invaded by cancer cells.
7. The body takes _____ days to form antibodies to fight invading microorganisms.
8. Memory cells persist for the _____ of the person.
9. The process of warding off disease through antibodies is known as _____.
10. Sometimes the body has misdirected attacks, which we call _____ diseases.

3.5 Special Concerns Regarding the Circulatory Systems of the Young and Elderly

Section Objectives

- Define the terms atherosclerosis, arteriosclerosis, hypertension, and myocardial infarction.

A. Defects and Diseases of Blood Circulation

Hypothermia is a condition that can result if the body temperature falls more than 2° C (4° F) below the healthy normal value of 37° C (98.6° F). Death is a real possibility if hypothermia persists for more than a few hours. Anyone whose body temperature drops below 32° C (90° F) has a 17 to 33 percent chance of dying. The elderly are more susceptible to hypothermia because of poor blood circulation, especially during the cold of winter. There are two reasons for this. First, the aging body becomes less able to maintain an even temperature when subjected to external cold. Second, the body mechanism that normally detects a drop in its own

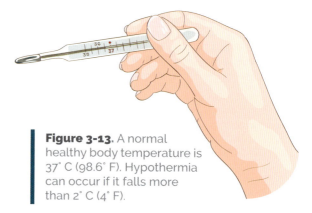

Figure 3-13. A normal healthy body temperature is 37° C (98.6° F). Hypothermia can occur if it falls more than 2° C (4° F).

temperature gradually loses its sensitivity as it ages. As a result, some older people do not realize that they are dangerously cold in frosty air and may fall asleep, become unconscious, and die. The aged body is also less likely to detect an increase in body temperature and, therefore, is also more prone to heat stroke in the hot summer months.

B. Changes in Arteries

Atherosclerosis refers to the presence of fatty streaks in the blood vessels, which may develop into masses of fatty tissue that can harden the artery wall. The fatty tissue is known as **atheroma**, and a large mass of atheroma is called a **plaque**. These tissue plaques can diminish the elasticity of the artery wall, or its ability to expand, contract, and stretch. As a result, the narrowed passageway interferes with the flow and delivery of blood. Atherosclerosis is more commonly referred to as "hardening of the arteries." This condition is made worse by **arteriosclerosis**, which is the gradual loss of elasticity in the arterial walls, due to aging. The combination of atherosclerosis and arteriosclerosis can lead to the formation of blood clots, which can diminish the amount of blood flowing through the arteries.

Most people in the United States and Canada eat large amounts of fatty foods and cholesterol-rich foods, such as meat, eggs, and butter. Research on atherosclerosis has suggested that a diet with much fat and cholesterol contributes to the development of **atherosclerotic plaques**. If there is a genetic disorder in the family that causes a person to have high cholesterol levels, diet modifications and cholesterol-lowering drugs can help to bring the cholesterol level down.

In European countries, including France, Greece, Italy, Portugal, and Spain, people enjoy a

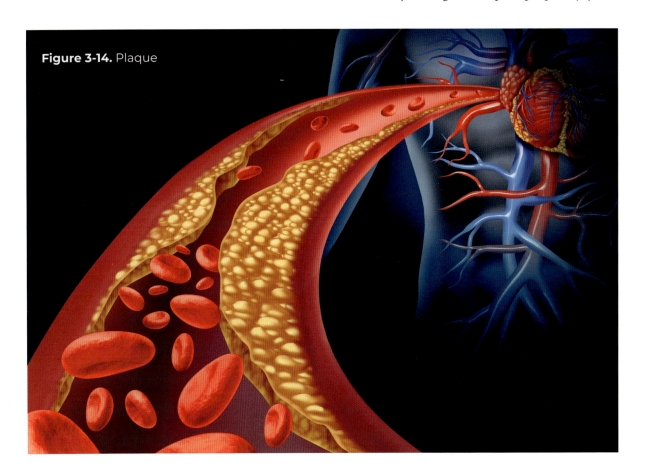

Figure 3-14. Plaque

well-balanced, healthier diet. They eat generous amounts of fruits and vegetables, small portions of nuts, and lean meats; and they eat fish and shellfish at least twice a week. The pace of life in Europe is slower than in the United States, and eating dinner with family and friends is an important event – much like it used to be in the United States. Today, most Americans are "in a hurry," and eating is often viewed as an obstacle to surmount before moving on to another activity. Recall that Our Lord knew that He faced His passion, crucifixion, and death before he sat down with His disciples on Holy Thursday. He knew what he faced, but he set an example for us. Dining with family and friends is important and should not be dismissed lightly.

C. Strokes and Heart Attacks

Plaques occasionally form in the vessels that supply the brain with blood. When this happens, the person with this condition has an increased risk of **stroke**. A stroke is a condition by which the person has a loss of brain function due to a disturbance in the blood supply to the brain. A heart attack, or, as it is known in the medical field, a **myocardial infarction**, is most commonly caused by a **thrombosis**, or blockage, of one of the **coronary arteries**. You will recall that coronary arteries supply the heart muscle with blood. If a **thrombus**, or blood clot, forms, it can cut off the blood supply to specific regions of the heart. This will result in tissue damage and degraded function in that region.

Heart attacks generally occur only if a person's coronary arteries are already narrowed by atherosclerosis. If the size of the infarct, or damaged area, is small and does not impair the pacemaker, the heart attack may not prove fatal. For such individuals, there is a good chance of recovery, which often depends on surgery.

D. Hypertension

As the heart pumps blood through your arteries, the force of the blood flow exerts pressure on the arterial walls. If the heart pumps blood through the circulatory system with a force much greater than is needed to maintain a steady blood flow, high blood pressure results. **Hypertension** is the medical term for high blood pressure. If left untreated, hypertension can eventually cause damage to the arteries of the body, especially if a person has atherosclerosis and arteriosclerosis. Hypertension is dangerous because the increased pressure within the circulatory system forces the heart to work harder to keep the blood moving. This extra work can damage the lining of the coronary arteries, and fatty plaques are more likely to form at the areas where damage has occurred. Medication, dietary habits, and lifestyle changes to reduce stress may be needed to bring blood pressure back to normal levels.

CHAPTER 3: **THE CIRCULATORY SYSTEM**

Section Review — 3.5 A, B, C & D.

1. What is the condition called if the body temperature falls more than 4 degrees F below the healthy body temperature of 98.6 degrees F?

2. One reason why the elderly are more susceptible to hypothermia is that the aging body is less able to _____ an even temperature when subjected to cold.

3. Another reason is that the aging body loses its ability to detect a drop in _____.

4. The aged body is less likely to detect an _____ in body temperature during summer months.

5. Hardening of the arteries is the definition of _____.

6. Fatty tissue, or atheroma, which forms in artery walls is called _____.

7. The gradual lessening of elasticity in artery walls is called _____.

8. Research shows that people who eat large amounts of fatty food are likely to develop _____ in their arteries.

9. In European countries, people eat more fruits and _____, nuts, lean meats, and fish.

10. A _____ is the result of plaque in the blood vessels which supply the brain with blood.

11. A heart attack is often the result of blockage, or thrombosis, in one of the coronary arteries which supply the _____ muscle with blood.

12. _____ is the medical term for high blood pressure.

13. Diet, medication, and _____ changes can help bring high blood pressure to normal levels.

E. The Circulatory System of the Newborn

Of all the wonders of the human body, perhaps none is so astonishing as the change that takes place when a baby is born. During this brief moment, the newborn child must start to breathe on his own. Here God's gentle hand changes the child's circulatory system from non-air breathing to air breathing in an instant. It is a wondrous example of God's workmanship.

Since implantation, the baby has depended on the mother's blood supply for oxygen and nutrients, which leave the mother's blood vessels and enter the placenta. The baby also has blood vessels in the placenta which pick up the oxygen and nutrients from the mother. The blood vessels of the mother and baby do not actually mix directly, but only indirectly through the placenta's tissues. Waste products from the baby are also transferred through the placenta, where they are picked up by the mother's blood and taken to her liver and lungs for removal.

Since the oxygen is supplied by the mother, and carbon dioxide and other wastes are removed by the mother's lungs, the baby's lungs do not need to work while inside the mother, so they are dormant and mostly filled with fluid. The mother's liver is also acting as a filter for the baby, so the baby's liver is not very active at this time. Therefore, the baby does not need much circulation to these organs, and they are mostly bypassed.

The circulation of the unborn baby has been designed to insure that the most vital organs receive

Biology For Life

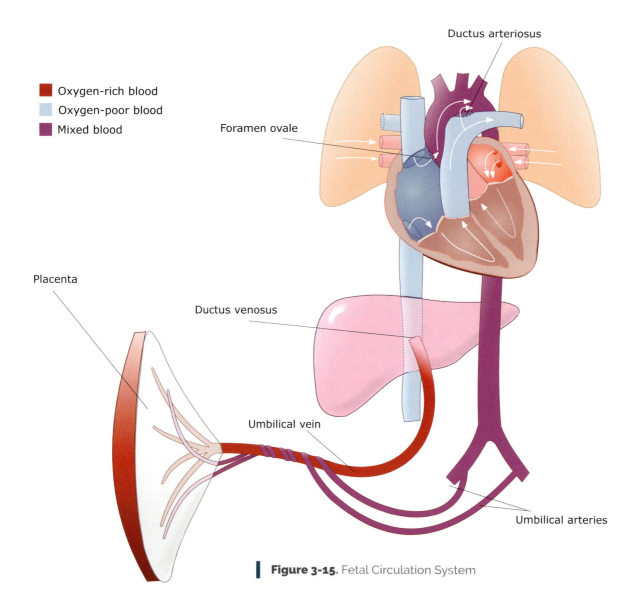

Figure 3-15. Fetal Circulation System

the greatest concentration of oxygenated blood. The brain of the fetus requires the most oxygen.

Blood that carries oxygen and nutrients enters the fetal circulation through the large umbilical vein via the placenta. The blood then travels to the liver, where it branches or divides. Part of the blood enters the circulation around the liver, and the remainder travels up towards the fetal heart. Blood enters the inferior vena cava by way of the **ductus venosus**, a special blood vessel found only while the baby is a fetus, and then goes to the heart.

Blood enters the right atrium but does not go to the right ventricle to be pumped to the lungs, since the fetal lungs do not function at this time. Instead, the blood goes directly from the right atrium through a tiny opening, the **foramen ovale**, which leads to the left atrium. This shunting of blood from the right to the left atrium allows more oxygenated blood to enter the left ventricle. At this point, the blood is pumped through the aorta to the head and the upper extremities (shoulders, arms, and neck).

CHAPTER 3: THE CIRCULATORY SYSTEM

Blood returning from the head enters the right atrium from the superior vena cava and is directed downward through the tricuspid valve and into the right ventricle. The blood is then pumped to the pulmonary artery, which, instead of sending the blood to the lungs, shunts most of it to the descending aorta via the **ductus arteriosus**, another special, temporary, fetal blood vessel.

A small amount of blood flows to and from the lungs. Blood from the descending aorta is returned to the placenta via the two fetal umbilical arteries. The transition from the circulation of the unborn to the circulation of the newborn is a gradual change. It results from pressure changes in the lung, heart, and major vessels at the time of a newborn's birth. This transition involves the functional closure of the three fetal shunts:

1. The **ductus venosus**: oxygenated blood from the placenta first goes to the baby's liver and is then shunted through the ductus venosus directly to the inferior vena cava and finally to the baby's heart.
2. The **foramen ovale**: blood from lower part of the body travels through the foramen ovale which is inside the fetal heart, which is shunted directly from right atrium to left atrium, bypassing the right ventricle and the lungs.
3. The **ductus arteriosus**: deoxygenated blood that comes from the fetal head is shunted by the ductus arteriosus directly from the pulmonary artery to the descending aorta to be returned to the placenta and the mother.

Once the newborn begins to cry, its lungs expand. The newborn begins to breathe on its own, which dilates the blood vessels of the lungs. This is followed by increased blood flow to the lungs. As the lungs receive blood, the pressure in the right atrium, right ventricle, and pulmonary arteries decreases. At the same time, there is a blood pressure increase in the left side of the heart. Since blood flows from an area of high pressure to an area of low pressure, the circulation of blood through the three fetal shunts is reversed. The most important factor controlling ductal (fetal shunt) closure is the increased oxygen content of the blood of the newborn. During birth or shortly thereafter, the foramen ovale closes due to compression of the atrial septum, or the wall that divides the right from the left atrium. The ductus arteriosus partly closes, and stops working four days after birth. Its complete closure will take considerably longer.

Congenital heart defects result from the failure of these three fetal shunts to close at birth. Because of the reversible flow of blood through

Biology For Life

the ducts (fetal shunts) during the early newborn period, heart murmurs can occasionally be heard. In conditions such as crying or straining, the increased pressure can shunt deoxygenated blood from the right side of the heart across the ductal opening, which causes a momentary bluish color to the newborn.

Jaundice is a very common disease among newborn babies, but it is more common among premature babies. It is estimated that as many as 50% of newborn babies become jaundiced in the first three days of life. This is a condition which makes the skin of the baby turn yellow, especially on its face. If action is not taken, the yellow coloring spreads over the body of the baby. The condition is caused by too much bilirubin in the blood. **Bilirubin** is produced by the normal breakdown of red blood cells and normally passes through the liver. Sometimes the baby's liver is not working normally and cannot process the bilirubin. Newborn babies with jaundice are treated by a process called phototherapy. They are placed under blue lights, which convert bilirubin into a form that can be removed from the body in the urine.

F. Conclusion

We close our chapter on the circulatory system with the observation that just as the heart and lungs are necessary for bodily life, prayer and the Eucharist are necessary for spiritual life. The heart pumps blood through our vessels that carry oxygen and nutrients and then carry away carbon dioxide and waste products. In this very complex and almost incomprehensible process, the heart sustains our physical life.

The Eucharist is the Body and Blood of Jesus, our Lord and Savior. Communion with the flesh of the risen Christ preserves, increases and renews to life the grace we received at Baptism (CCC #1393). It is Our Lord Himself who reminds us of the necessity of the Eucharist in our lives when he said, "Amen, amen, I say to you, unless you eat the flesh of the Son of Man and drink his blood, you shall not have life in you. He who eats my flesh and drinks my blood has life everlasting, and I will raise him up on the last day" (Jn. 6:54).

In the raising of our minds and heart to God in prayer, we procure for ourselves immeasurable blessings that help strengthen and sustain our spiritual life. St. Alphonsus, "the Doctor of Prayer," wrote much about prayer and fervently practiced it. One of his famous quotes is, "He who prays will be saved, and he who does not pray will be lost."

Section Review — 3.5 E & F.

1. Blood enters the inferior vena cava by way of the _____ _____, a special blood vessel found only while the baby is a fetus, and then goes to the heart.
2. What is the purpose of the Foramen Ovale?
3. What is the purpose of the Ductus Arteriosus?
4. Why are the three fetal shunts important?
5. What is a common disease among newborns? What is the cause? How is it treated?

Chapter 3 Supplemental Questions

Answer the following questions.

1. Identify the parts of the human heart on Figure 3.1.
2. Which parts of the human heart carry oxygenated/deoxygenated blood?
3. Which blood vessels carry oxygenated/deoxygenated blood?
4. Describe the five kinds of blood vessels.
5. What are the two main loops (or circuits) of the circulatory system? How are they different?
6. What are the differences in structure of the arteries, veins, and capillaries?
7. What are the differences in function of the arteries, veins, and capillaries?
8. At what point in the circulatory system is the blood pressure highest/lowest?
9. How are erythrocytes and leukocytes different?
10. How is the heart rate controlled?
11. What are the major components of blood plasma?
12. Describe the function of red blood and white blood cells.
13. What are the two major types of white blood cells and what are their functions?
14. What is the function of blood clots?
15. Describe the blood clotting mechanism.
16. What are the differences among the four major blood types and why are these important?
17. What is the Rh factor and why is it important?
18. What are the two functions of the lymphatic system?
19. Describe the body's immune system.
20. Define the terms atherosclerosis, arteriosclerosis, hypertension, and myocardial infarction.
21. Study all diagrams in this chapter. Be able to draw and label all diagrams.

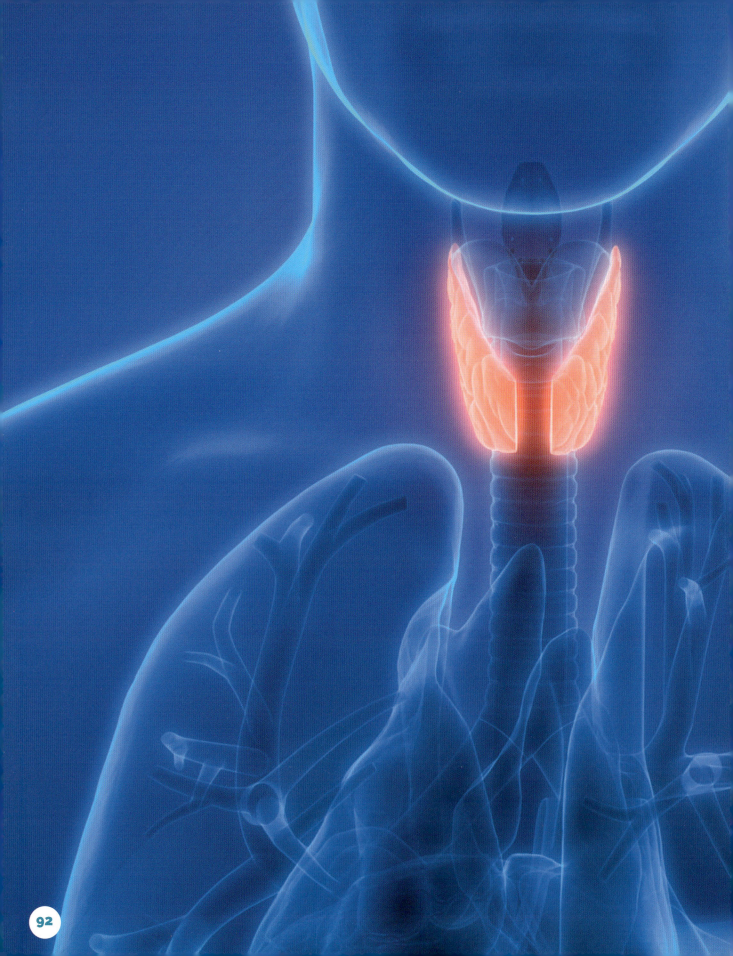

CHAPTER 4
The Endocrine System

Chapter 4
The Endocrine System

Chapter Outline

4.1 Introduction

4.2 The Endocrine System
- A. How Hormones Work
- B. Adrenal Glands
- C. Thyroid and Parathyroid Glands
- D. Islets of Langerhans
- E. Gonads
- F. The Pineal Gland
- G. The Thymus Gland
- H. The Pituitary Gland
- I. The Hypothalamus

4.3 Regulation of Hormone Secretion
- A. Tropic Hormones and Negative Feedback
- B. Interaction of the Nervous and Endocrine Systems
- C. Endocrine Glands and Major Hormones Secreted

4.4 Special Concerns Regarding the Endocrine Systems of the Newborn and Elderly
- A. The Newborn
- B. The Elderly

4.1 Introduction

The endocrine [en-də-krən] system is responsible for all the functions of your body. It works through chemicals or chemical messengers called **hormones**. They are released from specialized groups of tissues known as **glands**. Hormones regulate each and every process of your body. These are produced when your body needs to grow, when you need sleep, when you are hungry, and when you face emergencies. Hormones are produced for many other functions, which we will discover in this chapter.

God, in His omniscience, knew that we would face dangers requiring us to react faster than our ability to reason. Consider for a moment the firefighters, paramedics, and policemen who face dangers every day. Think of a time that you faced danger. Perhaps it was an unleashed dog chasing you when you were walking or riding a bicycle. You may recall in great detail how your body responded. You were scared. Your heart took action by pumping blood faster, which allowed extra blood and glucose to rush to the muscles and brain. These changes prepared your body to work more vigorously and efficiently. Ask your parents about emergencies they have faced in their lifetimes, and it is guaranteed that they can describe them precisely.

The endocrine system is one of the components of our human nature that God has made to help us survive, grow, reproduce, and regulate the activities of our bodies. There are processes so intricate that, if they depended upon our knowledge and will to be accomplished, much of our time would be spent in mastering these basic

Biology For Life

functions rather than considering those things that God desires us to contemplate: doing His will, keeping His commandments, helping others, and working out our salvation. We would do well to consider the intricacies of God's handiwork, which reflect His goodness, His creative power, and His wisdom. Such a meditation will allow us to grow in the knowledge of God. We should give thanks for all of these as we learn about the endocrine system.

4.2 The Endocrine System

The endocrine system is made up of specialized, ductless tissues called glands. These glands produce chemicals called **hormones**. Hormones act on specific cells to regulate processes such as growth, development, chemical and physical processes within cells, and responses to crises. Figure 4-1 shows most of the endocrine glands in the body.

The endocrine system does not work by itself and depends on connections with the nervous system. The nerves extend out to the structures that they control, including organs, tissues, and cells. The nerves provide feedback to the brain, which, in turn, causes the endocrine glands to respond. The endocrine glands may be very far from the structures that are affected by the secretion of a particular hormone. Although each nerve may be connected to one specific structure, hormones are carried throughout the body.

The body responds more quickly to nerve impulses than to hormones, but the effects of the chemical messengers are longer lasting than nerve impulses. This is part of God's wonderful foresight that can be described as cause and effect. The cause is a very short-lived electrical signal in the form of a nerve impulse; the effect is a longer-lived chemical substance.

A. How Hormones Work

Although each hormone comes in contact with many kinds of cells, only specific types of cells, called **target cells**, respond to specific hormones. Target cells contain proteins called **receptors** that are located on the surface of cell membranes. Each kind of cell has its own specific receptor that is located on its cell membrane. For this reason, each hormone can bind only to a particular type of receptor. When a hormone comes in contact with the cell, it binds to the cell, but only if the specific receptor is located on the cell's membrane.

When a hormone binds to a receptor, it triggers a series of reactions in the cell. These reactions are initiated by an increase of a molecule called **cyclic AMP** (c-AMP). It is often referred to as the **secondary messenger**. Cyclic AMP is important in many biological processes and causes cells to produce enzymes. Enzymes are protein catalysts

> **Section Objectives**
>
> - Distinguish between the conditions of diabetes mellitus and diabetes insipidus in terms of which endocrine glands and hormones are involved, as well as the symptoms of each condition.
> - Explain how receptors, enzymes, cyclic AMP, and prostaglandins are involved in the action of hormones.
> - State the location of each gland, the hormone that each gland secretes, and the actions of the hormone for the following glands: thyroid, parathyroid, thymus, adrenal cortex, adrenal medulla, islets of Langerhans, posterior pituitary, anterior pituitary, and pineal
> - Know which hormones are affected in the conditions of hypothyroidism, hyperthyroidism, cretinism, goiter, dwarfism, and gigantism.

THE BODY

CHAPTER 4: **THE ENDOCRINE SYSTEM**

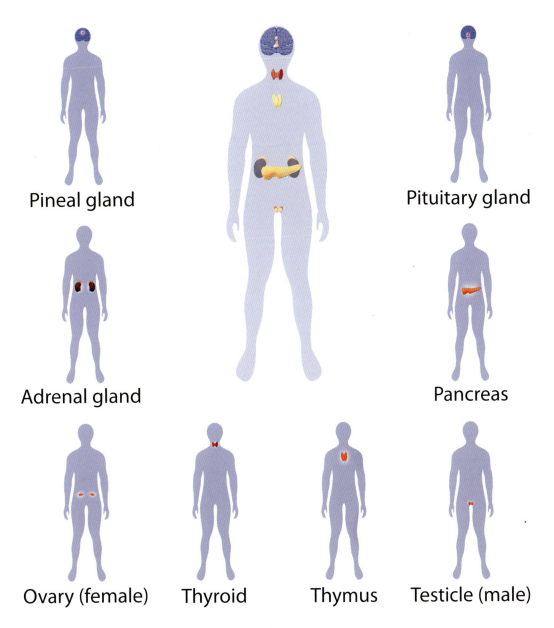

Figure 4-1. The Endocrine Glands are an integrated system of small organs that involve the release of extracellular signaling molecules known as hormones.

that assist chemical reactions by lowering the **activation energy** for the reaction. These have the effect of speeding up the reaction, though afterward they remain unchanged.

Other types of hormones, called **steroid hormones**, can enter the cell and travel into the nucleus. These hormones may activate inactive genes in the nucleus. The active genes can direct changes in the cell's chemistry, such as those associated with maturation.

Prostaglandins [präs-tə-glan-dən] are hormones but play important roles as mediators. They are made

Biology For Life

from a fatty acid known as **arachidonic [ar-a-kə-dä-nik] acid** and are produced by all cells in the body, except lymphocytes. There are ten unique receptors for prostaglandins in the body. Although they have a variety of functions, prostaglandins are known for their mediating responses to injuries, wounds, allergens, and pain.

B. Adrenal Glands

An adrenal gland is located on the top of each kidney. Each adrenal gland has two parts, the adrenal cortex and the adrenal medulla. The adrenal cortex produces **corticoid [kȯr-ti-kȯid] hormones**. Some of the corticoid hormones maintain water and salt balance by regulating the absorption of the salt and water in the kidney. Other corticoid hormones are involved in the conversion of proteins to glucose, which helps maintain proper blood sugar levels during times of physiologic stress.

The adrenal medulla secretes the hormones **epinephrine** (also known as adrenaline) and **norepinephrine** (also known as noradrenaline). The adrenal medulla produces both of these hormones during periods of extreme fear, anger, pain, or cold. Both of these hormones work together to increase the heart rate and the flow of blood to the muscles and brain. They accomplish this by dilating or widening the blood vessels in the liver, heart, and skeletal muscles. Epinehrine helps convert glycogen to glucose, so that more energy is made available to the body. The body's response to epinephrine and norepinephrine prepares a person to either fight danger or flee to avoid danger. This response is known as the "fight-or-flight" response. It can cause the rate of metabolism to increase by as much as 100%.

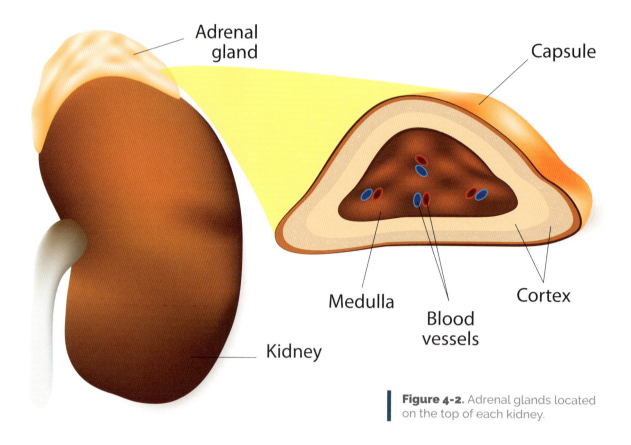

Figure 4-2. Adrenal glands located on the top of each kidney.

Section Review — 4.2 A & B.

1. What do we call the chemicals produced by the endocrine system?
2. The endocrine system depends on connections to the _____ system.
3. The nerves provide feedback to the _____, which then causes the endocrine glands to respond.
4. Hormones can travel throughout the _____.
5. Only _____ cells respond to specific hormones.
6. Target cells contain proteins called _____.
7. A hormone triggers a series of _____ in a cell.
8. _____ causes cells to produce enzymes.
9. Enzymes assist _____ reactions.
10. Steroid hormones enter a cell and travel into the _____.
11. _____ are known for their responses to injuries, wounds, allergens, and pain.
12. Adrenal glands are located on the top of each _____.
13. Each adrenal gland has two parts, the cortex and the _____.
14. The adrenal cortex produces _____ hormones.
15. Some corticoid hormones maintain water and salt balance in the _____.
16. During periods of extreme fear, anger, pain, or cold, the adrenal medulla produces _____.

C. Thyroid and Parathyroid Glands

The **thyroid gland** lies over the top part of the trachea in the neck and is shaped like an "H." It produces a hormone called **thyroxine** that regulates the metabolism of the body. In most cells, thyroxine controls the speed at which glucose is oxidized, or "used up." In addition, it influences protein production and affects the growth rate of children.

If too much thyroxine is present, a condition called **hyperthyroidism** results, and the metabolic rate increases. A rise in the metabolic rate increases the heart rate, blood pressure, and body temperature. People with hyperthyroidism may sweat heavily, become nervous, and develop bulging eyes. This condition may be treated with drugs that will reduce thyroxine secretions. Another treatment for hyperthyroidism consists of surgically removing a part of the thyroid gland.

A condition known as **hypothyroidism** results from a deficiency of thyroxine. Persons with this condition have a very low metabolic rate, usually lack energy, and may be overweight. If hypothyroidism occurs during childhood, it can cause stunted growth and severe mental retardation. This condition is known as **cretinism**. Hypothyroidism cannot be cured, but it can be corrected with thyroxine medication.

Thyroxine contains the element **iodine**. If there is not enough iodine in a person's diet, the thyroid gland becomes enlarged. This condition is known as **goiter**. Goiter can be prevented through diet by ingesting iodine in the form of iodized salt. For this reason, salt manufacturers add small amounts of iodine to salt that is sold in grocery stores. It is probably worthwhile to keep iodized salt at the dinner table. Chances are that your family eats a well-balanced diet, but the inexpensive, small amounts of iodine in table salt cannot hurt you. Seafood, especially saltwater fish and shellfish provide adequate amounts of iodine. In addition, dairy products and breads and cereals possess iodine.

The **parathyroid glands** are four tiny glands located on the posterior (back) surface of the thyroid gland. Each of these glands is less than eight millimeters long and is the smallest endocrine gland in the human body, about the size of a grain of rice. Taken together, the four parathyroid glands come to about the same size as the pineal gland. The parathyroid glands produce **parathyroid hormone**, which regulates the levels of calcium and phosphate in the blood. The bones, muscles, and nerves require both calcium and phosphate ions to grow and function properly. Bones are made mostly of calcium **phosphate** ($Ca_3(PO_4)_2$).

The thyroid gland secretes a hormone called **calcitonin [kal-sə-tō-nin]**. Calcitonin works in opposition to parathyroid hormone. Calcitonin ensures that the intestines do not absorb too much calcium, and it prevents loss of calcium from the bones of the mother during pregnancy and lactation.

D. Islets of Langerhans

The **pancreas** has two important roles. It secretes enzymes through ducts that are used for digestion, and it has special cells called endocrine cells that secrete hormones. These special cells are located throughout the pancreas in areas referred to as **islets of Langerhans**.

The islet cells secrete two different hormones into the bloodstream. One of these hormones is called **glucagon**. It converts glycogen into glucose within the liver and muscles. It does this when the concentration of glucose in the blood is low. Around mid-morning, somewhere between breakfast and lunch, you may feel a little jittery, perhaps hungry. These sensations occur because your blood sugar is low. Your body is releasing glucagon to make glucose available.

The other hormone secreted by islet cells is called **insulin**. Insulin lowers the levels of glucose in the blood and promotes the uptake of glucose by the body cells. Some of the glucose is used by the cells for cellular respiration, and some of it is converted to glycogen. The glycogen is then stored in the liver and muscle tissues. Most people become aware of the effects of insulin after lunch, during the afternoon. The body begins to slow down to allow this conversion to take place. The next time you eat Thanksgiving dinner you may observe many of those around you becoming sleepy after eating the large meal. Most of this "tiredness" is due to the effects of insulin.

In many places, such as Latin America and the Mediterranean countries of Europe, lunch is followed by a "siesta," or quiet time, when people take a nap. They wake up refreshed and return to their work. Americans work through this time and are expected to be productive, even though their bodies may be telling them that it is time for a nap. There are some who suggest that this early afternoon tiredness is partly due to a daily, natural body rhythm.

A condition called diabetes mellitus results when a person's body does not either produce enough

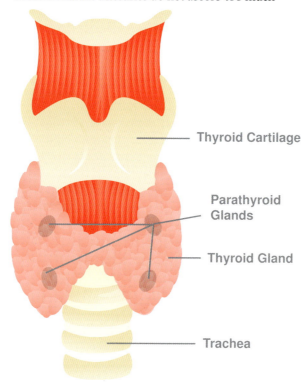

Figure 4-3. Thyroid and parathyroid glands located in the neck.

insulin or has built up a tolerance to it. The former condition is commonly known as Type 1 diabetes, and people who have it are called diabetics. In type 2 diabetes, the body produces enough insulin but cannot use it effectively. This condition is known as insulin resistance, which leads to the decreased production of insulin. The pancreas continues to make more insulin but eventually cannot keep up with the demands of the body. Type 2 diabetes tends to occur in older people.

People with Type 1 diabetes accumulate glucose in the blood, which is excreted in the urine. Additional amounts of water are also eliminated with the excess glucose. The person urinates excessively, and this leads to excessive thirst. Coma and death may result if diabetes remains untreated for too long.

Most diabetics require medical treatment that involves injections of insulin. However, if too much insulin is taken, a problem known as "insulin shock" may result. In this condition, the levels of glucose in the blood become very low, which results in the brain not receiving enough glucose for its cells to function properly. Diabetics who are on insulin treatments must carry sugary foods to eat in case insulin shock should occur.

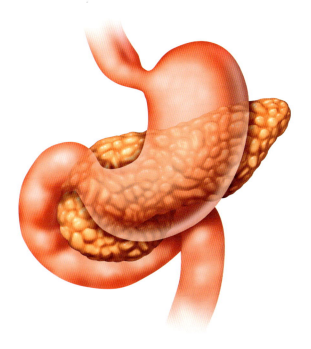

Figure 4-4. The pancreas is located near the stomach. It secretes enzymes, and has endocrine cells that secretes hormones.

Section Review — 4.2 C & D.

1. Where is the thyroid gland located?
2. What is the name of the hormone it produces? What is the function of this hormone?
3. How does this hormone affect children?
4. If too much of the hormone is present, what is the condition which results?
5. What condition is the result of not enough of this hormone?
6. What is the name of the element contained in thyroxine?
7. What is goiter?
8. How can we obtain a daily diet of iodine?
9. What are the parathyroid glands? Where are they located?
10. What does the parathyroid hormone regulate?
11. Where are the islets of Langerhans located?
12. What is the function of the hormone glucagon?
13. What is the function of the hormone insulin?
14. What is diabetes mellitus? What is the common name for it?
15. How is diabetes treated?

CHAPTER 4: THE ENDOCRINE SYSTEM

E. Gonads

The **gonads** produce sex hormones. In females, the gonads are called **ovaries**, and in males, they are called **testes**.

The ovaries produce hormones called **estrogens** [es-trō-jen]. These cause the beginning of the fertility, or **menstrual cycle**. In addition, estrogens cause the development of **secondary sex characteristics**, which develop during **puberty**. The time of puberty occurs in the early teenage years. During this time, a girl's breasts enlarge due to the development of the **mammary glands**. Her hips widen; she grows additional body hair; and she acquires an additional layer of body fat. This body fat often precedes a growth spurt. In our society, where glamour and thinness are highly valued, many girls go on diets during puberty because of this natural gain in weight. God has designed our bodies to mature in specific ways. If she eats a balanced diet, obtains sufficient sleep, and exercises, a girl has no practical need to go on a diet during puberty. Diets, especially starvation diets, can cause harm to a teenage girl's health.

The testes produce male sex hormones known as **androgens**. One of the androgens is known as **testosterone** [tes-täs-tə-rōn], which during puberty causes the development of the male secondary sex characteristics. At this time, a boy's voice will deepen, and there is an increase in the growth of body hair. A rapid spurt of growth in the muscular and skeletal systems also occurs, and boys will become taller and bulkier. Like girls, boys may put on weight in advance of a growth spurt. Unlike girls, boys lose fat under the skin during puberty. The increase in weight for boys is due to the increase in bone density and muscle mass. Boys like to lift weights at this age but they should be discouraged from powerlifting as there is a chance they can damage their developing skeletal system.

F. The Pineal Gland

The **pineal gland** is a pea-sized, pinecone-shaped gland that is located at the center of the brain. The pineal gland is the second smallest endocrine gland in the body. It produces a hormone called **melatonin**. Melatonin is involved in the sleep-wake cycle and is believed to cause drowsiness and lowering of the body temperature, thus preparing the body for sleep. It is also thought to regulate daily rhythms of the body. For example, if you wake up daily at 6:00 AM you may find yourself waking up at this time even when you do not need to. The pineal gland is also involved in the regulation of reproductive and fertility cycles.

G. The Thymus Gland

The thymus gland lies in the upper part of the chest and directly behind the sternum, or breastbone. It provides an area for lymphocytes to mature and does this through the action of the hormone **thymosin**. The thymus gland is important in protecting the body against disease and in protecting it against attacking itself through autoimmune diseases. An autoimmune disease is an illness which occurs when the body tissues are attacked by the body's own immune system. The evidence for the importance of the thymus gland in immune disorders comes from both the young and the elderly. When the thymus is not functioning at an early age, it results in **immunodeficiency** or a weakened immune system. In the elderly, the thymus may shrink with age, causing susceptibility to infection and cancer.

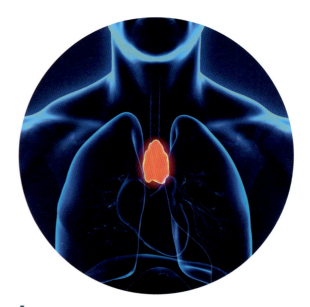

Figure 4-5. The Thymus Gland

Section Review — 4.2 E, F & G.

1. What do we call the gonads in females? In males?
2. What are the hormones produced by the ovaries?
3. What are the hormones produced by the testes?
4. Describe the pineal gland? Where is it located.
5. What is the hormone produced by the pineal gland?
6. What is the function of melatonin?
7. Where is the thymus gland located?
8. What is the name of the hormone produced by the thymus gland?
9. What is the function of the thymus gland?
10. If the thymus gland does not function at an early age, it can result in what?
11. What results in the elderly if the thymus shrinks?

The Pituitary Gland

The **pituitary [pə-tü-ə-ter-ē] gland** is located at the base of the brain. Nerves and blood vessels connect the pituitary gland to the hypothalamus, which is also known as the regulator, or "master switch," of the entire endocrine system. Some of the pituitary hormones affect parts of the body directly, and others target specific endocrine glands which will then release their secretions.

The pituitary gland has an anterior and a posterior section. The **posterior pituitary** stores and releases two hormones that are produced in the hypothalamus. One of these hormones is called **oxytocin**, and the other is called **antidiuretic hormone** (ADH), which is also known as **vasopressin**. Oxytocin has several functions among which include stimulate contraction of the muscles of the womb during labor, promote the flow of milk from the mammary glands, and to promote maternal caring behavior. Within minutes following the birth of the child, milk is readily available for the newborn. When the baby tries to nurse, the sucking action stimulates the secretion of oxytocin, which makes the milk flow (letdown), so that the baby is nourished. Oxytocin stimulates the smooth muscle cells in blood vessels and mammary glands and helps to regulate blood pressure.

ADH regulates the reabsorption of water in the structures of the kidney called **tubules**. If a person does not have an adequate amount of ADH, very little water will be reabsorbed, and the kidneys will produce an abnormally large volume of urine. This condition is known as **diabetes insipidus**. ADH also regulates blood pressure by making the muscles in the walls of the arterioles contract.

The **anterior pituitary** produces at least six hormones. Four of these affect the secretions of other glands and are known as **tropic** [trō-pik] **hormones**; tropic means to stimulate or stimulating. Tropic hormones specifically target other endocrine glands: the thyroid, the adrenal cortex, and the gonads. **Thyrotropic hormone**, or thyroid stimulating hormone (TSH), causes the thyroid gland to secrete thyroxine. The level of **adrenocorticotropic hormone**, or ACTH, determines the amount of corticoids secreted by the adrenal glands. Gonadotropic hormones, luteinizing hormone (LH) and follicle-stimulating hormone (FSH) control the production of the sex hormones by the testes and ovaries. Both of these hormones operate in both men and women, but differently. In men both LH and FSH are released simultaneously, which causes the release of testosterone and regulation of the production of sperm. In women both hormones are released alternately for about two weeks at a time and they are responsible for the

CHAPTER 4: THE ENDOCRINE SYSTEM

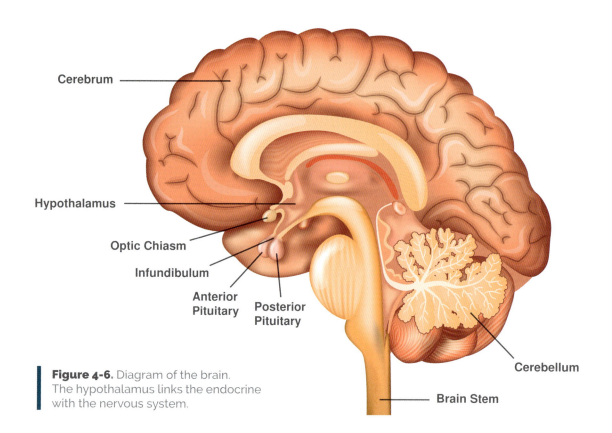

Figure 4-6. Diagram of the brain. The hypothalamus links the endocrine with the nervous system.

woman's fertility/menstrual cycle. A fifth hormone is called **somatotropin** or **growth hormone** (GH), which controls bone growth. A deficiency in growth hormone can lead to extreme shortness, a condition which is known as **dwarfism**. An excess of growth hormone during childhood can cause **gigantism**, which is an abnormally large stature. Another hormone that is secreted by the anterior pituitary gland is **prolactin**. This hormone stimulates the production of milk in the mammary glands after a woman has given birth and works in conjunction with oxytocin.

I. The Hypothalamus

The **hypothalamus** or master switch of the endocrine system connects directly to the brain and receives feedback about the entire body through the nervous system. Based on this information, the hypothalamus then sends signals to stimulate or inhibit hormone secretion by the pituitary gland. At least nine such hormones have been identified. The hormones that stimulate secretions are called releasing hormones; the hormones that slow down secretions are called inhibiting hormones. In this way, the hypothalamus controls body temperature, thirst, hunger, and salt and water balance, as well as our emotional behavior; so you can see why the term "master switch" is quite appropriate.

Section Review — 4.2 H & I.

1. Where is the pituitary gland located?
2. To what is the pituitary gland connected?
3. What is the function of the posterior pituitary section?
4. What is the function of the hormone oxytocin?
5. What is the function of the antidiuretic hormone (ADH)?
6. What is the function of the anterior section of the pituitary gland?
7. What is the name of the growth hormone?
8. What do we call a deficiency of the growth hormone? An excess?
9. What is prolactin?
10. List several things the hypothalamus controls.

4.3 Regulation of Hormone Secretion

Section Objectives

- Define the term tropic hormone and explain the concept of negative feedback.
- Give an example of the interaction between the nervous and endocrine systems.
- State the role of the hypothalamus in linking the nervous and endocrine systems.
- Know the function of a releasing factor and an inhibiting factor.
- Describe the differences in the endocrine systems of the newborn and the elderly.

The body needs different hormones at various times; therefore, an endocrine gland does not continuously release hormones. Certain signals within the body tell each endocrine gland when to secrete its hormones. Other signals stop hormone secretion. The pituitary gland and the central nervous system are involved in many of these control mechanisms.

A. Tropic Hormones and Negative Feedback

In a healthy individual, the concentration of each hormone remains within a narrow and definite range by means of a process called **negative feedback**. This process is similar to the way in which a thermostat regulates a household furnace. When the temperature falls below the thermostat setting, the furnace switches on and begins producing heat. When the temperature reaches the thermostat setting, the furnace switches off.

Due to the negative feedback system, the hormone level can be kept in its proper concentration. A cycle of actions takes place in a **negative-feedback system**, whereby the final event inhibits the first event. This inhibiting effect is the reason for the use of the word "negative." For example, an increase in "substance A," will cause an increase in "substance B." But an increase in substance B will cause a decrease in substance A. As a result, substance A controls the level of substance B and substance B controls the level of substance A.

The relationship between adrenocorticotropic hormone (ACTH) and the corticoid hormones shows how the negative-feedback system operates. The tropic hormone ACTH causes the secretion of corticoid hormones by the adrenal cortex. Corticoid hormones are produced until these reach a high level. In turn, high levels of corticoids inhibit the production of ACTH by the pituitary gland.

CHAPTER 4: THE ENDOCRINE SYSTEM

B. Interaction of the Nervous and Endocrine Systems

The endocrine and nervous systems work together to regulate all processes of the body with the help of the hypothalamus. For example, if blood is flowing at a reduced volume, the hypothalamus will send a nerve impulse to the posterior pituitary gland, which, in turn, secretes ADH. The end result is a rise in blood pressure that increases the flow of blood.

The hypothalamus communicates with the anterior pituitary gland by releasing hormones. These are called **releasing factors** and are secreted by the hypothalamus. These releasing factors stimulate the pituitary gland to secrete specific hormones. For example, suppose that the level of growth hormone is too low. In this case, the hypothalamus would secrete a hormone known as growth hormone-releasing factor (GHRF). When GHRF reaches the anterior pituitary, it causes the secretion of growth hormone. Some hormones, such as prolactin, are controlled by **inhibiting factors** secreted by the hypothalamus. If the level of prolactin in the blood is too high, the hypothalamus will secrete prolactin-inhibiting factor (PIF). The PIF causes the anterior pituitary gland to decrease the secretion of prolactin. This will result in the decreased production of milk.

The adrenal medulla and the pituitary glands are the only endocrine glands which have direct connections to the nervous system. The adrenal medulla receives nerve impulses from the **autonomic nervous system** (ANS), which is the control system for the body. The ANS works "automatically" and is not under our conscious control. The ANS functions like a computer, coordinating and regulating all processes of the body. For example, in a crisis, the ANS causes the adrenal medulla to secrete adrenaline for one of the most important reactions for our survival, the "fight-or-flight" response.

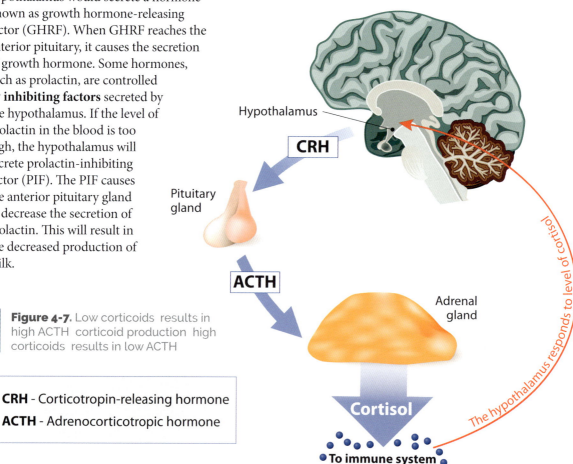

Figure 4-7. Low corticoids results in high ACTH corticoid production high corticoids results in low ACTH

CRH - Corticotropin-releasing hormone
ACTH - Adrenocorticotropic hormone

Biology For Life

Section Review — 4.3 A & B.

1. The negative-feedback system is a process whereby hormone levels can be kept in proper _____.
2. The endocrine system and the _____ system work together to regulate all processes of the body.
3. The _____ provides the link between the two systems.
4. To stimulate the pituitary gland to secrete specific hormones, the hypothalamus secretes hormones called _____ factors.
5. The hypothalamus can also secrete hormones that will decrease the production of other hormones; these are called _____ factors.
6. The adrenal _____ is the only endocrine gland other than the pituitary gland which has direct connections with the nervous system.
7. The autonomic nervous system (ANS) works automatically and is not under our _____ control.

C. Endocrine Glands and Major Hormones Secreted

As mentioned earlier, an endocrine gland does not continuously release hormones, because the body needs different hormones at different times. Accordingly, there are certain signals within the body to tell each endocrine gland when to secrete its hormones. Conversely, other signals tell each endocrine gland when to stop secretions. Below is a list of the endocrine glands and the major hormones they secrete.

Gland	Hormone	Function
Pituitary (anterior lobe)	Somatotropin or growth hormone	Regulates growth of bone and other tissues
	Gonadotropic hormone	Influences the development of sex organs and the production of hormones by the ovaries and testes.
	Adrenocorticotropic hormone (ACTH)	Stimulates hormone secretion by the adrenal cortex
	Prolactin	Stimulates secretion of milk by the mammary glands
	Thyroid stimulating hormone (TSH) or Thyrotropic hormone	Stimulates the activity of the thyroid gland

Continued

CHAPTER 4: THE ENDOCRINE SYSTEM

Gland	Hormone	Function
Hypothalamus (released by the posterior lobe of pituitary)	Oxytocin	Stimulates smooth muscles in the womb, the mammary glands, and blood vessels. Helps regulate child birth, milk production, and blood pressure.
	Antidiuretic hormone (ADH)	Controls water absorption in the kidneys and helps to regulate blood pressure by controlling the constriction of arterioles.
Pineal	Melatonin	Affects biological clocks and reproductive cycles; regulates the sleep-wake cycle
Thyroid	Thyroxine	Accelerates metabolic rate
	Calcitonin	Inhibits the release of calcium from bone
Parathyroids	Parathyroid hormone	Stimulates the release of calcium from bone; promotes calcium absorption by the digestive system
Adrenal cortex	Corticoids	Regulates metabolism, salt, and water balance; controls production of certain blood cells; influences structure of connective tissue
Adrenal medulla	Adrenaline	Causes constriction of blood vessels; increases pumping action of heart; stimulates liver to convert glycogen to glucose; stimulates nervous system
Pancreas	Glucagon	Stimulates glycogen conversion to glucose in liver and muscles
	Insulin	Controls sugar storage in liver and sugar breakdown in tissues
Thymus	Thymosin	Promotes the development of immune system cells; stimulates the immune system

Continued

Gland	Hormone	Function
Ovaries	Estrogens	Produce female secondary sex characteristics; begin growth of uterine (womb) lining; affect adult female reproductive functions
	Progesterone	Maintains growth of uterine (womb) lining
Testes	Testosterone	Produces male secondary sex characteristics, including muscle mass and reproductive system

Section Review — 4.3 C.

1. Review and study the chart listing the glands, the hormones, and their functions.

4.4 Special Concerns Regarding the Endocrine Systems of the Newborn and Elderly

A. The Newborn

While the endocrine system of the newborn is developed, its functions are often immature. For example, the posterior pituitary gland produces only limited quantities of ADH, which inhibits the passage of large amounts of urine. Because of the limited amount of ADH, the newborn does not reabsorb as much water as a child or adult; therefore, the infant will excrete more water in the urine. Consequently, infants are more susceptible to **dehydration**, a condition which occurs when the output of fluid is greater than the intake of fluid, resulting in too little water within the body.

B. The Elderly

The glands of the endocrine system, such as the thyroid, adrenal, pituitary, and hypothalamus, all slow down their output of hormones by varying degrees as a person ages. The loss of some hormone production can cause a reduction in the overall rate at which biological reactions occur in the body (metabolic rate). For this reason, it is often said that as a person ages, he or she slows down. Sometimes, however, a substantial loss of function in hormone-producing glands is not associated with the normal aging process. This can be an indication of a serious disorder, for example, diabetes mellitus or adult-onset diabetes. Diabetes mellitus is typically caused by obesity and lack of exercise. Adult-onset diabetes is also known as Type II diabetes. It is estimated that about 8% of people in the United States have diabetes; of these, it is estimated that 90% have Type II diabetes and 10% suffer from Type I diabetes.

> ### Section Review — 4.4 A & B.
>
> 1. Though the endocrine system of the newborn is adequately developed, its _____ are not fully working.
>
> 2. Infants are more susceptible to _____.
>
> 3. In the elderly, the glands tend to slow down their output of _____.
>
> 4. Diabetes mellitus is typically caused not by aging but by obesity and lack of _____.
>
> 5. Type 2 diabetes is also known as _____.

Chapter 4 Supplemental Questions

Answer the following questions.

1. Explain how receptors, enzymes, cyclic AMP, and prostaglandins are involved in the action of hormones.

2. State the location of the gland, the hormone(s) the gland secretes, and the action of the hormone(s) for the following glands: thyroid, parathyroid, thymus, adrenal cortex, adrenal medulla, Islets of Langerhans, posterior pituitary, anterior pituitary, and pineal.

3. Know which hormones are affected in the conditions of hypothyroidism, hyperthyroidism, cretinism, goiter, dwarfism, and gigantism.

4. Distinguish between the conditions of diabetes mellitus and diabetes insipidus in terms of which endocrine glands and hormones are involved, as well as the symptoms of each condition.

5. What are tropic hormones?

6. Explain the concept of negative feedback.

7. Give an example of the interaction between the nervous and endocrine systems.

8. State the role of the hypothalamus in linking the nervous and endocrine systems.

9. Know the functions of a releasing factor and an inhibiting factor.

10. Describe the differences in the endocrine systems of the newborn and the elderly.

11. Know the chart in section 4.3-C.

12. Study all diagrams in this chapter. Be able to draw and label all diagrams.

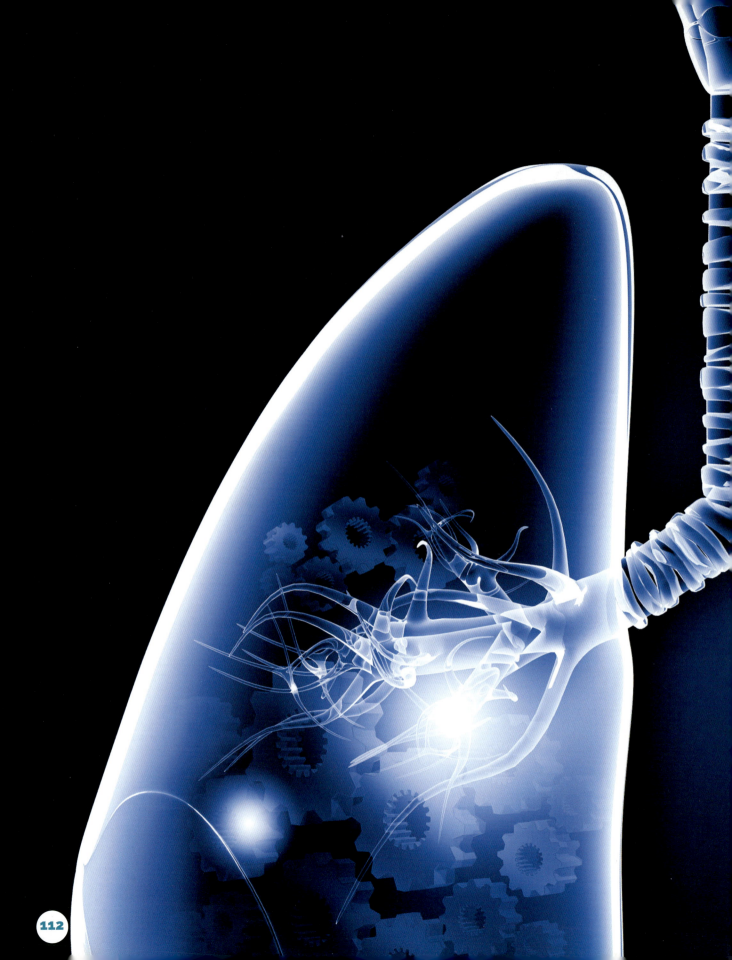

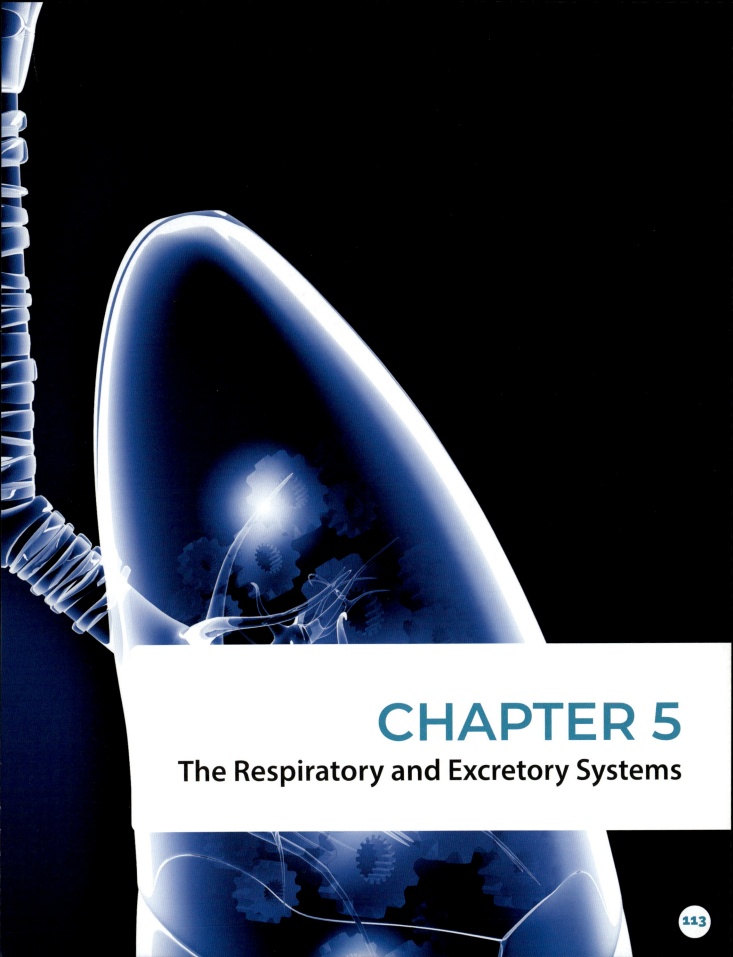

CHAPTER 5
The Respiratory and Excretory Systems

Chapter 5

The Respiratory and Excretory Systems

Chapter Outline

5.1 Introduction

5.2 Respiration
  A. Different Types of Respiration
  B. Structures of the Human Respiratory System
  C. Gas Exchange
  D. The Mechanism of Breathing
  E. Control of Breathing Rate

5.3 Cellular Respiration
  A. Anaerobic Step
  B. Krebs Cycle
  C. Electron Transport System
  D. The Finger of God

5.4 The Excretory System
  A. The Organs of Excretion
  B. The Structure of the Nephron
  C. The Function of the Nephron
  D. Homeostasis of Body Fluids

5.5 Special Concerns Regarding the Respiratory and Excretory Systems of the Newborn and Elderly
  A. The Newborn
  B. The Elderly

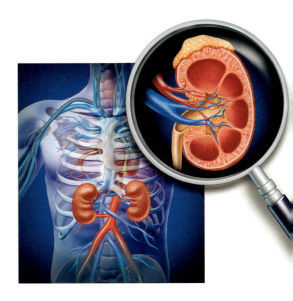

5.1 Introduction

> **Section Objectives**
>
> - Give the Catholic Church's perspective on prayer and its importance in life.

A proper concentration of oxygen in our bloodstream is one of the most basic needs of the human body. Every cell in our body uses oxygen. Just as our body uses oxygen to grow and mature, our soul uses prayer to grow and mature. Prayer is a channel of divine grace. When we lack oxygen, our senses become dull, the body lapses into a coma, and we eventually succumb to death. Without prayer, our spiritual life loses an important source of grace and may wither and die. Praying keeps our souls alive, because prayer is contact with God, and God is the life of the soul. Peter Kreeft says, in *Prayer for Beginners*, "Nothing but prayer can make saints, because we meet God in prayer." Prayer is the hospital for the soul where we meet God, our Doctor. God does not just ask us to pray; He commands us to pray, in fact, to "pray constantly" (1 Thess 5:17).

Praying constantly in our busy, hectic world can seem impossible, but, actually, anywhere there is oxygen to breathe, there is the ability to pray. Prayer can come in many forms and methods; it does not need to be vocal prayer. One can offer to Our Lord one's thoughts, actions, suffering, joys, or even one's sleep. If one is in a state of grace, everything can be a prayer if it is offered to God in a spirit of humble resignation to God's holy will.

As we study the respiratory system, let us come to an understanding that God loves us so much that He gives us every breath we take. The breaths we take seem to come automatically, without our even thinking about it. However, if for even one second God stopped thinking about us, stopped loving us, we could not take the next breath. Each life-giving breath in each of us is a gift from God. God gives us each second in our lifetime in order to give back to Him in some way. He wants us to give love to Him and to give love to others. God wants to give us a chance to show our love for Him by giving love, kindness, and respect to those whom God has placed in our own home. Take your breaths for life, for eternal life.

CHAPTER 5: THE RESPIRATORY AND EXCRETORY SYSTEMS

5.2 Respiration

Section Objectives

- Describe the processes of external respiration, internal respiration, and cellular respiration.
- Explain the relationship between cellular respiration and gas exchange in the lungs.
- Identify the major structures of the human respiratory system.
- Explain how oxygen as well as carbon dioxide are transported in the blood.
- Describe how the diaphragm and rib cage move air into and out of the lungs.
- Explain how the breathing rate is determined.

We cannot exist without **oxygen**. Our cells need oxygen to carry out all life processes of the body. With each breath, we take in oxygen that is vital for life. Oxygen travels through a precise series of steps before it is used by the cells. As the cells use oxygen during the process of respiration, they produce waste products which must be removed from the body. In this chapter, we will learn how the oxygen of the air is absorbed into the body's cells through the **respiratory system**, and how wastes are removed from the cells by the excretory system.

A. Different Types of Respiration

The term respiration, in its most general definition, refers to the act of breathing. However, respiration can be divided into three types of processes. **External respiration** is the term given to the exchange of gases between the atmosphere and the blood. This part of respiration takes place in the **lungs**. During external respiration, oxygen passes from the air into the lungs, then into the bloodstream; then carbon dioxide, the body's waste gas, passes from the bloodstream into the lungs and released into the air.

The term **internal respiration** refers to the gas exchange between the blood and the cells. Oxygen is transported by the bloodstream into cells of the body, and carbon dioxide diffuses out of the cells and into the bloodstream.

The third type of respiration refers to the chemical process that takes place within the mitochondria of the cells. This type of respiration is called **cellular respiration**. During cellular respiration, glucose is metabolized, or converted, by oxygen, and energy is released. The energy is then stored in the molecule known as **ATP** (adenosine triphosphate). **Carbon dioxide** (CO_2) and **water** (H_2O) are the waste products of cellular respiration that are removed by the bloodstream through internal respiration, which returns these waste products to the lungs. The lungs then release the waste CO_2 and H_2O during breathing, which is external respiration.

B. Structures of the Human Respiratory System

Locate the structures on the diagram of the respiratory system in Figure 5-1 as they are described.

How does air that contains life-sustaining oxygen enter our bodies? Certainly all of us have breathed through our mouths during our lifetime.

Oxygen O_2

CHAPTER 5: **THE RESPIRATORY AND EXCRETORY SYSTEMS**

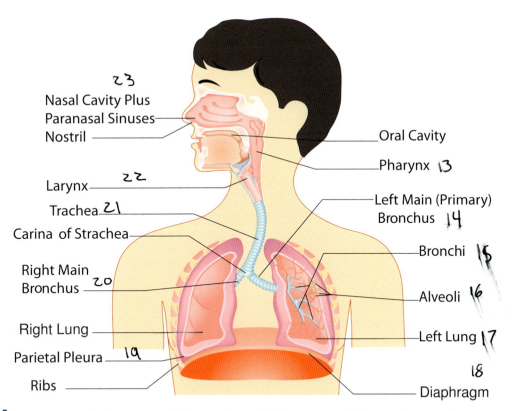

Figure 5-1. The human respiratory system. The human respiratory system can be subdivided into an upper respiratory tract and a lower respiratory tract based on anatomical features. The upper respiratory tract includes the nasal passages, the pharynx, and the larynx, while the lower respiratory tract is comprised of the trachea, the primary bronchia, and the lungs, all contained in the thoracic cavity (the area contained by the ribs).

For example, when we play a competitive sport, our muscles have an increased demand for oxygen and we may obtain a larger volume of this precious gas by mouth breathing. Or, you may have noticed that when you have a bad cold that is accompanied by nasal congestion, you breathe through your mouth. However, under normal circumstances and when we enjoy good health, our loving Creator God intended for us to obtain the oxygen we breathe in through our noses, and he spared no details.

The **nasal cavity** contains many bony projections, which create a large surface area for air to pass over on its way to the lungs. These bony projections are covered with soft **epithelial tissue** that contains many **capillaries**. The capillaries add moisture to dry air and warmth to cold air in order to prevent damage to the delicate lung tissues. Cells that secrete mucus are also present in the nasal cavity. Most dust, bacteria, and other airborne particles stick to the mucus and are removed from the air before it enters the lungs. Cilia, which are tiny hairs, move the mucus back toward the throat, where the mucus can be expelled by coughing or swallowed.

Both the nasal passages and mouth open into the **pharynx** (throat area). If the nasal passages are blocked, which happens when a person has a cold, the air can reach the lungs via the mouth. However, the mouth is not as effective in filtering and warming the air. In addition, breathing through the mouth increases the chances of airborne particles entering the lungs.

Biology For Life

CHAPTER 5: THE RESPIRATORY AND EXCRETORY SYSTEMS

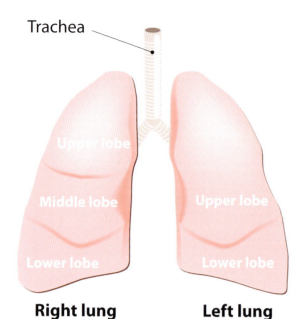

Figure 5-2. The trachea and lungs

The **trachea** [trā-kē-a], also called the windpipe, extends downward from the pharynx. When you swallow, the epiglottis prevents food and liquid from entering the trachea. The **larynx**, or voice box, is located just below the epiglottis in the upper part of the trachea. In the larynx, a pair of **vocal cords** stretches across the trachea. Speech and other sounds we make are made possible by vibrations that are conducted along the vocal cords.

From the trachea, air is carried to the right and left **bronchi** [bräŋ-kī] (**bronchus**-singular), the large air tubes which lead to the lungs. Bronchi branch into finer and smaller tubules within the lungs known as **bronchioles** [bräŋ-kē-ōl]. At the end of each bronchiole, there are clusters of balloon-like microscopic air sacs, called **alveoli** [al-vē-ō-lī].

The linings of the trachea, bronchi, and bronchioles contain cells that secrete mucus. Particles of dust or other foreign substances, which may have traveled past the nasal passages, will stick to this mucus. Large particles that are trapped in mucus are expelled from the lungs by coughing. Many adaptations of the body serve to filter and improve the quality of the air that finally reaches the alveoli.

The lungs are large sack-like organs where external respiration takes place. The lungs are located in the thoracic cavity, which is the area enclosed by the ribs, vertebral column, and breastbone. The thoracic cavity is separated from the abdominal cavity by the diaphragm. Also located within the thoracic cavity are the heart and major blood vessels.

Each lung is surrounded by a thin, soft double membrane known as the **pleura**. The outer pleura is attached to the chest wall, and the inner pleura is attached to the lungs. A lubricating fluid called the pleural fluid is contained in the space between the two thin membranes. The pleural fluid allows the lungs to move freely.

C. Gas Exchange

The walls of the alveoli, are very thin and moist, and are surrounded by tiny blood vessels known as capillaries. Gases easily pass across the alveolar membrane by the process of diffusion, moving from one area to another, following what is known as a concentration gradient: high concentration to low concentration. External respiration takes place across the alveoli.

Air is a mixture of gases, consisting mostly of nitrogen (about 78%) and oxygen (about 21%). During external respiration, carbon dioxide leaves the bloodstream and exits the lungs, and simultaneously, oxygen from the air travels to the lungs and enters the bloodstream. These gaseous exchanges, called diffusion, occur because carbon dioxide and oxygen tend to move from areas of high concentration to areas of low concentration. Before the blood enters the lungs, it has given up its oxygen to the cells of the body and has absorbed carbon dioxide, a waste product of the cells. At this point, the blood is **deoxygenated**, or low in oxygen. That is, the concentration of carbon dioxide is higher in the deoxygenated blood than in the air surrounding the alveoli. According to God's plan, carbon dioxide, diffuses out of the blood and into the air of the alveoli.

Since the concentration of oxygen is higher in the air surrounding the alveoli than in the blood, oxygen immediately diffuses from the air into the blood. The blood that returns to the heart after it leaves the lungs is low in carbon dioxide but rich in oxygen. Oxygen-rich blood is called **oxygenated**.

Internal respiration occurs when blood reaches the cells of the body. Cells continuously use oxygen and produce carbon dioxide during cellular respiration. The concentration of oxygen is always less in the cytoplasm of body cells than in the bloodstream; therefore, oxygen diffuses from the bloodstream into the cells. Conversely, carbon dioxide accumulates in the cytoplasm of the cells, and there is more carbon dioxide in the cytoplasm than there is in the bloodstream. Carbon dioxide moves out of the cells and into the bloodstream. Blood leaving the body's cells is rich in carbon dioxide but low in oxygen. This blood returns to the lungs to repeat the entire cycle.

Red blood cells can carry oxygen because oxygen molecules bind to a blood protein known as **hemoglobin** [hē-mō-glō-bin], which is contained in the red cells. Hemoglobin molecules are rich in iron. When oxygen is attached to the hemoglobin, the blood is a bright red color. When oxygen is given off to the body cells, the blood is a dull, purplish red. The dull color is the color of the hemoglobin without oxygen attached.

D. The Mechanism of Breathing

The body must constantly receive a fresh supply of oxygen, and air in the lungs must be constantly replaced. When you breathe, the air is continuously being renewed in the lungs. The term **inspiration**, or **inhaling**, is used to describe the process of air being taken into the lungs. During **expiration**, or **exhaling**, the air is expelled, or exhaled, from the lungs.

A large, dome-shaped muscle called the **diaphragm** [dī-a-fram] separates the chest cavity from the abdomen. In order to inhale, the diaphragm contracts. The diaphragm flattens out, and, at the same time, other muscles move the ribs up and out. Both of these muscle actions serve to increase the volume of the thoracic cavity (which is enclosed by the ribs). This increase in volume will reduce the pressure in the pleural fluid in the lungs, which results in air being pulled into the lungs. As air comes in, the walls of the alveoli stretch.

The opposite process occurs when you exhale. The diaphragm relaxes and returns to its dome shape, and the rib muscles relax. These combined movements serve to reduce the volume in the thoracic cavity, which increases the pressure on the pleural fluid. Air is pushed out of the lungs, and the walls of the alveoli return to their smaller size.

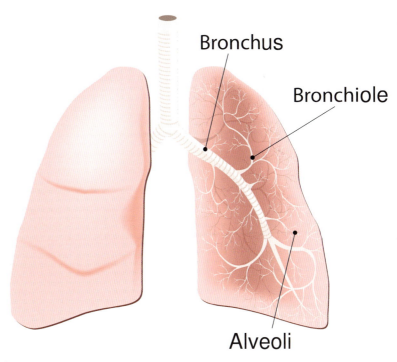

Figure 5-3. The structure of the bronchi

E. Control of Breathing Rate

The rib muscles and diaphragm are **striated muscles** (grooved skeletal muscles) and can work under voluntary control. That is, a person can choose to take deep or shallow breaths. One may also choose to stop breathing for a short time; however, automatic control eventually takes over, causing the body to gasp for air. God has made breathing an unconscious response, yet every breath we take is a gift of love and life from our Creator. The rib muscles and diaphragm are automatically stimulated by several nerves. These nerves bring impulses from a **breathing center** that is located deep in the brain.

Your breathing rate slows when you sleep and increases during exercise or other strenuous activities. Your breathing rate depends on your activity level and is controlled by a complex mechanism. This complex mechanism involves a group of nerve cells in the brain that is connected to the breathing center. These cells constantly monitor the carbon dioxide content in the bloodstream. When the level of carbon dioxide increases, nerve cells send messages to the breathing center, which, in turn, sends nerve impulses that increase the breathing rate. This leads to a decrease in the level of carbon dioxide.

Section Review — 5.2 A, B, C, D & E.

1. What are the three types of respiration?
2. The windpipe is also known as the _____.
3. What prevents food and liquid from entering the trachea when you swallow?
4. The thoracic cavity and the abdominal cavity are separated by what?
5. Oxygen molecules bind to a blood protein known as what?
6. When the diaphragm contracts, you _____. When the diaphragm expands, you _____.
7. Your breathing rate _____ when you sleep and _____ during exercise.

5.3 Cellular Respiration

Section Objectives

- Describe the processes of cellular respiration.
- List the steps in the process of cellular respiration.

Cellular respiration is the process in cells by which the chemical energy of "food" molecules is released and used to make ATP. Carbohydrates, fats, and proteins can all be used as fuels in cellular respiration; however, glucose is the primary fuel and is used below as an example to examine the reactions and pathways involved.

The combustion of glucose is represented by the chemical reaction:

$C_6H_{12}O_6$ (glucose) + $6O_2$ (oxygen) → $6CO_2$ (carbon dioxide) + $6H_2O$ (water) + **energy** (ATP)

As we discussed in section 2.7.C, each and every chemical reaction within living organisms is regulated by enzymes, or protein molecules that work as catalysts. Catalysts serve several purposes, including lowering the activation energy and speeding up the reaction. When an enzyme binds to a molecule, it is a specific fit, much like a lock and key. The enzyme easily breaks bonds to start the reaction, which proceeds at a very fast rate. As long as there is a compound for the enzyme to bind, the reactions keep moving forward. The

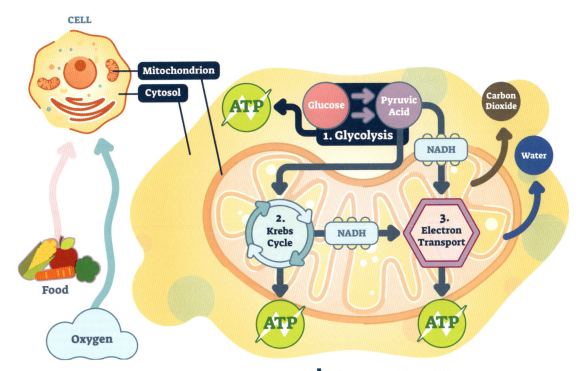

Figure 5-4. The cellular respiration process.

catalyst is never used up and is continuously recycyled for additional reactions.

Cellular respiration occurs in a three-stage process: the anaerobic stage (anaerobic means without oxygen), or **glycolysis**; the Krebs cycle; and the electron transport system. In the anaerobic stage, there is a net gain of two ATP molecules. The Krebs cycle produces two more ATPs from the ADP and phosphate in the mitochondrion. Finally, the electron transport phase results in an additional 32 ATPs within the mitochondrion, giving a net gain of 36 ATPs for the cell.

A. Glycolsis-Anaerobic Step

In order for a cell to "burn," or metabolize a glucose molecule to use for energy, the glucose must first be transported into the cytoplasm of a cell. It is in this location that enzymes assist in the reaction by causing the glucose molecule to lose some of its hydrogen atoms. Two molecules of a substance called **pyruvic acid** (pyruvate; $C_3H_4O_3$) are formed in a reaction that is called **glycolysis**:

$C_6H_{12}O_6$ (glucose) → $2C_3H_4O_3$ (pyruvate) + 4H + **energy** (ATP)

There is a net gain of 2 ATPs during this step. This process does not require any oxygen and is therefore called the anaerobic step in cellular respiration.

An easy way to recall that anaerobic means "no oxygen" is to note the presence of "n" in the term; the "n" stands for "no." Aerobic literally means "with air," but since oxygen is the active element in these reactions, aerobic really means "with oxygen".

B. Krebs Cycle-Aerobic Step

The two pyruvate molecules, created in the anaerobic step travel to a mitochondrion in the cell. These react with oxygen to make hydrogen and carbon dioxide in a multi-step reaction that is controlled by enzymes. The Krebs cycle is called the **aerobic step** because it requires oxygen.

$2C_3H_4O_3$ (pyruvate) + $3O_2$ → 8H + $6CO_2$ + **energy** (ATP)

ATP is known as the "universal currency" of biological energy for living organisms. That is, it is involved in nearly every cellular process that requires energy. Cells use this energy for important reactions and processes that keep them alive. During cellular respiration, ATP breaks down into ADP and water, and thus energy is released. ADP can be changed into ATP by the addition of water and energy. This whole process takes place in the mitochondria.

C. Electron Transport System

Within the electron transport system, the hydrogen that was produced in both the anaerobic step and the Krebs cycle reacts with oxygen to make water. This is accomplished through a series of enzyme-controlled reactions. It might be helpful to recognize that a teaspoon of glucose placed over an open flame yields the same amount of energy as a teaspoon of glucose that is broken down during cellular respiration. The open flame reaction releases all of its energy in the form of heat. Since cellular respiration is controlled by enzymes, some energy is lost as heat and some of the energy is used by the cell.

Enzymes cause chemical reactions in the cell to proceed more rapidly. They actually do this by lowering the activation energy by binding to a particular molecule, called the substrate, and breaking bonds, thus enabling a turnover of bound substrate into product. These reactions occur without wasting electrons, though some heat is produced as a byproduct. The reaction stops during feedback inhibition-too much product available, or you run out of substrate, or some other process directs it to stop. For example, if I have fasted over a 36 hour period and eat a tablespoon of honey, the honey (glucose, fructose) will be broken down by glycolysis/Krebs Cycle quickly until none of it is left. All of these electrons will be captured and some heat will be produced as a byproduct. Similarly, take a firefly. The enzyme luciferase catalyzes the protein luciferin in the presence of glucose. All electrons are conserved and the light is produced without heat – cold light is a mystery.

D. The Finger of God

Cellular respiration within eukaryotic organisms results in about 39% of the energy stored in the "food" molecule becoming captured by the cell. The remaining 61% is lost as heat energy. However, the heat energy is not completely lost, and the body regulates its use. That is, rather than being lost to the surrounding environment, the heat is used by our bodies to maintain a normal temperature of 98.6° F - the perfect temperature for all chemical reactions or "life processes." It is an amazing design which only an all-intelligent Creator God could produce.

> ### Section Review — 5.3 A, B, C & D
> 1. What are the three stages in cellular respiration?
> 2. What is the name of the substance that is formed in glycolysis? How many molecules are formed?
> 3. What is another name for the Krebs cycle?
> 4. What is the "universal currency" of biological energy?

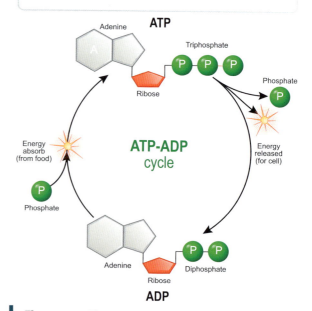

Figure 5-5. The ATP - ADP Cycle

5.4 The Excretory System

> **Section Objectives**
>
> - Identify the source of nitrogen waste products in the body.
> - Describe the major structures of the human excretory (urinary) system.
> - Explain how a nephron functions.
> - Label the parts on a diagram of the nephron and on a diagram of the respiratory system.
> - Explain the role of the kidneys in controlling water balance.

Cellular respiration produces **carbon dioxide** as one of its waste products. This waste gas must be removed from the cell through the circulatory system and then exhaled from the body. Metabolic processes produce other chemicals that are of no use to the cell and would be harmful if they built up. The process of excretion rids the cell and thus the organism of these other harmful molecules.

A. The Organs of Excretion

The **liver** is one of the most important organs in the human body, and it has several functions. It removes the buildup of harmful substances, or toxins, from the body. Amino acids (the building blocks of proteins) are also broken down, or metabolized, in the liver. When the liver **metabolizes** [mĕ-tăb-ō-līz] amino acids, it produces the molecule **urea**, which contains nitrogen. In high concentrations, urea can be poisonous to cells, so the urea must be eliminated from the system.

In addition, the liver removes the products that result from the breakdown (phagocytosis) of red blood cells, which occurs in the spleen, liver, and bone marrow. These products include iron and bilirubin. Iron is stored by the liver and will be used by the bone marrow for the production of new red blood cells. **Bilirubin** is excreted from the liver and is transported to the gall bladder in a substance known as **bile**. The **gall bladder** secretes bile into the digestive system, where it helps digest fats. Finally, the other digestive function of the liver includes its role in processing food, water, and other substances that enter the digestive system.

The kidneys filter urea and excess salts from the blood, and help maintain proper water balance in the body. The **kidneys** are the main excretory organs

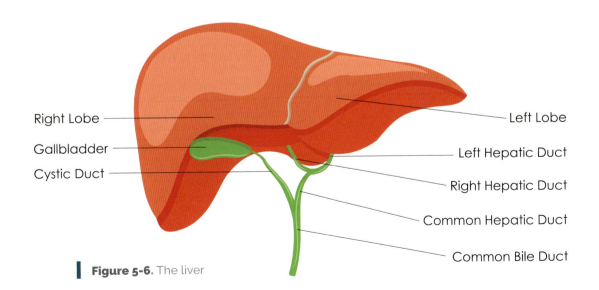

Figure 5-6. The liver

Biology For Life

of the body. About 180 liters of fluid (enough to fill a 50 gallon container) are filtered through the kidneys daily. Water that is not put back into the blood is excreted from the body in the form of urine, which is mostly made up of water.

The kidneys are a pair of bean-shaped organs located near the lower part of the vertebral column. The **renal arteries** carry blood into the kidneys, and purified blood leaves the kidneys through the **renal veins**.

Urine, which is produced by the kidneys, flows down through a tube called the **ureter** to a storage organ called the **urinary bladder**. It empties to the outside of the body through a tube called the **urethra**.

The internal structure of the kidney is shown in Figure 5-8.

The outermost layer of the kidney is called the **renal cortex**. The **renal medulla** lies below the **cortex**. The filtration of blood and the formation of urine take place in the cortex and medulla, and urine collects in a large cavity called the renal pelvis. The ureter drains urine from the renal pelvis to the bladder.

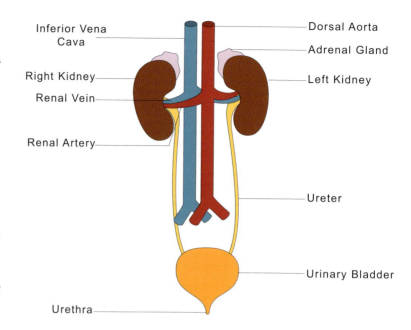

Figure 5-7. The excretory system

B. The Structure of the Nephron

In each kidney, the renal artery branches into many smaller vessels, each of which ends in a filtering unit called a **nephron**. Over one million microscopic units called nephrons make up the renal cortex and medulla of the kidney.

A nephron consists of a group of tiny blood vessels, as well as the **glomerulus** [glō-mě-rū-lus]. The glomerulus is located inside a cup-like structure known as **Bowman's capsule**.

Each Bowman's capsule is the beginning of a long, continuous tube, which has sections that differ in structure and function. The twisted **proximal tubule**, or **tube**, is located just behind the Bowman's capsule. This tubule leads to the **loop of Henle**. The loop of Henle dips into the medulla and returns back to the cortex. When the loop returns to the cortex, it becomes the **distal tubule**. The distal tubule then drains into the collecting duct, which goes through the medulla and opens into the **renal pelvis**.

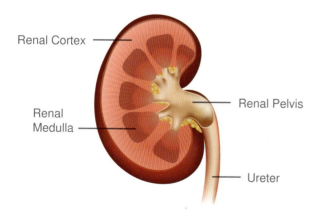

Figure 5-8. Internal structure of the kidney

C. The Function of the Nephron

Pressure forces much of the blood plasma through the glomerulus of a nephron and into its Bowman's capsule. The membranes of the glomerulus and Bowman's capsule allow small molecules and ions (charged particles) to pass through. Water, ions, sugar, amino acids, and urea move from the bloodstream into the Bowman's capsules of the nephrons. As noted above, over 180 liters (47-1/2 gallons) of water and 600 grams (1-1/3 lb) of sodium are filtered by the Bowman's capsules every 24 hours. Much of the water that enters the Bowman's capsules is reabsorbed and not excreted.

As fluid travels throughout the tubules of the nephron to the collecting duct, the fluid's composition changes. The blood recovers many substances from the fluid in the nephron by **active transport**, where membranes use energy to pump substances against the direction in which they would normally move by diffusion. In diffusion, particles move from a higher concentration to a lower concentration. During active transport the membranes can "pump" substances back to an area of greater concentration. Active transport requires energy but returns glucose, amino acids, and sodium to the blood. (This is similar to a process called reverse osmosis about which you may have heard, a process in which fresh water is extracted from salt water in the oceans.)

Cells in the proximal tubule, distal tubule, and collecting ducts pump sodium ions back into the capillaries. Water molecules also return to the capillaries. In this manner, the body recovers over 99% of the water and sodium that is filtered into the Bowman's capsules. The urea, however, remains in the tubules to be dumped into the renal pelvis area.

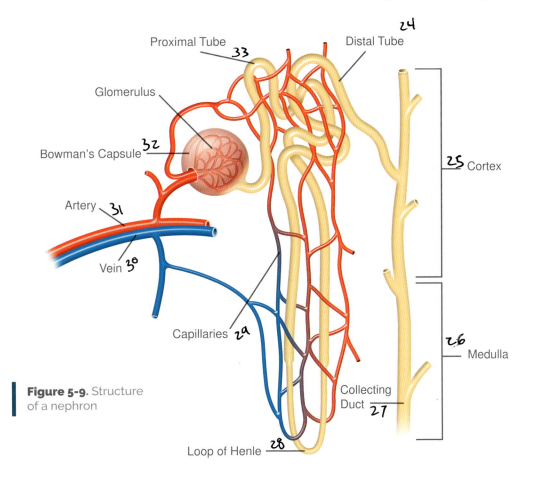

Figure 5-9. Structure of a nephron

After the re-absorption process has been completed, the fluid, which reaches the renal pelvis, is now called **urine**. It consists mainly of the water that has not been reabsorbed, some salts, and urea. The ureter drains the urine from the renal pelvis to the bladder. Healthy kidneys produce from 1 to 2 quarts of urine daily.

D. Homeostasis of Body Fluids

A watery tissue fluid that contains dissolved salts and other complex substances surrounds all of the body's cells. For the cells to function properly, the concentration of salts in this fluid must be kept constant. The term **homeostasis** of body fluids refers to the constancy of this concentration. If the body loses too much water or takes in too much water, the salt concentrations in tissue fluids can change. Salt in the diet, as well as the loss of salt through excretion, can affect the concentration of tissue fluid. Homeostatic mechanisms keep the concentration of tissue fluid fairly constant. The kidneys are an important part of this process.

There are four main ways in which the body loses water. Air entering the lungs is moistened, and about one liter (about 1 qt.) per day of this moisture is lost when you exhale. You can only imagine the amount of water vapor that daily enters the atmosphere from every person in the world releasing one quart of water each day! Another 100 milliliters (1/2 cup) of water are lost each day in the fecal matter, the waste product of the digestive system. About 50 milliliters (1/4 cup) of water evaporate as perspiration, and the kidneys excrete about one liter of water. The body can balance water losses so that the concentration of tissue fluid is kept fairly constant. The kidneys are an important part of the homeostatic mechanisms used by the body to do this.

When you perspire, the membranes of the kidney tubules become more permeable (allow more water to pass through), so most of the water passing into the kidneys is reabsorbed into the blood. Little water is left for urine formation, and the urine becomes a darker yellow than usual, because the dissolved materials have become more concentrated. If you drink an excess of water, the cells of the distal tubule and collecting duct become impermeable to water (that is, water cannot pass through). Water then stays in these tubules, and the urine is higher in volume and lighter in color.

Doctors recommend that you should follow the "8 x 8 rule": drink eight 8-ounce (about 1.9 liters) glasses of water a day to maintain the proper amount of water in your body. The rule could be stated, "drink eight 8-ounce glasses of fluid a day."

Water and fruit or vegetable juices are the best ways to acquire these fluids. Many people drink soda during the course of the day. Soda has very high quantities of sugar and should be avoided or only occasionally enjoyed. Many other people drink soda that contains artificial sweeteners. These are a poor substitute for water and can have the tendency to make a person feel thirsty again within a short period after drinking them. In addition, many artificial sweeteners can be dangerous to your health if ingested too frequently.

In summary, the kidneys are complex organs with several functions, chief of which is to regulate fluid homeostasis. That is, the kidneys: (1) retain sugar, (2) control the amount of sodium that is pumped out of the tubules, and (3) regulate the amount of salt, which is excreted into the urine. Together, both of these processes serve to regulate the salt balance of the tissue fluids throughout the body.

> **Section Review — 5.4 A, B, C & D**
>
> 1. What are the functions of the liver?
> 2. What are the main excretory organs of the body?
> 3. What do the kidneys look like? Where are they located?
> 4. The _____ carry blood into the kidneys, and purified blood leaves the kidneys through the _____.
> 5. What is a nephron?
> 6. The ureter drains the urine from the _____ to the _____.
> 7. The more water in your system the _____ in color urine will be; the less water in your system the _____ in color it will be.

5.5 Special Concerns Regarding the Respiratory and Excretory Systems of the Newborn and Elderly

> **Section Objectives**
>
> - Define the following terms: apnea, urinary, incontinence.

A. The Newborn

The most critical and immediate physiologic change required of the newborn is the onset of breathing. Fortunately, for most babies, chemical and thermal (relating to heat) stimuli are the primary stimuli that initiate respiration in the newborn. Chemical factors in the blood, such as the low concentration of oxygen and high concentration of carbon dioxide, initiate impulses that excite the breathing center in the medulla of the brain. The abrupt change in temperature as the infant leaves a warm environment and enters a relatively cooler atmosphere excites sensory impulses in the skin that are transmitted to this respiratory center. Descent through the birth canal and normal handling during delivery, such as drying the skin, also may have some effect on the initiation of respiration. Slapping the newborn's heel or buttocks, as seen in the movies, has no beneficial effect and wastes valuable time in the event of respiratory difficulty. Unfortunately, some babies, especially premature ones, have trouble initiating breathing. They can sometimes be jump-started by massage (which replaced the old tradition of smacking a baby's bottom to make it cry) or with a neonatal respirator.

The initial entry of air into the lungs is opposed by the presence of fluid that has filled the baby's lungs

CHAPTER 5: THE RESPIRATORY AND EXCRETORY SYSTEMS

and alveoli. However, this lung fluid is removed by the pulmonary capillaries and lymphatic vessels. During labor and delivery, as the baby's chest emerges from the birth canal, fluid is squeezed from the lungs through the newborn's nose and mouth. Air enters the upper airway to replace the lost fluid.

In a cesarean birth, where the baby is surgically removed from the mother's womb and does not travel through the birth canal, the chest is not compressed, and the newborn may need additional respiratory support. The normal respiratory rate of the newborn is 30 to 60 breaths per minute. Periods of crying will increase the respiratory rate. The respiratory rate will decrease during sleep. Periods of **apnea** (when breathing stops) lasting less than 15 seconds are normal.

All structural components are present in the newborn's renal (kidneys and bladder) system. However, due to immaturity, there is a deficiency in the kidney's ability to concentrate urine and to cope with conditions of fluid and electrolyte (salts and acids) fluctuations, such as dehydration (loss of too much water) or a high concentration of salts or sugars in the body.

The total volume of urinary output in 24 hours is about 200 to 300 ml (about one cup) by the end of the first week of life. However, the bladder of the newborn will automatically empty when stretched by a very small volume of 15 ml (about three teaspoons), resulting in as many as 20 urinations in one day. The first urination should occur within 24 hours.

B. The Elderly

The lung tissue loses some elasticity as we age. The lungs gain volume but lose some of their inner surface area. This loss of surface area results in a decreased ability to absorb oxygen from the air. As a result, the elderly person may become out of breath much more quickly than a younger person after a long walk or walking up a flight of stairs. The ribs and breastbone (sternum) may fuse in the elderly, which forces the diaphragm to work harder. However, the loss of lung function is not necessarily an inevitable part of aging. Research has shown that exercise can increase the ability of the lungs to do their job. The diaphragm does not seem to be adversely affected by normal aging.

Smoking, however, will speed up the deterioration of the lungs and produces serious, life-threatening illnesses, such as heart disease, lung disease, and emphysema.

In women, loss of tone in muscles that control urinary function can result in **urinary incontinence** (the inability to hold urine in the bladder). It is not an inevitable part of aging, but rather incontinence is associated with other disorders, such as **chronic neurologic disease**, an illness which involves the nervous system in both men and women. Lost muscle tone can almost always be regained with prescribed exercise. About 85% of women who are not confined to bed and are not neurologically impaired can be cured of urinary incontinence or dramatically improved. Both women and men with incontinence can be helped. Many communities have incontinence assessment and treatment programs that can help people whose physicians do not have adequate experience in this area.

> ### Section Review — 5.5 A & B
>
> 1. What is the normal respiratory rate of a newborn?
> 2. The first urination of a newborn should occur within _____.
> 3. What are some examples of life-threatening illnesses that can occur from smoking?
> 4. Loss of tone in muscles that control urinary function can result in _____.

Chapter 5 Supplemental Questions

Answer the following questions.

1. Describe the processes of external respiration, internal respiration, and cellular respiration.

2. Explain the relationship between cell respiration and gas exchange in the lungs.

3. Identify the major structures of the human respiratory system.

4. What is the function of mucus?

5. Describe how the diaphragm and rib cage move air into and out of the lungs.

6. Explain how breathing rate is determined.

7. Identify the source of nitrogen waste products in the body.

8. List and describe the major structures of the human excretory (urinary) system.

9. Explain how a nephron functions.

10. Label the parts on a diagram of the nephron and on a diagram of the respiratory system.

11. Explain the role of the kidneys in controlling water balance.

12. Define the following terms: apnea, and urinary incontinence.

13. Know the diagrams in section 5.1, 5.2, and 5.3. Be able to draw and label all diagrams.

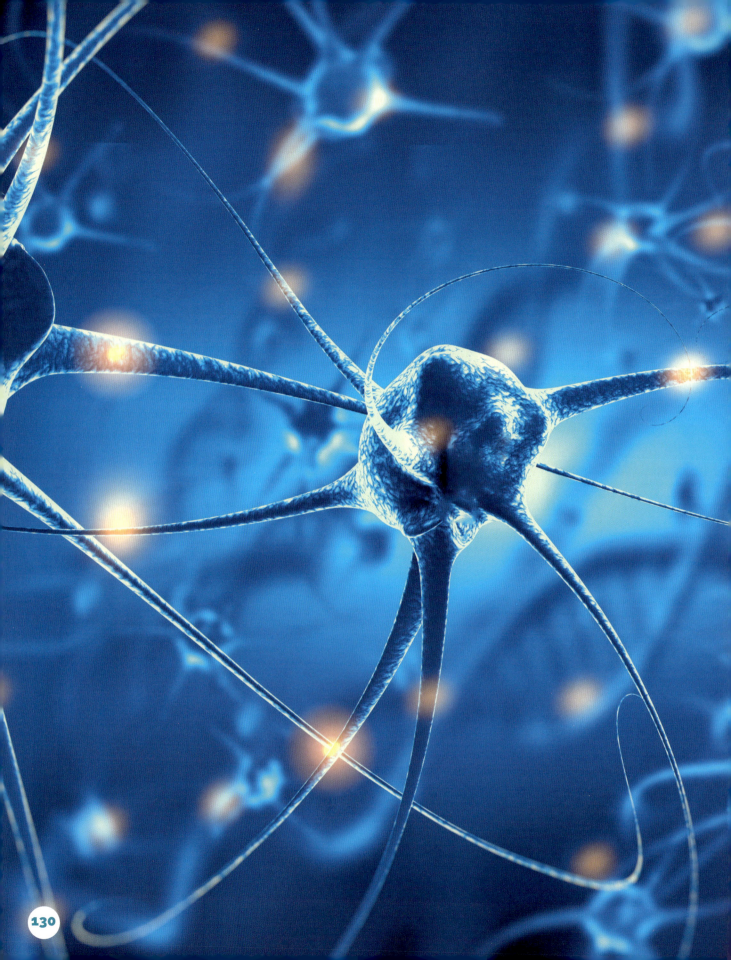

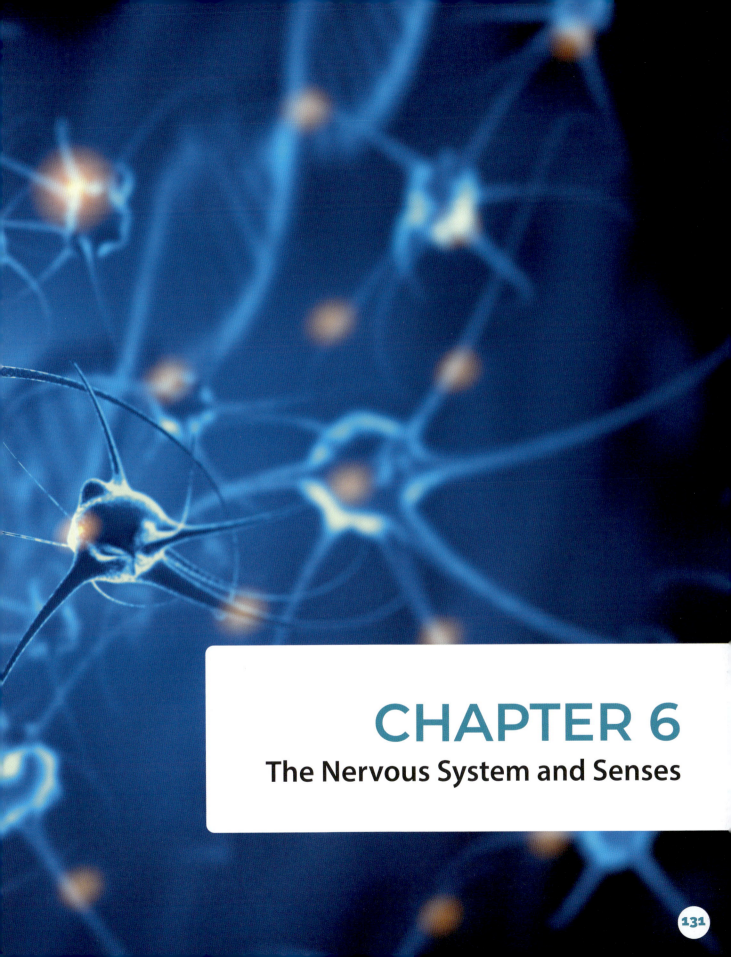

CHAPTER 6
The Nervous System and Senses

Chapter 6

The Nervous System and Senses

Chapter Outline

6.1 Introduction

6.2 Transmission of Information
  A. Neurons and Nerves
  B. The Nerve Impulse

6.3 The Central and Peripheral Nervous Systems
  A. Brief Overview of the Central Nervous System
  B. The Brain
  C. Spinal Cord
  D. Peripheral Nervous System

6.4 The Senses
  A. Touch
  B. Taste
  C. Smell
  D. Hearing and Balance
  E. Vision

6.5 Special Concerns Regarding the Nervous System and Senses of the Newborn and Elderly
  A. The Newborn
  B. The Elderly

6.1 Introduction

What happens when you touch something hot? You react immediately by pulling your hand away and saying something suddenly, such as, "Ouch!" The reason your body is able to respond as quickly as it does is because of the **nerves**, or **neurons**, in your skin. These neurons are part of your nervous system. Touch receptors in your skin transmit the pain signal to your nerves, which stimulate the muscles in your arm, so that your hand pulls away from the hot object. Other nerves then convey the pain sensation to your brain, where it is interpreted as signaling something that is too hot to touch. The whole process takes place so quickly that it seems to happen all at once.

The nervous system is responsible for coordinating all of the actions described above, as well as coordinating our ability to reason, problem solve, and have feelings and emotions. In this chapter, you will learn how the nervous system functions to control the actions of the body.

6.2 The Transmission of Information

Section Objectives

- Describe the structure and function of the neuron.
- Explain the transmission of a nerve impulse down an axon.
- Discuss the polarization of the nerve cell membrane.

The **nervous system** is composed of the brain and spinal cord, as well as all of the nerves that connect these two structures with the rest of the body. There are three major functions of the nervous system. First, the nervous system receives information about the environment and about the other parts of the body. Second, it interprets the information that it receives. Finally, the nervous system causes the body to respond to this processed information.

A. Neurons and Nerves

The nerve cell, or **neuron**, is the basic unit of structure and function in the nervous system. A neuron carries information from one location to another. There are three different types of neurons. A **sensory neuron** picks up information from the environment. A **motor neuron** carries information to muscles or glands, which causes them to act. The **interneurons**, or associative neurons, carry

CHAPTER 6: THE NERVOUS SYSTEM AND SENSES

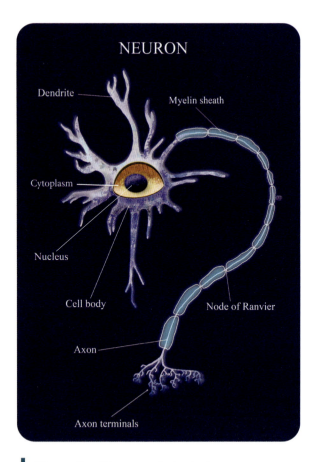

Figure 6-1. Diagram of a typical neuron.

information between the sensory and motor neurons.

Look at Figure 6-1, which is an illustration of a typical neuron. The many-branched dendrites carry nerve impulses *toward* a neuron cell; a single axon carries impulses *away* from a neuron cell. Axons and dendrites in the central nervous system are typically only about one micrometer thick (it takes 25,400 micrometers to make an inch; human hair is between 20-180 micrometers wide), while some in the peripheral nervous system are much thicker. The longest axon of a human motor neuron can be over a meter long, reaching from the base of the spine to the toes. Sensory neurons have axons that run the length from the toes to the dorsal column, or **spinal cord**; these axons measure over 1.5 meters in adults. Giraffes have single axons several meters in length running along the entire length of their necks!

In Figure 6-1, notice that the cell body of the neuron contains the nucleus and most of the cytoplasm, the fluid-like material outside of the nucleus. Many thread or root-like extensions, called **dendrites**, extend outward from the cell body. A long, thin fiber, known as an **axon**, also extends out from the cell body. Axons are sometimes called nerve fibers and are arranged in bundles called **nerves**. Axons are often coated with a white, fatty material known as **myelin**. It is a protective covering, much like rubber insulation that covers electric wires and has the same function.

The neurons respond to a stimulus by working together as a group. The dendrites receive stimulation from the environment outside the body or from the body. If this stimulation is strong enough, a nerve impulse, or message, is generated in the axon. The impulse then travels, or is propagated, along the axon. When the impulse signal reaches the end of the axon, it passes to a muscle, to a gland, or to the dendrites of another neuron. The axon, however, does not physically touch any of these structures. The signal passes across a tiny gap called the **synapse**. This synapse is only about 20 nanometers (0.2% of the width of a cell) in length; nevertheless, the signal must jump across this gap in order to be transmitted to the next neuron, gland, or muscle.

B. The Nerve Impulse

How does a nerve impulse start? What makes a nerve impulse travel along an axon? Although electric current is involved in generating and conducting nerve impulses, a nerve impulse is more than a simple electric current. It can be likened to a set of switches or gates that open and close due to concentrations of ions. You may recall that ions are atoms that have either gained or lost electrons, resulting in an imbalance of electrical charges. Ions with more positive than negative charges will have a net, or overall, positive charge. Conversely, ions with more negative than positive charges will have a net negative charge. For example, if you dissolve ordinary table salt in water, each salt molecule disassociates into positive sodium ions and negative chlorine ions. The

Biology For Life

sodium atoms lose one electron and the chlorine atoms gain one electron.

Tissue fluid that is found outside the cells of the body contains many dissolved ions. The cell membrane of the neuron is selectively permeable, which means that only certain materials are allowed to pass through the membrane. Due to the selective permeability of the cell membrane, the concentration of specific ions inside and outside of the cell membrane is different. During the neuron's resting state (when it is not being stimulated), there are slightly more positively charged sodium ions on the outside of the membrane. Inside the membrane, there are more positively charged potassium ions and negatively charged molecules than on the outside of the cell. As a result, there is a slight difference in electrical charge between the inside and the outside of the cell membrane. The outside of the cell is positively charged and the inside of the cell is negatively charged. It is during this state that the cell membrane is called **polarized**.

A neuron receives input from the senses or other neurons. If the stimulation is strong enough, it changes the permeability of the cell membranes. Sodium ions rush into the cell, and potassium ions move out. The result of this ion movement is a change in the polarization of the membrane. It has become reversed, or **depolarized**; that is, more positive charges have accumulated inside and more negative charges have accumulated outside. Restated, there is a change in the permeability of the cell membrane and a reversal in electrical charges as the impulse moves along the axon. This marks the beginning of the nerve impulse, which travels very quickly, more than several meters per second.

After the impulse has passed, the cell quickly begins to reestablish its original distribution of ions. Higher concentrations of positive charges are built up on the outside of the membrane, which is followed by higher concentrations of negative charges on the inside of the cell membrane. The nerve cell is now back in its resting state.

Neurons have an all-or-none response, which means that either, if the stimulus is strong enough, the impulse will travel the entire length of an axon, or, if the stimulus it is too weak, no impulse is generated. In addition, each neuron always conducts impulses at the same rate, neither faster nor slower at any time. After an impulse reaches the end of an axon, it must cross the synapse to reach another neuron, gland, or muscle. The electrical current does not move across the synapse. The gap is bridged by the releasing of chemicals called **neurotransmitters**. There are more than a dozen different neurotransmitters, which diffuse, or flow, across synapses and attach to receptor sites on neurons, glands, or muscle cells.

This is how it works. When a nerve is stimulated sufficiently to exceed the threshold of the all-or-nothing response, an impulse is generated that travels the length of the nerve cell to the axon at the end of the cell. When the impulse reaches the end of the axon, it causes the release of a neurotransmitter. The neurotransmitter crosses

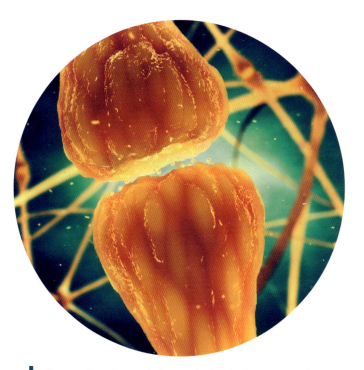

Figure 6-2. Synaptic transmission is the process by which neurotransmitters are released by a neuron and activate the receptors of another neuron or cell.

the synapse, binds to receptors that are located on the dendrites of another nerve cell, which stimulates it and allows further propagation of the nerve impulse.

When the target is a muscle cell, the neurotransmitter that is released is called **acetylcholine (ACh)**. The ACh molecules diffuse across the synapse and bind to receptors on the membrane of the muscle cell, causing a contraction in the muscle.

God designed the body to permit the control of this system. It would not be good for the muscle to remain in the contracted state forever. Consequently, there is an enzyme found in the synapse called **acetylcholinesterase (AChE)**. This AChE enzyme inactivates the acetylcholine by degrading it into its precursors acetic acid and choline molecules. AChE works extremely quickly and can process 5,000 acetylcholine molecules per second. Once the ACh molecules have been broken down, the muscle cell can relax. The acetic acid and choline molecules are then transported back to the axon nerve endings of the nerve from which they came, where they are turned into new acetylcholine molecules to be used by the next impulse. The muscle cell is now ready to contract again, if it receives sufficient additional stimulation.

> **Section Review — 6.2 A & B**
>
> 1. What are the three structures that an axon can synapse with?
> 2. Describe the process by which a nerve impulse starts and continues to travel along the neuron.
> 3. A chemical which is released from an axon and causes an impulse in another neuron is a _____.
> 4. Name and be able to identify the six major features of the neuron as identified in Figure 6-1.
> 5. How do neurons cause muscles to contract?

6.3 The Central and Peripheral Nervous Systems

> **Section Objectives**
>
> - Name the parts of the central nervous system.
> - Describe the location, structure, and function of the following: cerebrum, cerebellum, thalamus, hypothalamus, medulla oblongata, and spinal cord.
> - Explain the role of the spinal cord in controlling simple reflex behaviors.
> - Differentiate between the somatic and autonomic parts of the peripheral nervous system.
> - Contrast the actions of the sympathetic and parasympathetic nervous systems.

The nervous system is made up of neurons and is divided into two main parts: the **central nervous system** and the **peripheral nervous system**. The central nervous system consists of the brain and spinal cord, and the peripheral nervous system contains all of the nerves that lie outside the central nervous system.

A. Brief Overview of the Central Nervous System

The brain and spinal cord are covered by three protective membranes, which are called **meninges** [mĕ-nĭn-jēz]. The outermost layer of the meninges is called the **dura mater**. The middle layer is called the **arachnoid mater**, and the innermost protective layer is called the **pia mater**. Between the arachnoid mater and the pia mater is an area known as the subarachnoid space,

CHAPTER 6: THE NERVOUS SYSTEM AND SENSES

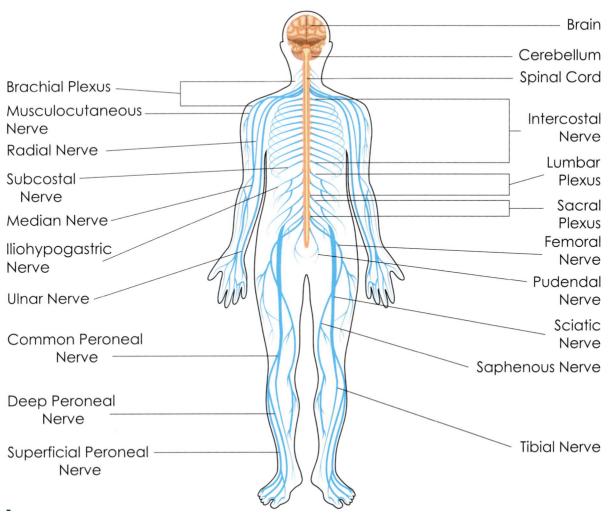

Figure 6-3. The body's nervous system. It is a network of specialized cells that communicate information about a person's surroundings and about himself. It processes this information and causes reactions in other parts of the body.

which contains a watery fluid called **cerebrospinal fluid (CSF)**. This fluid is produced in cavities within the brain, and from there the fluid drains into the brain's blood vessels. The purpose of the cerebrospinal fluid is to cushion the brain and spinal cord against shock and to remove waste materials from the brain.

The nervous tissue of the brain and spinal cord lies beneath the meninges. The outer portion of the brain tissue is gray, and the layer beneath the gray matter is white. The spinal cord, in contrast, has an outer layer that is white and a gray inner layer.

B. The Brain

The human brain is very large. It is composed of three major structures: the cerebrum [sĕr-ē-brum], the cerebellum [sĕr-ē-bĕl-um], and the brain stem. The brain contains about 100 billion neurons connected with one another in complex networks. All physical and mental functions depend on the establishment and maintenance of these neuron networks.

Approximately 75 billion neurons are located in the cerebrum. The **cerebrum** is the large upper region of the brain. The surface of the cerebrum is known as the **cerebral cortex**, and it is folded into ridges and

Biology For Life

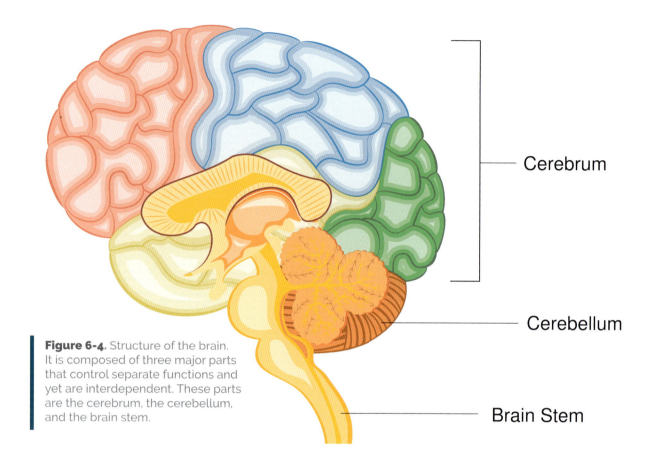

Figure 6-4. Structure of the brain. It is composed of three major parts that control separate functions and yet are interdependent. These parts are the cerebrum, the cerebellum, and the brain stem.

depressions called **convolutions**. These convolutions greatly increase the surface area of the cerebral cortex. The cerebral cortex controls many functions that are associated with intelligence, including memory, creativity, and reasoning.

The cerebrum is divided into right and left halves called **cerebral hemispheres**. A bundle of axons known as the **corpus callosum** connects the right and left hemispheres, allowing them to communicate with each other. Deep folds in the cerebral cortex divide each hemisphere into four lobes, which are known as the frontal, temporal, parietal, and occipital lobes.

The cerebrum receives information about the environment from the sensory neurons, and then it sends information to the muscles and glands by means of the motor neurons. Messages that begin on the left hemisphere of the cerebrum cross over to neurons that control movement on the right side of the body. Similarly, messages that start on the right hemisphere control movement on the left side of the body. The cerebrum is a "two-way street." Not only does it send messages, but it receives information too. The left hemisphere of the cerebrum processes information received from the right side of the body. Conversely, the right hemisphere of the cerebrum processes information received from the left side of the body.

For each and every one of us, one hemisphere of the brain is dominant over the other. The left hemisphere is dominant in most people, which explains why most people write with their right hand. Also, in most people, an area in the left hemisphere controls speech. The left side of the brain is thought to have been designed for analytic thought, especially mathematics and logic. The right hemisphere of the brain is believed to have been created for art and music.

The **cerebellum** lies beneath the back of the cerebrum. Like the cerebrum, the cerebellum has a convoluted surface of ridges and folds. It also has right and left hemispheres, which are joined by a finger-shaped structure called the **vermis**. The left cerebral hemisphere connects to the right half of the cerebellum, and the nerve pathways extend to the right side of the body. Likewise, the right cerebral hemisphere is connected to the left half of the cerebellum, which extends to the left side of the body.

When you decide to go for a brisk walk, your legs are controlled by signals from the motor neurons. The cerebrum passes the signals to the cerebellum, which in turn passes it on to the muscles. The cerebellum is what allows us to walk with coordinated steps. Only unrefined, jerky movements would be possible if were not for the cerebellum. The cerebellum also helps to maintain balance within the body.

The **brain stem** contains all of the nerves that connect the spinal cord with the cerebrum. The **medulla oblongata** [ŏb-lŏng-gā-ta] is the enlarged portion of the brain stem that extends down from the center of the brain and connects to the spinal cord. This part of the brain controls basic life functions, such as the rate of breathing and the heartbeat. Even if a person's cerebrum and cerebellum are severely damaged in an accident, the medulla oblongata may enable life to continue.

At the upper end of the brain stem are the **thalamus** and the **hypothalamus**. There are actually two thalami, one on each side of the brain stem. Each thalamus directs nerve impulses through synapses with other neurons and to the parts of the cortex where they will be interpreted. Also, the thalami receive and screen stimuli and prevent the brain from being overwhelmed with information. Consider the thalamus as the central processing unit of a computer, which allows it to operate the system with countless processes going on in the background.

The hypothalamus lies beneath the thalamus. The hypothalamus controls sensations involved in maintaining the internal environment of the body and gives rise to sensations such as hunger and thirst. In addition, body temperature, water balance, and blood pressure are controlled by the hypothalamus. Recall from Chapter 5 that the hypothalamus is the master switch for the endocrine system as it regulates the release of many hormones.

> **Section Review — 6.3 A & B**
>
> 1. Where is the cerebrospinal fluid located?
> 2. Name and be able to identify the three major features of the brain as identified in Figure 6-4.
> 3. The cerebral cortex is folded into ridges and depressions called _____.
> 4. The outermost protective layer of the meninges is called the _____.
> 5. What are the functions of the hypothalamus?

C. The Spinal Cord

The **spinal cord** extends downward from the medulla oblongata. The spinal cord passes sensory and motor information between the brain and other parts of the body. All sensory and motor nerves located below the neck must pass through the spinal cord on the way to and from the brain. See Figure 6-5 for a cross section of the spinal cord.

The spinal cord runs down the back through holes in the vertebrae known as foramen. The vertebral column or backbone protects the spinal cord from injury. If the spinal cord becomes crushed or severed, the person will lose the ability to move the parts of the body below the area of injury. The injured person will also have no sensation in those parts of the body.

The spinal cord also controls most reflex behavior. A **reflex** is an **automatic response** to a

Biology For Life

CHAPTER 6: THE NERVOUS SYSTEM AND SENSES

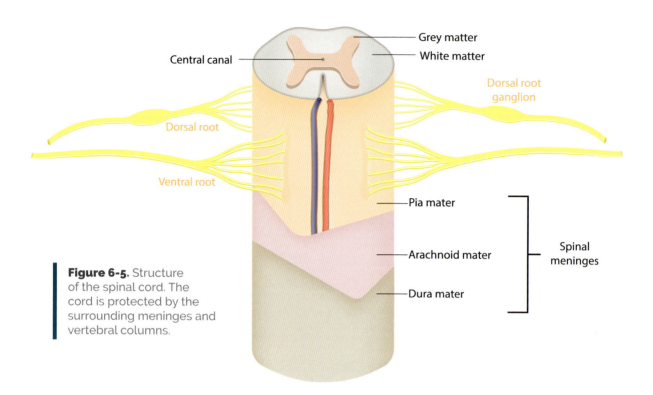

Figure 6-5. Structure of the spinal cord. The cord is protected by the surrounding meninges and vertebral columns.

stimulus, without the person consciously thinking about the response. The brain does not control many simple reflex responses.

As an example of a automatic reflex response to a stimulus, consider what happens when a sharp object touches your skin. Suppose you are walking barefoot along a beach, and you step on a broken shell. Your foot will quickly pull up and off of the shell. The sharp edge of the shell will activate dendrites in a sensory neuron in the foot. This sensory neuron synapses with an interneuron, and the interneuron synapses with a motor neuron that sends an impulse to a leg muscle. In short, when the sensory neuron is activated by the sharp edge of the shell, an impulse is sent to your leg muscle. The muscle contracts, and you pull your leg away from the sharp object on the sand. The impulse from the sensory neuron continues to travel upward through the spinal cord to the cerebral cortex in the brain. However, this impulse does not reach the brain until after you have already pulled your foot away. The brain does not interpret the sensation of pain until after you have moved your foot from the object.

Amazingly, this entire process is accomplished in less than a second. It protects us from danger even before we have become aware of it!

D. Peripheral Nervous System

The **peripheral nervous system** includes everything in the nervous system except the brain and the spinal cord. It includes the cranial and spinal nerves. The **cranial nerves** consist of 12 pairs of nerves that emerge from the brain. These nerves primarily connect with areas in the head and face. Muscles that activate the eyes and tongue are controlled by cranial nerves. Cranial nerves also carry sensory information to the brain.

There are 31 pairs of **spinal nerves** in humans. Each spinal nerve divides into two branches before entering the spinal cord. The sensory branch enters the gray matter of the spinal cord toward the back portion of the cord. The motor branch enters the gray matter at the front side (the abdominal side of the body) of the spinal cord. The cell bodies of the sensory neurons are located just outside the spinal cord in swellings called **ganglia** [găng-glē-ă]. The

Biology For Life

cell bodies of the motor neurons are contained within the spinal cord itself and not in the ganglia.

The motor neurons of the peripheral nervous system are classified into two groups: the somatic nervous system and the autonomic nervous system. The **somatic nervous system** consists of motor neurons that connect the central nervous system to the striated, or voluntary, muscles. The autonomic nervous system consists of neurons that attach to glands, smooth muscle, and cardiac muscle. These structures are not under our conscious control and are involuntary.

The autonomic nervous system is divided into two parts: the sympathetic nervous system and parasympathetic nervous system. The **sympathetic nervous system** dominates in times of great stress. It is also known as the "fight or flight" system. For example, suppose you are running in a 100-meter race. Nerve impulses will increase your both your heart rate and blood pressure . The level of sugar in your blood will rise, so that extra energy is readily available. These combined nervous system responses enable you to move with speed.

The actions of the **parasympathetic system** are opposite to those of the sympathetic nervous system. After an emergency, the parasympathetic system returns the body to its normal state. It is also known as the "rest and digest" system. In the above example, the parasympathetic system would bring heart rate and blood pressure back to resting rates. The parasympathetic system dominates under normal, non-stressful situations. The **vagus** nerve, one of the cranial nerves, is the main nerve that is active in the parasympathetic nervous system.

Section Review — 6.3 C & D

1. An automatic uncontrolled response to a stimulus is called a _____.
2. The peripheral nervous system consists of 12 pairs of _____ nerves and 31 pairs of _____ nerves.
3. The _____ is dominant during times of great stress.
4. After an emergency, the _____ returns the body to its normal state.

6.4 The Senses

Section Objectives

- Explain how skin, taste, and smell receptors inform the brain about the environment.
- Describe the structure and function of the parts of the ear and label the parts of the ear on a diagram.
- Explain the structure and function of the parts of the eye and label the parts of the eye on a diagram.

In order for the nervous system to respond to events in the environment, it must obtain information about its surroundings. This information is provided by the five senses: vision, hearing, smell, taste, and touch. Each sense organ contains specific sensory receptors located on the cell membranes that are located at the ends of nerve fibers. These are responsible for the conversion of stimuli into electrical impulses. For example, nerve cell receptors in the eye convert light and color into electrical impulses that the brain can interpret.

The type of receptor in a given sensory neurons will be sensitive to a specific type of stimulus. For example, neurons containing mechanoreceptors in the skin are sensitive to tactile stimuli, while olfactory receptors in the nose are sensitive to odors. The electrical impulses generated in the sense neuron are sent to the spinal cord and thus to the brain. The brain then interprets the impulses it receives from these neurons as associated with sight, hunger, or other sensations.

A. Touch

The skin has receptors that are sensitive to touch, pain, pressure, heat, and cold. This makes the skin the largest sense organ of your body! Each of these sensations of touch, pain, pressure, heat, and cold has its own unique type of receptor and, consequently, a unique sense neuron. For example, touch receptors will inform the brain of even the lightest touch, since many of the touch sense neurons are located at the base of hairs. These hairs generate impulses when the hairs move even slightly.

Pressure receptors are buried deeper in the skin than the touch receptors. Firm pressure is needed to activate these receptors. Unlike the other sense of touch, the sense of pain has no specialized receptors. Pain receptors are free ends of nerve fibers that are not covered with myelin. Finally, there are specific receptors for detecting heat and cold.

Most of the time we do not recognize, or are not conscious of, many of the sensations around us. For example, we all have had the experience of walking along and becoming aware of a pebble that found its way inside one of our shoes. The pressure receptors in our feet made us aware of this sensation. Before this happens, we are usually not paying attention to how the shoes or socks feel around our feet. The sense receptors in our feet process this information too, but we are not alerted to it as we are to the pebble. After we remove the pebble, the pressure receptors no longer receive the

Figure 6-6. The sense of touch is found all over the human body. This is because your sense of touch originates in the bottom layer of your skin, called the dermis.

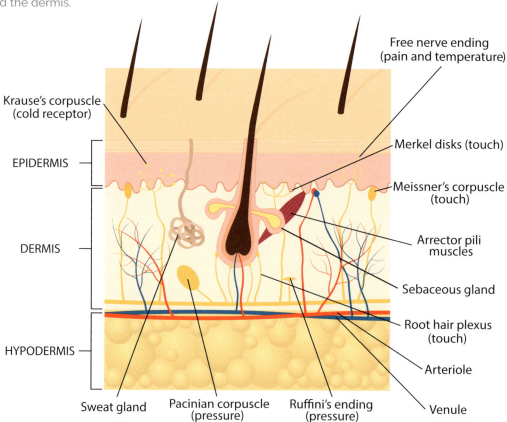

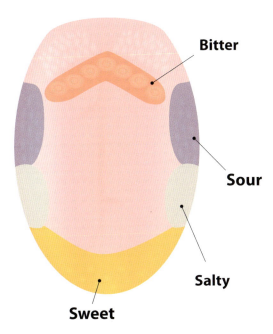

Figure 6-7. The taste buds can recognize four basic kinds of tastes: sweet salty, sour, and bitter. The salty/sweet taste buds are located near the front of the tongue; the sour taste buds line the sides of the tongue; and the bitter taste buds are located at the very back of the tongue.

stimulus, and we no longer feel the pressure. The pressure receptors return to a quiet state, ready for the next stimulus. As you sit and read this book, you are not thinking about how your clothing feels on your arms and legs or how cold or hot it is inside the room. Yet the sense receptors in your skin still process this information.

B. Taste

Taste is a sense which is directly stimulated by chemicals. The taste receptors are located in the **taste buds** and are found in the front, back, and sides of the tongue. Each taste bud consists of about 40 receptors, supporting cells, and an opening called the taste pore. The taste buds lie in bunches called **papillae**, which are visible as the bumps on your tongue.

Although most of a person's 10,000 taste buds are on the tongue, some taste buds are also located on the tissue that lines the throat. Food molecules activate the receptors in the taste buds, which, in turn, send nerve impulses to the brain. The brain interprets these impulses as taste. No one knows precisely how taste receptors function. Some researchers think that the receptor sites accept only specific chemical molecules that are present in the saliva. During the process of eating, these molecules constantly bathe the taste buds and eventually reach the receptor cells through the taste pores. Other researchers think that the shape of the molecule determines its taste.

There are many types of taste receptors, and each type senses a different taste. Most foods have a blend of several different tastes; therefore, many foods will stimulate more than one type of taste receptor. However, if you have a cold and your nose is blocked, your sense of taste does not function well, because much of what is tasted also depends on the sense of smell.

> **Section Review — 6.4 A & B**
>
> 1. What are the five types of receptors which are located in the skin?
> 2. Where are the taste receptors located?
> 3. Taste buds lie in bunches called _____.
> 4. What are the four major types of taste receptors?

C. Smell

The sense of olfaction or smell is another chemical sense. The olfactory, or smell sense, receptors are embedded in the lining of the nose and respond to molecules in the gaseous state. About 50 million of these specialized sense neurons are located in the human nose. When you smell a substance, most of the molecules must be in the form of a vapor. These dissolve in the mucus that covers the **olfactory receptors**. When an olfactory receptor is stimulated

Biology For Life

CHAPTER 6: THE NERVOUS SYSTEM AND SENSES

it produces impulses that travel through the olfactory nerve. The olfactory nerve is one of the 12 pairs of sensory cranial nerves.

Scientists know that information is sent from the olfactory nerve to the brain where it is interpreted as smell. However, how olfaction occurs, that is, how the olfactory receptors work is a mystery. There are many theories that attempt to explain how it happens, and not just in humans but other organisms as well.

Some scientists think that the perception of smell occurs when a specialized molecule on the receptor surface reacts with a specific chemical in inhaled air. The reaction generates an impulse that results in a particular smell. Other scientists believe that the outline, or shape, of a molecule is the cause of its particular odor. They think that a molecule of a specific shape fits into an olfactory receptor, which will accept only that specific shape, just as a lock

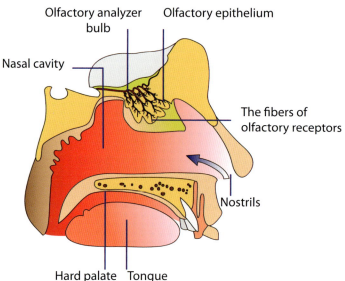

Figure 6-8. The olfactory receptors are located in the human nasal passage.

works with one specific key. Current researchers believe that there are about 400 kinds of scent receptors and that the human nose can distinguish 1 trillion different odors! We keep learning more about our senses every year and this is another example of our Creator's magnificent attention to detail.

D. Hearing and Balance

The ear is a complex organ that is responsible for both senses of hearing and balance. The ear is divided into three main areas: the outer ear, the middle ear, and the inner ear. The outer ear consists of the ear flap, or auricle, and the **auditory canal**. The auditory canal ends at the **tympanic membrane**, or eardrum. The middle ear begins on the other side of the tympanic membrane. The Eustachian tube connects the middle ear to the pharynx. The Eustachian tube contains air and functions to help equalize the pressure between the outer ear and the middle ear.

There are three tiny bones in the middle ear known as the **malleus** (hammer), **incus** (anvil), and **stapes** (stirrup). These bones connect the

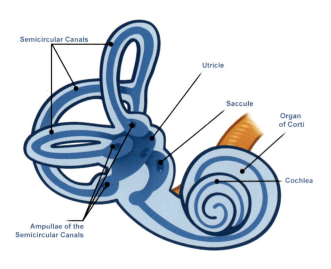

Figure 6-9. Structure of the cochlea. Sound waves must pass through the outer, middle, and inner ears for hearing to take place. The cochlea is located in the inner ear and contains the receptors for sound.

Biology For Life

CHAPTER 6: **THE NERVOUS SYSTEM AND SENSES**

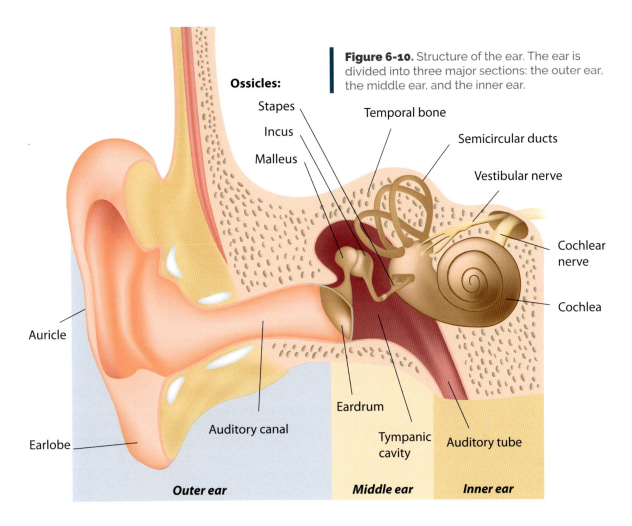

Figure 6-10. Structure of the ear. The ear is divided into three major sections: the outer ear, the middle ear, and the inner ear.

tympanic membrane with the **oval window**, which is a membrane-covered opening between the middle and inner ear.

The **cochlea** [co-klee-uh] is a fluid-filled, coiled structure in the inner ear that resembles a snail's shell. Receptor cells in the cochlea contain nerve endings. The **semicircular canals**, which are also in the inner ear, are involved in the senses of balance and motion.

Sound travels through the air as waves of vibrating air molecules. These vibrations are captured by the auricle and travel through the auditory canal. The vibrating air causes vibrations on the tympanic membrane, which, in turn, passes the vibrations to the three small bones of the middle ear. The last of the bones, the stapes, transmits the vibrations to the oval window. These vibrations pass through the fluid within the cochlea, and the motion of the fluid stimulates receptor cells in the **organ of Corti**. The organ of Corti contains sensory neurons that make up the auditory nerve and is responsible for sending impulses to the brain. These impulses are then interpreted as sound by the brain.

The ears are also involved in the sense of balance. Three semicircular canals are positioned at right angles to one another, much like the two walls and ceiling which meet at the corner of a room. The canals are arranged in this manner so that a person can sense motion in different planes.

The semicircular canals are filled with fluid and also contain sensory hair cells upon which

Biology For Life

CHAPTER 6: THE NERVOUS SYSTEM AND SENSES

rest tiny calcium crystals called **otoliths**. When the head changes position, the crystals move from their position on the tiny hairs. Sensory neurons detect the strength of the pressure on each hair, and, if stimulated, send impulses to the brain. These impulses communicate the position of the head to the brain.

The function of the semicircular canals explains the phenomenon of "dizziness." When the body is subjected to repeated circular motion, such as when spinning on an amusement park ride, the person often feels dizzy after getting off of the ride. The reason for this is because the fluid in the semicircular canals also spins, in an attempt to adjust the position of the head. The fluid remains spinning for a short time after the ride stops, and this spinning is interpreted as a feeling of dizziness by the brain. It feels as if you are still spinning, when actually, what is still spinning is the fluid in the canals, not your body.

> ### Section Review — 6.4 C & D
>
> 1. What are the two chemical senses?
> 2. What is another name for the smell receptors?
> 3. The _____ separates the outer ear from the middle ear.
> 4. Changes in the position of the head are detected by pressure exerted on receptors in the ear in the _____.

E. Vision

Humans rely very heavily on the sense of sight, and the eye is well designed for receiving light. A tough outer layer of tissue called the **sclera** covers the eye. The front portion of the sclera is called the **cornea**. The cornea is transparent, which means that all light can easily pass through it. The cornea is more curved than the rest of the eye. Its curved surface causes the incoming light rays to bend, thus helping to form an image of the object.

The **iris**, which is the colored portion of the eye, is located behind the cornea. The iris contains smooth muscle tissue that can dilate (enlarge) or constrict (make smaller) the pupil. The **pupil** is the black opening in the center of the iris. The adjustment of the muscles of the iris will change the size of the pupil. In strong light, such as on a bright sunny day, the pupil will become smaller, so that less light is admitted into the eye. In dim light, such as at night, the pupil becomes larger, so that the maximum amount of light can enter the eye.

The **lens**, which is the eye's light-focusing structure, is behind the pupil. Muscles are attached to the lens, and when these muscle contract, the lens changes shape. This process allows the lens to bend light and to form sharp images of objects that are at varying distances from the eye.

The lens of the eye focuses light rays on the retina. The **retina** is a thin membrane on the back of the eye that contains light-sensitive receptors. These receptors are called cones and rods. The **cones** are sensitive to bright light, and the **rods** are sensitive to dim light. The cones can distinguish form and color very well, whereas the rods cannot distinguish color. In dim light, when the rods are

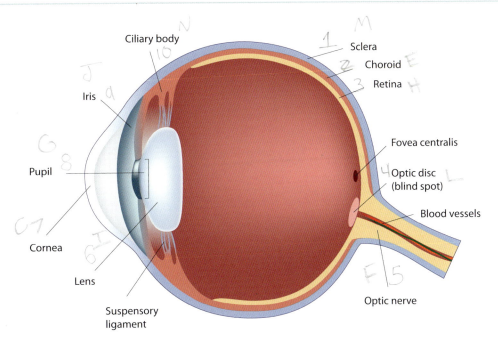

Figure 6-11. The human eye. The eye is much more complex than a simple camera. Our two eyes give us the ability to see objects in three dimensions - height, width, and depth.

activated, you can see only shades of gray. Rods are primarily used for night vision.

Think of the eye as a camera. In fact, cameras have been built on the principles of the human eye, yet our most expensive and high quality cameras cannot compare with the remarkable abilities of the human eye. Vision is made possible when light that is emitted from an object is directed towards the eye. Light enters the eye through the cornea and lens. These two structures bend the light rays and form an image on the retina. This image appears upside down and backwards. However, the brain processes this image correctly, so that you see it as it is supposed to appear.

The Human Mind - A Most Wonderful Creation of God

It is important to note that we cannot compare how the senses work (structure and function) with sense awareness (perception). Descriptions of the structures and functions of the senses cannot adequately explain how the brain processes and interprets this information. For example, an object emits light, which passes through the cornea and lens and to the retina. This light energy is then converted to chemical energy, which generates an electrical impulse. The electrical impulse is sent to the brain, and then the brain processes this image (or sees it).

Take another example of the difference between structure and function and perception, or sense awareness. If you are lighting a candle but the wick does not light quickly enough, the flame on the match travels upwards towards your fingertips. The receptors on your skin detect the warmth. Within one second, you notice that you will not be able to light the candle with this match. Instinctively, your arm recoils and you blow out the flame and drop the match. Does this sound familiar?

The two examples above describe mechanisms and processes, but do little to help us understand how the brain processes this information. No matter how deeply we probe into the nervous system, in the end, we need to envision an "inner man" that transforms the electrical signals into a sensation, because what a person perceives never appears in the brain as such. It is a mystery. As everything in itself is a mystery, we must accept that God designed our minds and bodies in a

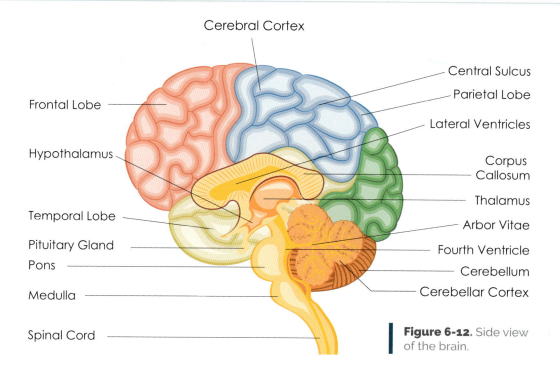

Figure 6-12. Side view of the brain.

marvelous way in order that we know Him, love Him, and serve Him.

The brain itself is shrouded in complete silence, even while the person hears the deafening roar of a jet aircraft engine. Likewise, the brain, encased as it is in the skull, is covered with darkness and produces no light, even when the person perceives the brilliance of the sun's glare. Our brains do not become colder when we touch snow or harder when we touch iron. The brain is chemically and physically isolated from the odors, sounds, flavors, textures, temperatures, and colors that exist outside the skull. Not a single sugar molecule passes from the chocolate candy in the mouth to the gustatory, or taste, region of the cerebral cortex, and yet we perceive the sugar's sweetness in our mouths. The brain tissue itself takes on none of the sourness of a tasted lemon or the pungent odor of the skunk's spray that we smell. Even if we know the particular series of impulses that correlate with these sensations, we can neither perceive nor comprehend the activities of the mind.

Yet, despite the non-material aspect found in sensory perceptions, a bodily organ is still indispensable for all the activities of a sensory being. Without an eye and an optic nerve and visual cortex, there is no sight. The electrical and chemical activities are not themselves sensations; yet the "inner man" uses them to make the sensations possible. This explains why brain damage can temporarily or permanently inhibit sensory capabilities.

Section Review — 6.4 E

1. The tough outer layer of tissue that covers the eye is called _____.
2. The _____ is the colored area of the eye that contains smooth muscles to adjust the amount of light which enters the eye.
3. The retina is a thin membrane in the back of the eye that contains light-sensitive receptors. These receptors are called _____ and _____.
4. What is the difference between cones and rods?

6.5 Special Concerns Regarding the Nervous System and Senses of the Newborn and Elderly

Section Objectives

- List one difference in the nervous system and senses of the newborn and elderly.

A. The Newborn

At birth, the newborn's nervous system is incompletely integrated, but sufficiently developed to sustain life outside of the mother. Most of the newborn's nervous system functions are primitive reflexes. The autonomic nervous system is important during this time, and the conditions outside of the womb stimulate the infant's breathing, help maintain acid-base balance, and partially regulate its temperature control. The newborn has acute, well-developed senses of taste, smell, and hearing, as well as the perception of pain.

Many of the neurons of the newborn are not yet covered with myelin. Myelin is necessary for rapid and efficient transmission of some, but not all, nerve impulses along the neural pathway. Impulses travel along myelinated axons 50 times faster than they do along unmyelinated axons, sometimes as fast as 100 meters/second (224 miles/hour). The neurons become myelinated in the head area and then down toward the toes. Also, the neurons at the center of the body become myelinated first before the peripheral nerves. As more neurons become myelinated, the infant develops a greater mastery of motor skills.

The newborn's senses are well developed to have a significant effect on a baby's growth and development, including the process of attachment (bonding). At birth, the eye is structurally incomplete. The ciliary muscles are immature, which limits the ability of the eyes to accommodate and fixate on an object for any length of time. The pupils react to light, and the blink reflex is responsive to a minimal stimulus. The corneal reflex is activated by a light touch. Tear glands usually do not begin to function until two to four weeks of age.

The newborn has the ability to briefly focus on a moving object that is within eight inches of his face and in the middle of his visual field. The clarity of an infant's vision is reported to be between 20/100 and 20/400. (The designation 20/400 means that an object 20 feet away from this person appears as it would be to a person with normal sight 400 feet away from the object.) Normal vision is 20/20, so

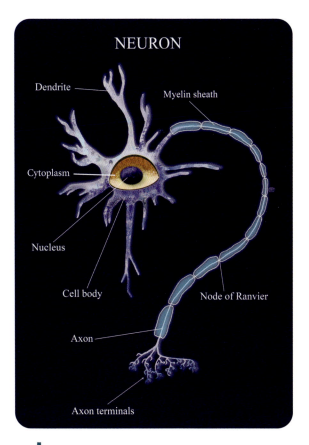

Figure 6-13. Myelin is a protective covering around many nerves. It is necessary for the rapid and efficient transmission of nerve impulses.

an infant is nearsighted and cannot distinguish objects that are farther away. The infant also has visual preferences: medium colors such as yellow, green, and pink over dim or bright colors such as red, orange, and blue. Other visual preferences are black and white contrasting patterns, especially geometric shapes and checkerboards. Infants also have a tendency to look at large objects with medium complexity, rather than small complex objects, and prefer reflecting objects to dull ones. This nearsightedness helps the infant to bond with its mother, who holds it closely.

After the amniotic fluid has completely drained from the newborn's ears, their sound reception is similar to that of an adult. The newborn can detect a loud sound of about 90 decibels (sound of a lawn mower) and will react with a startle reflex.

The newborn responds to low frequency sounds, such as a metronome, heartbeat, or lullaby, by decreasing both motor activity and crying. High frequency sounds tend to elicit an alerting reaction in the newborn. There is an early sensitivity to the sound of human voices and to specific speech sounds. For example, infants younger than three days old can tell the voice of their mother from that of other women. As young as five days old, the newborn can differentiate between stories repeated to them during the last trimester of pregnancy by their mother and the same stories recited after birth by a different woman. This recognition of the mother's voice is another way bonding is advanced.

The internal and middle ear structures are large at birth, but the external ear canal is small. A bone near the ear, called the mastoid process, and the

Figure 6-14. A newborn can focus on a moving object that is within eight inches of his face and in the middle of his visual field.

bony part of the external canal have not yet developed. Consequently, the tympanic membrane (eardrum) and facial nerve are close to the surface and can be easily damaged. Loud noise or shouting very close to the infant's ears should always be avoided, as should any sudden jerking movement of the neck.

Newborns react to strong odors, such as alcohol and vinegar, by turning their heads away. Infants who are breast-fed are able to smell breast milk and will cry for their mothers when the breasts are engorged and leaking. Infants can differentiate by smell the breast milk of their mother from the breast milk of other women. Maternal odors are thought to influence the process of attachment and successful breast-feeding. Again, these odor and taste recognitions aid the bonding process with the mother.

The newborn is able to distinguish between tastes, and various types of solutions will elicit different facial expressions. A sweet solution elicits an eager suck and a look of satisfaction; a sour solution causes puckering of the lips; and a bitter liquid results in an angry, upset expression. The newborn has a preference for glucose and water over plain water, just like his elder siblings.

The newborn perceives tactile (touch) sensation in any part of the body, although the face, especially the mouth, as well as the hands and soles of the feet, seem to be the most sensitive areas. Research has shown that touch and motion are essential to normal growth and development of the infant. Gentle patting of the back or rubbing of the abdomen usually elicits a calming response from the infant. However, painful stimuli, such as even a slight pinprick, will elicit a loud, upsetting response.

B. The Elderly

With advancing age, there is a slowdown in the rate at which nerves are able to transmit signals to one another. This can affect some types of thought, as well as some reflexes, and can include a slower

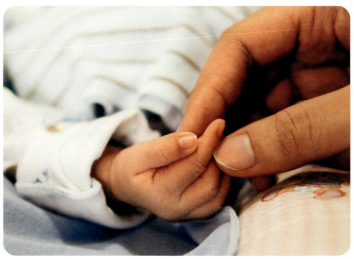

Figure 6-15. A newborn's mouth, hands, and soles of the feet seem to be the most sensitive areas for tactile sensations.

response to new information. Learning may take more time, but the elderly can still learn new things.

Changes in sleep patterns may occur with the elderly person. Forgetfulness may increase slightly, although people may simply become more aware of forgetfulness as they age. There is a "use it or lose it" mechanism to mental acuity. Just as those who stay physically active tend to stay physically vigorous, those who stay mentally active by taking courses, keeping up with current events, or reading books, tend to remain mentally sharp.

For the most part, despite the association of senility with aging, the profound loss of mental acuity is a symptom of a disorder, such as Alzheimer's disease, and is not a normal part of the aging process.

The ability of the elderly to see, hear, taste, feel, and react to stimuli changes with time. These changes are not great and happen gradually. For example, the pupils of our eyes do not rapidly contract and dilate as they once did. These changes are due to a "stiffening" and loss of flexibility that occurs with age. Many of the disorders that result from changes, such as cataracts and hearing loss, may be treated quite effectively.

CHAPTER 6: THE NERVOUS SYSTEM AND SENSES

> "Amen I say to you, as long as you did it to one of these my least brethren, you did it to me."

An elderly person's mental and physical reaction time may diminish, and many states now require seniors to undergo driving tests more frequently than younger people. Losing driving privileges may be difficult for some seniors to accept. It causes a loss of freedom, and they need to begin to depend on others. However, the loss of driving privileges hardly compares to the potential for loss of life due to automobile accidents that result from diminished reaction time.

With a positive attitude and some help from family, friends, and local agencies, healthy seniors may maintain a considerable amount of freedom. If you have time, consider volunteering your time in a retirement or assisted living center. In today's world, the families of the elderly often live far away from them. Many of these people have no one to talk to or spend time with, except the paid staff who look after them. When we spend time with the elderly, we are adhering to the commands of our Lord (Matthew 25:40), "Amen I say to you, as long as you did it to one of these my least brethren, you did it to me."

Section Review — 6.5

1. Why is myelin important?
2. What happens to the nervous system and senses as a person ages?

Chapter 6 Supplemental Questions

Answer the following questions.

1. Explain the transmission of a nerve impulse down an axon.//
2. Discuss the polarization of the nerve cell membrane.
3. What is the function of a neurotransmitter?
4. How are acetylcholine and acetylcholinesterase related?
5. Name the parts of the central nervous system.
6. Describe the location, structure, and function of the following: cerebrum, cerebellum, thalamus, hypothalamus, medulla oblongata, and spinal cord.
7. Explain the role of the spinal cord in controlling simple reflex behaviors.
8. Differentiate between the somatic and autonomic parts of the peripheral nervous system.
9. Contrast the actions of the sympathetic and parasympathetic nervous systems.
10. Explain how skin, taste, and smell receptors inform the brain about the environment.
11. Describe the structure and function of the parts of the ear, and label the parts of the ear on a diagram. Refer to the two diagrams in section 6.4 D.
12. Explain the structure and function of the parts of the eye, and label the parts of the eye on a diagram. Refer to section 6.4 E.
13. Study all diagrams in this chapter. Be able to draw and label all diagrams.

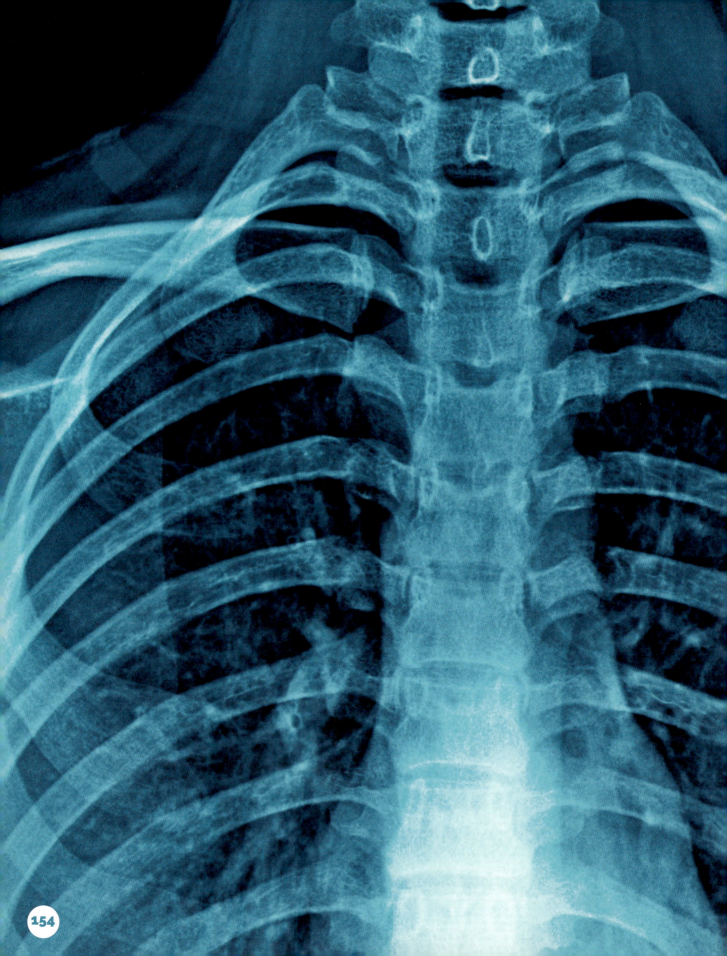

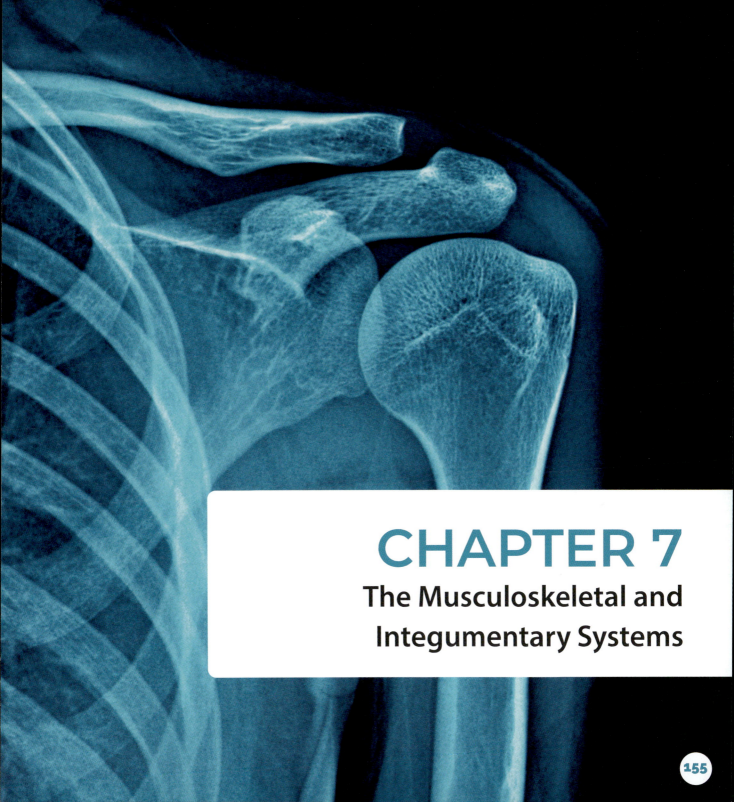

CHAPTER 7
The Musculoskeletal and Integumentary Systems

Chapter 7
The Musculoskeletal and Integumentary Systems

Chapter Outline

7.1 Introduction

7.2 The Human Skeleton
    A. Bone and Cartilage
        A.1 Compact Bone
        A.2 Spongy Bone
        A.3 Cartilage
        A.4 Broken Bones
    B. Joints
    C. The Human Skeleton
        C.1 Axial Skeleton
        C.2 Appendicular Skeleton
    D. Special Concerns Regarding the Bones of the Young and Elderly
        D.1 Congenital Defects of Bone
        D.2 Effects of Aging on Bones

7.3 Muscles
    A. Skeletal Muscles
    B. Chemistry of Skeletal Muscles
    C. Joint Movement
    D. Cardiac Muscle and Smooth Muscle
    E. Special Concerns Regarding the Muscles of the Young and Elderly

7.4 The Human Skin
    A. Structure and Function of Human Skin
    B. Special Concerns Regarding the Skin of the Young and Elderly

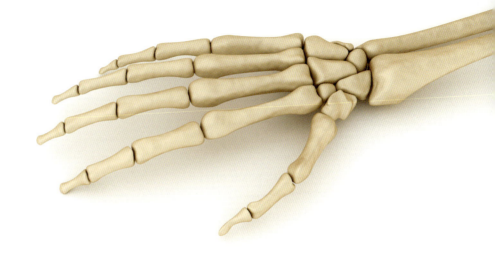

7.1 Introduction

Why were we created? *The Baltimore Catechism* answers this profound question simply and clearly: made in the image and likeness of God, we were created to know Him, to love Him, and to serve Him in this world and in the next. It is for this end that man was created. *The Catechism of the Catholic Church* explains that man, although made of body and soul, is a unity. This unity of body and soul allows man to raise his voice in praise to the Creator. "For this reason, man may not despise his bodily life. Rather, he is obliged to regard his body as good and to hold it in honor, since God has created it and will raise it up on the last day" (CCC 364). The *Catechism* explains that our bodies need to be used as God intended. In this chapter, the emphasis lies on specific parts of the body, namely bones, muscles, and skin.

Bones, muscles, and skin enable man to move and function. As you study how and why bones, muscles, and skin function in unity, you may want to think of what God intended us to do with our movements. Movement allows us to perform works of charity for our neighbors, to labor for the greater honor and glory of God, and to clasp our hands and bend our knees in reverent prayer to our glorious Creator and King in heaven.

Consider what Scripture tells us to do with our bodies. Saint Paul, in 1 Corinthians 10:31, commands, "Therefore whether you eat or drink, or whatsoever you do, do all things for the glory of God." In Colossians 1:10, St. Paul says, "That you may walk worthy of God, in all things pleasing; being fruitful in every good work, and increasing in the knowledge of God."

Our Lord Jesus teaches us in Saint Matthew's Gospel, "Come ye blessed of My Father, possess the kingdom prepared for you from the foundation of the world. For I was hungry, and you gave Me to eat; I was thirsty, and you gave Me to drink; I was a stranger, and you took Me in; naked, and you clothed Me; sick, and you visited Me; I was in prison, and you came to Me" (Mt. 25:34-36).

Have you ever wondered why movement of the body seems so simple? In reality, the ability to move our bodies is very complex. Bodily movements are possible because we have muscles that move parts of our bones. Our skin surrounds and protects the delicate framework of the body, which consists of our bones and muscles. After completing this chapter, you will appreciate the many activities occurring in our bodies that make motion possible.

CHAPTER 7: THE MUSCULOSKELETAL AND INTEGUMENTARY SYSTEMS

7.2 The Human Skeleton

Section Objectives

- List the type of tissue that makes up the skeletal system.
- Identify the tough membrane that covers the bones.
- Name the channels in bone through which blood vessels run.
- Know what type of cell secretes bone matrix.
- Know what type of protein gives strength to bone.
- Know the term for the lubricating fluid of the joints.
- Give an example of a joint from each of the following categories: ball and socket, gliding, hinge, pivot, immovable.
- List the major bones discussed in the chapter and be able to identify these bones on a drawing of the human skeleton.
- Describe a congenital and an age-related problem with the bones.

attach to bone. Contained within the periosteum are two important types of cells: **osteoclasts**, a special type of cell that constantly move throughout your bones to remove old materials making room for new bone growth, and **osteoblasts**, another special cell in the bone tissue responsible for the formation of new bone. These osteoblast cells are responsible for bone growth, which makes our bones grow longer as we mature from baby to adult. Beneath the periosteum is a hard, dense bone material known as **compact bone**. **Spongy bone** is located beneath the hard compact bone, particularly at the knobby ends of bones.

Look at Figure 7-1 to see an example of one of the many bones in our bodies. A layer of **marrow** lies within the hollow spaces in the spongy bone. Most spongy bone contains **red marrow**, which manufactures red blood cells for the body. The long bones of the arms and legs have a cavity that contains **yellow marrow**, blood vessels, and nerves, in addition to red marrow. Yellow marrow is made up primarily of fatty tissue. Yellow marrow can change into red marrow when the red blood cell count gets low.

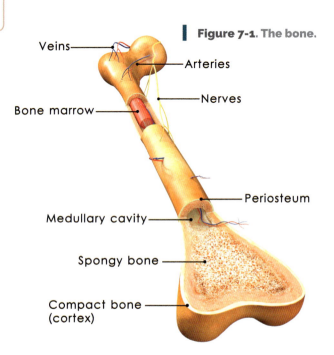

Figure 7-1. The bone.

The human skeleton has many functions. It supports the body, protects our internal organs, and enables the body to move by its attached muscles. Like the other parts of the body, skeleton is made up of tissues. The type of tissue in the skeleton is called **connective tissue** because this tissue connects and supports parts of the body. Connective tissues include **bone**, cartilage, tendons, ligaments, **fat**, and **blood**.

A. Bone and Cartilage

Bone is one of the types of connective tissue. Look at Figure 7-1, a typical long bone in the body.

All bones are surrounded and protected by a tough outer membrane known as the **periosteum** [pĕr-ĭ-ŏs-tē-um]. It is the location where muscles

CHAPTER 7: THE MUSCULOSKELETAL AND INTEGUMENTARY SYSTEMS

A.1 Compact Bone

Compact bone is composed of bone cells called **osteocytes** [ŏs-tē-ō-sīts]. These osteocyte cells secrete a nonliving material called the **matrix** [mā-trĭks], which surrounds the osteocytes. Osteocyte cells are arranged in concentric, cylinder-shaped groups of bony material that surround a central canal known as the **Haversian canal**. The Haversian canal provides a route for microscopic channels to provide tissue fluid between the Haversian canal and osteocytes of the larger Haversian system. Haversian canals ensure that the bone is able to retain nutrients and maintain blood vessels and nerve fibers. The absence of Haversian canals would lead to the decay and destruction of compact bone. A Haversian system contains a central Haversian canal surrounded by concentric rings, like tree rings, of matrix and osteocytes. Groups of Haversian systems together make up the bulk of compact bone.

We tend to think of bone as a lifeless tissue. However, it is composed of many living cells and must continuously receive food and oxygen and remove wastes. Bone can repair itself and even strengthen itself. It conducts these processes by way of the Volkmann's canals. The blood vessels of the periosteum (a dense fibrous membrane covering the surface of bones) connect with the Haversian canals by the **Volkmann's canals**. The Haversian canals are situated along the long axis of the bone, while the Volkmann's canals go across the bone, between the Haversian canals.

The process of bone formation is called **ossification** [ŏs-ĭ-fĭ-kā-shun]. This ossification process is complete by the time a person reaches the age of twenty.

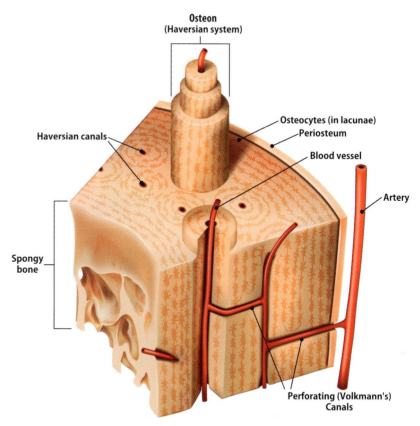

Figure 7-2. Bone structure. A typical long bone has a shaft of compact bone surrounding a center of yellow marrow.

Ossification is best understood if we compare it with the building of a modern-day concrete bridge. During construction of a bridge, iron bars are positioned inside of a supporting wooden frame. Concrete is poured into the wooden frame and allowed to harden. The wooden frame is then removed, and the hard concrete becomes part of the developing bridge structure.

Bone formation can be likened to this process if we substitute collagen for the iron bars and calcium crystals for the concrete. Bone matrix, the non-living material, contains large amounts of a fiber-like protein known as **collagen**. The collagen gives bone matrix its strength and durability. While bone is in the process of forming, the osteocyte cells deposit crystals of calcium phosphate and

Biology For Life

CHAPTER 7: **THE MUSCULOSKELETAL AND INTEGUMENTARY SYSTEMS**

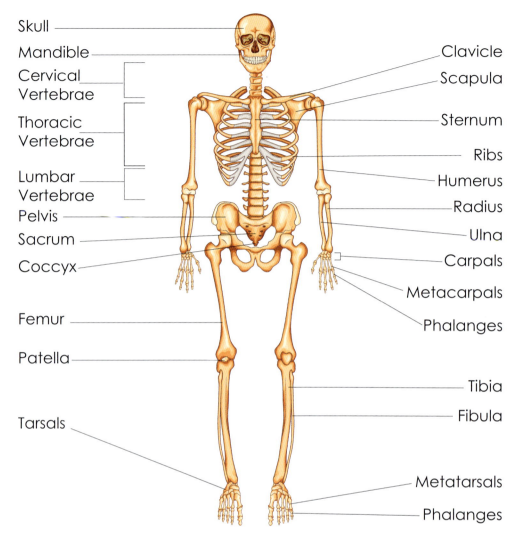

Figure 7-3. The Human Skeleton. The skeleton makes possible a wide range of movements while supporting the body and protecting vital organs. It consists of 206 bones.

calcium carbonate around the collagen fibers. These crystals of calcium provide the bone with "hardness."

The osteocyte cells and bone matrix of non-living material are both involved in the important functions of storage and release of calcium and phosphorus. These two minerals, calcium and phosphorus, are essential for the proper functioning of many of the body's tissues. It is important, therefore, that our diets contain adequate amounts of calcium (from foods like milk, cheese, and yogurt) and phosphorus (from dairy products, meat, and fish).

A.2 Spongy Bone

Spongy bone is also known as **cancellous bone**. In comparison with compact bone, spongy bone has a higher surface area, but it is less dense, softer, weaker, and less stiff. It typically occupies the interior region of bones. Spongy bone has many blood vessels and frequently contains red bone marrow. Red bone marrow is the region where red blood cells are produced, by a process known as hematopoiesis [he-ma-to-poi-e-sis]. As part of its structure, spongy bone contains many tissue elements in the form of small beams and

Biology For Life

struts, or rods, called **trabeculae** [tră-bĕ-kŭ-lā]. The trabeculae form as lightweight support beams that strengthen the compact sides of the bone, without adding too much weight. When stress is put on the bone by physical work, the bone grows new trabeculae to add strength where the work increased the stress on the bone. So, when you work out, you not only strengthen your muscles, you also strengthen your bones.

A.3 Cartilage

Cartilage is another type of connective tissue that makes up part of the skeleton. Unlike bone, cartilage is very flexible and is lacking in minerals. Cartilage is found in the ears and at the tip of the nose. When developing in an embryo, most of the skeleton is made of cartilage. With continued growth and development, the cartilage is replaced by osteocytes, which secrete calcium and other minerals.

A.4 Broken Bones

Some of us will experience a broken bone in our lifetime. Our Creator has designed our bones to heal automatically, which is a wonderful gift. Our bodies are self-healing! The assistance of a physician can help the healing process, of course. It is important to "set" the bone, that is, to make sure that the two broken ends line up to each other correctly. The physician will take an X-ray picture of the bone and then place a cast on the broken bone, or insert pins and plates, so that it will heal properly.

Following a break or fracture, a blood clot forms around the break to prevent further internal bleeding. In a few days, the clot begins to dissolve, and osteocytes produce new matrix material, beginning with collagen. In turn, the collagen is hardened by the addition of calcium crystals. At this point, the newly-made bone does not have a well-defined structure. It takes up to six weeks for a callus to develop. A callus is a temporary formation of cells which forms at the area of a bone fracture. The callus is an indication that a new "woven bone" is forming. Then it will take another six weeks or longer for the woven bone to form mature bone.

Bone breaks are very painful and make a person very uncomfortable and panicky. What would you do if someone close to you broke a bone? The first thing to do is to remain calm, or to at least appear calm. Quickly make the other person as comfortable as possible. Keep the person from moving the broken bone and help keep pressure off of it. If it seems to be a broken leg, then make the person lie down on the side opposite of the fracture. You can then contact emergency rescue services by calling 911, or have someone quickly take the person to a nearby emergency room.

Though relatively few of us will experience a broken bone in our lifetime, it is nevertheless a good idea to know what to do if it happens to you or to a friend or family member. The American Red Cross offers free first aid classes and can provide you with training on how to respond to injuries, including broken bones. Sometimes it is the right person at the right time who can provide aid and comfort to an injured person, help with a quicker recovery and positive outcome, and perhaps even save a life. Rely on God and St. Luke, the patron of doctors, to help you in an emergency situation.

> **Section Review — 7.2 A**
>
> 1. What are the different types of connective tissues?
> 2. Draw a long bone and label each part.
> 3. Bone-forming cells are known as _____.
> 4. What is the purpose of red marrow?
> 5. Compact bone is composed of bone cells known as _____.
> 6. What is ossification?
> 7. What is another name for spongy bone?

CHAPTER 7: THE MUSCULOSKELETAL AND INTEGUMENTARY SYSTEMS

B. Joints

The central column of the human skeleton is made up of the skull, the **vertebra** (or **vertebral column**), and the rib cage. The rest of the bones, including the arms and legs, are attached to this central column of bones. **Joints** occur when two or more bones come together. Joints are classified according to the type of movement they allow.

Immovable joints do not allow any movement. Most of the skull bones are fused together at immovable joints.

All other joints allow some movement to occur. In these **movable joints**, the bones are attached by tough, fibrous, connective tissues called **ligaments**. Since ligaments can stretch somewhat, the bones are allowed to move but not to separate. There is a thin layer of cartilage on the ends of bones in a movable joint. This layer provides for a smooth contact surface. A special lubricating fluid, known as **synovial** [sĭ-nō-vē-al] **fluid**, helps reduce friction. Synovial fluid helps keep the joint ends from wearing out.

There are six types of freely movable joints: **hinge joints, ball-and-socket joints, pivot joints, ellipsoid joints, gliding joints, and saddle joints.**

Hinge joints, such as those found in the knee, elbow, and joints of the fingers, allow bending in a single direction, like a door swinging on its hinges.

The most freely movable joint in the body is the **ball-and-socket joint**, in which the rounded head of a bone fits into a hollow socket of another bone. The two shoulder joints and the two hip joints are the only ball-and-socket joints in the body. Complex movements, such as throwing a baseball or swimming, are made possible by the ball-and-socket joint of the shoulder because it allows the bone to rotate in two planes.

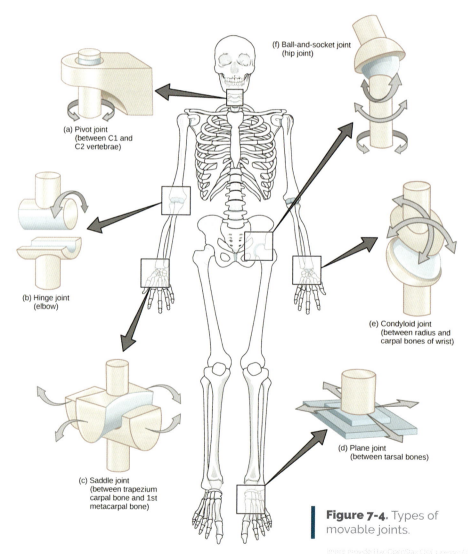

Figure 7-4. Types of movable joints.

A **pivot joint** is where the bone rotates in place against another bone. The joint between the atlas and the axis in the neck allowing you to swivel your head left and right is an example of a pivot joint.

The joints between the metacarpals and phalanges allowing you to move your fingers up, down, left, and right are examples of **ellipsoid joints**, also known as condyloid joints. **Ellipsoid joints** are bones with a convex (bulged) surface that fits into a concave (indented) surface of another bone.

A **gliding joint**, or plane joints, are located between carpal bones in the wrist and the vertebrae in the backbone and allows limited twisting, turning, and sliding between vertebrae. Pads of cartilage called **vertebral disks** are found between each vertebra in the backbone.

The last type of joint is the **saddle joint**. The saddle-shaped articular surfaces are convex in one direction and concave in another which permits movements in all directions except axial rotation. This joint is found only in the hands where your thumb attaches to the carpal bone.

> **Section Review — 7.2 B**
>
> 1. What are the six types of movable joints?
> 2. Give an example of each type of movable joint.

C. The Human Skeleton

There are 206 bones in the human skeleton. New bones are constantly built up; our skeletons are replaced every two years.

The skeleton is divided into two major portions: the **axial skeleton** and the **appendicular** [ăp-en-dĭk-ū-ler] **skeleton**.

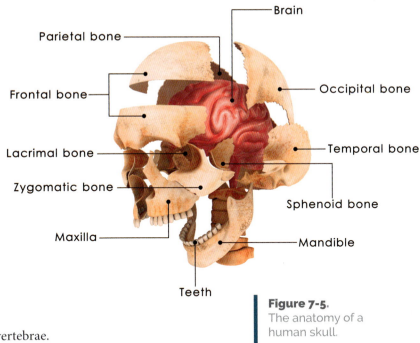

Figure 7-5. The anatomy of a human skull.

C.1 The Axial Skeleton

The **axial** [ăk-sĭ-al] **skeleton** consists of bone and cartilage that support and protect the organs of the head, neck, and trunk. The four parts of the axial skeleton are discussed in detail below.

Skull

The skull is composed of the brain case, or **cranium**, and the **facial bones**. The cranium protects the brain from injury. The facial bones form the basic shape of the face and provide attachments for various muscles that move the jaw and control facial expressions.

Hyoid bone

The hyoid bone is located in the neck, between the lower jaw and the **larynx** (voice box). This bone is not connected with any other bones but is fixed in position by muscles and ligaments. The hyoid bone supports the tongue and serves as an attachment for certain muscles that help to move the tongue and function in swallowing.

Vertebral column

The **vertebral column**, which is also known as the spine or backbone consists of 33 bones known as vertebrae. The vertebrae[1] are separated by intervertebral disks, the latter are composed of cartilage and serve as a cushion, in order to absorb shocks between bones. The function of the vertebral column is to keep the human posture erect and support the body when standing or sitting upright.

Twenty-four vertebrae are articulated and allow limited bending and twisting motions. Seven cervical vertebrae are at the top of the column and support or hold the head. Then come 12 thoracic vertebrae which anchor the ribs. These are larger than the cervical vertebrae and are followed by the 5 largest vertebrae, the lumbar vertebrae. Near the bottom of the vertebral column, 9 vertebrae are fused together into two formations, five are fused to form the **sacrum**, which is the large triangular bone at the bottom of the spine or backbone; four more form the **coccyx** or tail bone. The sacrum connects with the coccyx below it and the lumbar vertebrae above to form part of the pelvis, which is a ring of bones at the bottom of the tailbone.

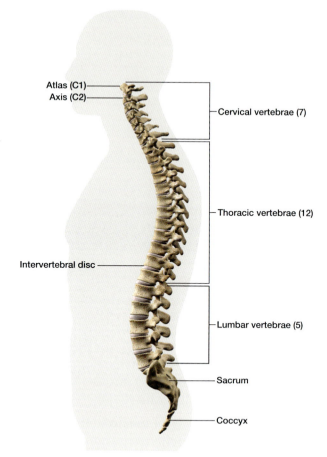

Figure 7-6. Vertebral column. The vertebral column, or backbone, is the feature that distinguishes vertebrates from other animals.

Thoracic cage

The thoracic cage, or the rib cage, protects the organs of the **thorax** (chest area) and upper abdomen. The rib cage is composed of twelve pairs of ribs, 7 pairs of true ribs, 3 pairs of false ribs, and 2 pairs of floating ribs. The true ribs are connected to the vertebrae in the back and to the **sternum** (breastbone) in the front of the body. The false ribs are connected to the sternum by joining with the costal cartilages of the ribs above. The floating ribs are so-called because they are attached to the vertebrae only, and not to the sternum or cartilage coming off the sternum. These ribs are relatively small and delicate, and are capped by a cartilaginous tip.

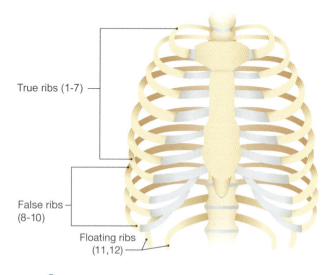

Figure 7-7. The anatomy of a rib cage.

---

[1] Note the spelling for singular and plural: one vertebra, many vertebrae. The adjective is vertebral.

C.2 The Appendicular Skeleton

The **appendicular skeleton** consists of the bones of the arms and legs (limbs) and those bones that attach the limbs to the **axial skeleton**. The axial skeleton is the skull, the vertebral column, the ribs, and the breastbone. The parts of the appendicular skeleton are discussed in detail below.

Pectoral girdle

The **pectoral girdle** is formed by the **scapula** (shoulder blade) and **clavicle** (collarbone) on both sides of the body. The pectoral girdle is attached to the axial skeleton (the skull, vertebrae, ribs, breastbone) by muscles. The pectoral girdle not only connects the bones of the arms to the axial skeleton, but it also aids in arm movements.

Upper limbs

Each upper limb of the arm consists of a **humerus**, or upper arm bone, two lower arm bones, a **radius**, and an **ulna**. Below the radius and ulna, there are eight **carpals**, or wrist bones. The bones of the palm of the hand are called **metacarpals**, and the fourteen little finger bones of the hand, three per finger and two on the thumb, are called **phalanges**. Phalanges is also the name given to the small bones of the feet, the toes. In the illustrations of the arm and leg, you can see all the scientific names of the various bones.

Biologists have important reasons for using these terms, so we should learn them. However, on most occasions, it is certainly proper to call them finger bones rather than phalanges, even when the doctor asks where it hurts!

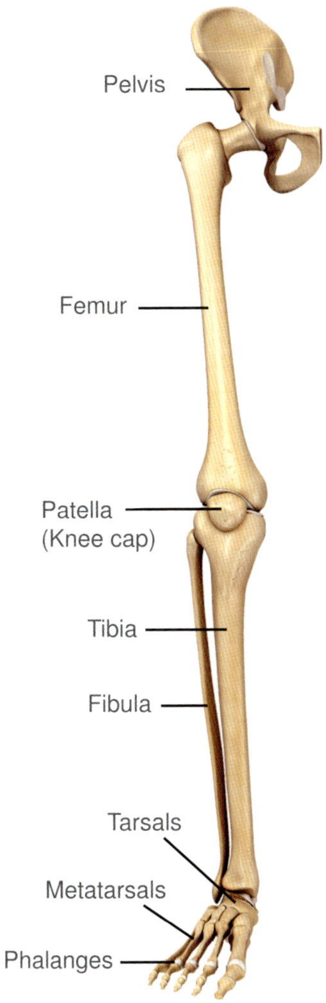

Pelvic girdle

The **pelvic girdle** is formed by two **innominate** bones, or hip bones. These are attached to each other in the front of the body and to the sacrum in the back of the body. They connect the leg bones to the axial skeleton and, along with the sacrum and coccyx, form the pelvis. The pelvis protects the lower abdominal and internal reproductive organs. In women, the pelvis forms the cavity in which the developing baby is contained during pregnancy.

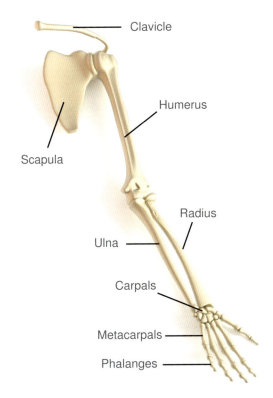

Figure 7-8. The pectoral girdle. These are bones of the shoulder area that provide support for the arms, allowing them a wide range of movement.

Figure 7-9. The pelvic girdle and leg.

CHAPTER 7: THE MUSCULOSKELETAL AND INTEGUMENTARY SYSTEMS

Lower limbs

These are the legs. Each leg consists of a **femur** (thigh bone) and two lower leg bones, the large **tibia** (shinbone), and the slender **fibula** (calf bone). These three bones connect with each other at the knee joint, where the **patella**, or kneecap, covers the front surface. At the bottom of the tibia and fibula, there are seven **tarsals**, or ankle bones. The bones of the foot are called **metatarsals**, while those of the toes are called **phalanges**, like the finger bones.

> **Section Review — 7.2 C**
>
> 1. How many bones are there in the human body?
> 2. What are the four parts of the axial skeleton?
> 3. Where is the hyoid bone located?
> 4. How many vertebrae make up the spinal cord?
> 5. The rib cage is composed of how many pairs of ribs?

D. Special Concerns Regarding the Bones of the Young and Elderly

Certain conditions may arise during pregnancy, due to defective genes, that can cause the bones of a baby to form incorrectly. There are also conditions that cause bones to deteriorate as we age.

D.1 Congenital Defects of Bone

The term *congenital* means an illness or defect that is present from the time of birth. The defects mentioned here are just a few samples of the possible types of congenital birth defects.

Scoliosis is a condition where the vertebral column (backbone) has a lateral curve. In congenital scoliosis, one side of the backbone grows faster than the other, thus producing a curvature. Other clinical conditions involving abnormal curvatures of the spine are **kyphosis**, excessive thoracic curvature, and **lordosis**, excessive lumbar curvature.

Congenital scoliosis may be surgically corrected. Sometimes, young children or teenagers need to wear a brace to help correct the curvature. However, a brace worn on the back is of limited value for congenital scoliosis.

Spina bifida is a disorder in which the backbone and spinal canal do not close before birth. Because the backbone has failed to close, the spinal cord and the membranes that cover the cord protrude from the infant's back. Most of the defects occur in the lowest area of the back, known as the **lumbar region** of the vertebral column. The lumbar region is the last part of the backbone to close. The protrusion of the spinal cord and its surrounding protective tissues damages the spinal cord and nerve roots. This, in turn, causes a decrease or lack of function or a complete paralysis of body areas controlled by the nerves in the spinal cord at or below the defect.

In spina bifida, the goals of initial treatment are to reduce the amount of damage to the nervous system caused by the defect, to minimize complications of infection, and to aid the family in coping with the disorder. Early surgical repair of the defect is important. Before surgery, the infant must be handled carefully to reduce damage to the exposed spinal cord. Spina bifida has been linked to diets that are deficient in folic acid, which is one of the B vitamins. Folic acid is contained in eggs, orange juice, and dark green leafy vegetables. Folic acid helps the body make new cells. Accordingly, it has been recommended that women who are likely to become pregnant take folic acid as a supplement before they are pregnant.

In **congenital hip dislocation**, the baby is born with the ball-shaped head of the thighbone lying outside its cup-like socket in the pelvis. Usually, the socket is not formed correctly. Congenital hip dislocation can be corrected with splints placed on the infant's thighs. If corrected early, the defect is often completely cured, and the child is able to walk normally.

A baby born with **clubfoot** has one or both of its feet bent, either downwards and inwards or upwards and outwards. Many babies with normally formed feet persistently turn them inwards, but the feet can be pushed back to the proper position. In clubfoot, the feet cannot be pushed back into the proper position. For infants with clubfoot, the physician will show the parents how to manipulate the foot regularly until it settles into a normal position. In more severe cases, splints and plaster casts are placed on the foot, and the foot is manipulated when the casts are periodically removed. Surgery is used in only the most severe cases of clubfoot.

In a baby with a **cleft palate**, there is a gap in the bone that makes up the roof of the mouth, also known as the palate. The gap runs along the midline of the palate and extends from behind the teeth to the cavity of the nose. It makes eating and swallowing difficult, and, in newborns, regurgitated milk may come through the nose instead of through the mouth. An uncorrected cleft palate may cause speech difficulties. Surgery is needed to repair a cleft palate, though this is not done until the infant is older. Surgery generally improves the infant's appearance and allows the development of normal speech. Speech therapy may be needed in some cases.

D.2 Effects of Aging on Bones

Osteoporosis is the thinning of bone tissue and loss of bone density over time. Osteoporosis affects both men and women, but it is much more common in women who are over 50 years of age. Osteoporosis results when the bones no longer store calcium, due to dietary deficiencies. The loss of estrogen in older women may contribute to the problem. The bones in older people lose calcium, which is then replaced with fatty and fibrous tissues. Consequently, the bones become weak and brittle. The bones can no longer bear weight easily, and normal activities, such as carrying groceries or even coughing, can result in bone fractures.

Osteoporosis can cause vertebrae in the backbone to shrink with age, resulting in a loss of height, more commonly known as "shrinking." For some women, estrogen replacement therapy helps to prevent bone loss and reduces the risk of osteoporosis. For those who cannot have hormone therapy or choose not to have the hormone therapy due to side effects, there are medications which act directly on bone to stop breakdown, thus effectively maintaining bone mass. However, the best approach for the elderly is eating a balanced diet, with foods rich in calcium. Calcium supplements may be necessary. Older people, as well as people of all ages, should remain as active as possible to keep the

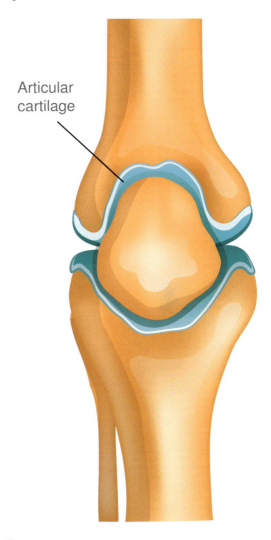

Figure 7-10. In the knee, articular cartilage covers the ends of the femur (the bone between the hip and the knee), the top of the tibia (the shin bone, the larger bone of the leg below the knee), and the back of the patella (the flat bone on the front of the knee).

bones strong. There is much evidence that weight training and/or strengthening exercises helps maintain strong bones in the elderly. Older people should avoid medications which increase the risk of osteoporosis, such as cortisone.

Osteoarthritis is a natural part of the aging process and involves flaking or cracking of the smooth lining of the joint, called the **articular** cartilage. Articular cartilage covers the ends of two bony surfaces that move against one another, or *articulate*. Articular cartilage is a strong, slippery substance that allows the joint surfaces to slide against one another without damage to either of the ends of bones. It varies in thickness and is about one quarter of an inch thick in large joints. In addition, it absorbs shock and provides an extremely smooth surface for producing motion.

The flaking and cracking of this cartilage is due to the normal wear and tear on these joints that occurs as one ages. The underlying bone may become thickened and distorted as the cartilage deteriorates. Movement becomes painful and restricted, causing the person to use the associated muscles less often. Inactivity of the muscles will cause these muscles to begin to waste away, or atrophy.

In osteoarthritis, occasional mild attacks of pain and stiffness in a joint are not a cause for concern. If symptoms are more severe, to minimize the wear and tear on the joints, a person should try to maintain a normal weight, since obesity contributes to the strain and pressure on the joints. Physicians recommend the use of a cane, taking frequent rests, sleeping on a firm bed, and keeping warm, since heat helps to ease most kinds of joint pains. Regular exercise helps to strengthen the muscles surrounding the affected joints and will minimize symptoms of arthritis. In more severe cases of arthritic pain, physicians may prescribe anti-inflammatory drugs. Surgery may be required for osteoarthritis of the hip. This surgery is known as a hip joint replacement.

Daily physical exercise is best for everyone during childhood, during early adulthood, during middle age, and even during old age. Simply walking can be a great exercise, but any other exercises recommended by a doctor can help prevent problems in older people. The current problem of obesity among the young, which in turn causes children to exercise less, will result, in middle age and later, in more and more serious problems, some of which are osteoarthritis, hypertension, and diabetes.

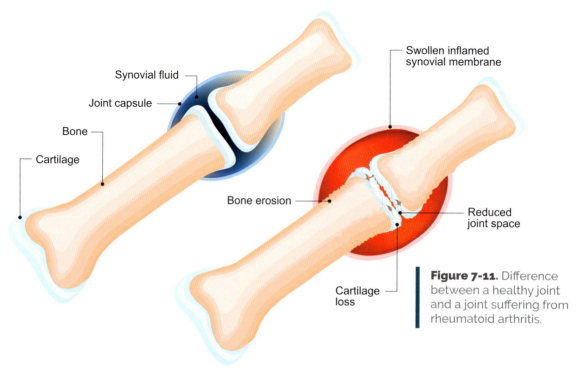

Figure 7-11. Difference between a healthy joint and a joint suffering from rheumatoid arthritis.

Rheumatoid arthritis is a long-term disease of the joints. In this disorder, the **synovium**, which is a thin membrane surrounding a joint, gradually becomes inflamed and swollen, and this leads to inflammation of other parts of the affected joint. The joints that are usually affected are the small ones in the hands and feet, mainly the knuckles and toe joints. Rheumatoid arthritis may occur in the wrists, knees, ankles, or neck. It occurs less often in the vertebrae or hips, which are more susceptible to osteoarthritis.

The best ways to manage the disease of rheumatoid arthritis are to obtain plenty of rest, exercise regularly and moderately, and sleep on a firm mattress with lightweight covers, so that very little pressure is placed on the joints. In the case of a severe attack of rheumatoid arthritis, hospitalization may be necessary. The affected joints will be placed in splints until the symptoms have subsided. Pain-relieving medication may be prescribed to treat very painful episodes of the disease. If a single joint is affected, sometimes surgery is necessary to remove the synovium of the badly inflamed joint. Rheumatoid arthritis varies in severity from person to person; however, only one in ten persons who have the disease are severely disabled by it.

> **Section Review — 7.2 D**
>
> 1. What is scoliosis?
> 2. What is the difference between kyphosis and lordosis?
> 3. What is spina bifida?
> 4. What are some problems that spina bifida can cause?
> 5. What is osteoporosis?
> 6. What joints are most affected by rheumatoid arthritis?

7.3 Muscles

> **Section Objectives**
>
> - Identify the type of tissue that makes up the muscular system.
> - Know the two major proteins found in sarcomeres.
> - Know the purpose of tendons and ligaments.
> - Define antagonistic pair.
> - Know what substance is broken down to provide energy needed for ATP synthesis during vigorous exercise.
> - Know the scientific term for heart muscle.
> - Compare and contrast cardiac, skeletal, and smooth muscles.
> - Describe a congenital and age-related problem with the muscles.

Muscle tissue is capable of contracting, or shortening. This contraction is what allows for motion to occur. When the body moves, muscles are doing the work. The muscles and other body tissues are covered and protected by the skin. Skin helps regulate the body's internal temperature and sense its external environment.

A. Skeletal Muscles

There are three main types of muscles in the body: **skeletal muscle, smooth muscle,** and **cardiac muscle**.

Most of the muscle in the body is skeletal muscle. The body mass of an average adult male is made up of 40–50% skeletal muscle, and the body mass of an average adult female is made up of 30–40% skeletal muscle. It is called skeletal muscle because the muscle is attached to the bony skeleton and helps move the bones. Skeletal muscle has a striped appearance when viewed with a microscope,

CHAPTER 7: THE MUSCULOSKELETAL AND INTEGUMENTARY SYSTEMS

Types of Muscle

Figure 7-12.

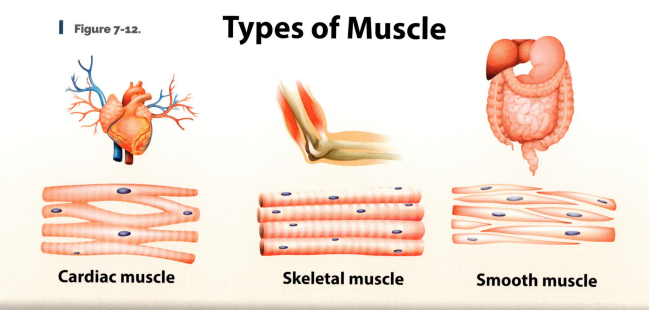

Cardiac muscle   Skeletal muscle   Smooth muscle

and it is therefore called **striated muscle**. You may have heard the word striated when you studied earth science, where it is used to describe the layering of sedimentary rocks.

Both skeletal muscle and smooth muscle contractions are controlled by nerve impulses that originate in the brain. Skeletal muscles are often referred to as **voluntary muscles**, and their contractions are under our conscious control. **Smooth muscles** are controlled without conscious thought and are called **involuntary muscles**.

Cardiac muscle is found only in the heart. It is similar to skeletal muscle, but is specially designed to contract over and over without tiring. This allows your heart to beat for a lifetime.

A muscle is known also as a **fiber**. See Figure 7-13. Many smaller units of fiber, known as **myofibrils** [mī-ō-fī-brĭls], make up each fiber. Dark

lines called **Z-bands** separate each myofibril into many identical-looking units known as **sarcomeres** [sär-kō-mērs]. A **sarcomere** is a segment of a fibril of a striated muscle. Within each sarcomere are many threadlike **protein filaments**, some thick and some thin. The thick filament is made of a protein known as myosin, and the thin filament is made of a protein called **actin**.

Each myosin molecule in the thick filament has a large head portion that forms a bridge to a thin actin filament. Using energy provided by ATP, myosin bridges pull the thin filaments toward the center

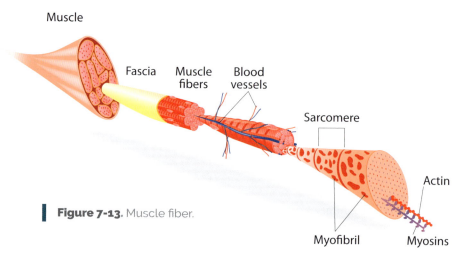

Figure 7-13. Muscle fiber.

Biology For Life

of the sarcomere. This results in the Z-bands being pulled closer together, which shortens the entire sarcomere, making the muscle fiber contract.

A muscle fiber can either contract fully or not at all. This is known as the all-or-none response. A *whole* muscle does not have an all-or-none response, since a whole muscle can contract at different strengths. The intensity of the entire muscle's contraction depends upon the number of its fibers that contract. More fibers contracting results in a stronger contraction of the whole muscle.

B. Chemistry of Skeletal Muscles

Muscles move by contracting, that is, by getting shorter. Just as a car or a plane requires energy to move, a muscle requires energy to contract. This energy comes from **ATP**. ATP is like the gas in the car's tank. The mitochondria provides ATP to the working parts of the muscle cell, which are the actin and **myosin filaments**. ATP breaks down, releasing energy so that the myosin fibers easily slide past the actin fibers, thus shortening the muscle. Both the actin fibers and the myosin fibers retain the same length, but they move past each other, like the two ends of an adjustable curtain rod. Each half of the curtain rod always remains the same length, but the total rod length is shortened when the one side slides over the other, until you attain the correct length for the window and the curtain. When the myosin fibers slide past the actin fibers, the muscle shortens and moves a bone.

Glucose is the primary fuel source for your muscles during day-to-day activities. The carbohydrate glycogen also serves as a fuel to meet the immediate demands of muscle cells during stressful emergencies or periods of strenuous exercise. **Glycogen** is stored in most muscles and the liver, and can be released to provide energy when needed. Upon release from its storage form it is quickly converted into glucose. The glucose molecules are rapidly broken down during the process of glycolysis. You read about this first step of cellular respiration in Chapter 5.3. During normal, anaerobic cellular respiration, the end product of glycolysis (glucose metabolism) is pyruvic acid. Pyruvic acid or

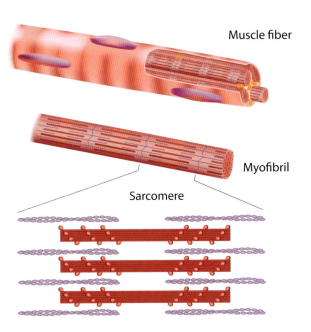

Figure 7-14. Structure of a myofibril.

pyruvate will be used by the Krebs cycle to make ATP in the mitochondria. It is worth noting that the ATP powers the muscle cell.

Sometimes during hard muscular work, there will not be enough oxygen to meet the demands of the cell. This is referred to as anaerobic metabolism and the end product is a buildup of lactic acid, instead of pyruvic acid. A person will breathe rapidly for awhile after stopping exercise because the increased oxygen intake is needed to convert the lactic acid into pyruvic acid. This causes a "burn" or pain in the muscle, which is also known as oxygen debt. The lactic acid must be transported to the liver by the blood where it is converted into pyruvate and then into glucose. These glucose molecules can be transported back to muscles cells for energy use, or they may be converted to glycogen.

In time, when a person becomes physically fit or accustomed to a specfic exercise, the heart rate and breathing become stronger, blood supply to the muscle cells increases, mitochondria become more abundant, the cells become supplied with more oxygen, and pyruvic acid will be broken down aerobically with oxygen. That is, the aerobic reactions can provide all of the energy needed and the cellular machinery prevents the buildup of lactic acid. No longer will the person be troubled again with sore muscles as long as he maintains the same level of physical fitness.

C. Joint Movement

Tendons, which are made up of connective tissue, attach muscles to bones. Most skeletal muscles are attached to two different bones. A **ligament** is a short band of tough, flexible, fibrous connective tissue that connects two bones or cartilages or holds together a joint.

When a muscle contracts, one bone usually acts as the anchor, and only one of the bones moves. The **origin of a muscle** is the place where the muscle attaches to the non-moving bone. The **insertion of a muscle** is the place where the muscle attaches to the moving bone.

When you bend your elbow, the biceps muscle does the work. The origin of the biceps is the top of the upper arm bone. The insertion, or place, of the biceps muscle is on a bone of the lower arm. When the biceps contracts, the lower arm is raised up. When you straighten the arm, the triceps muscle is used. The biceps then relaxes. Bending a joint is referred to as **flexion**, and straightening a joint is called **extension**.

The biceps and triceps muscles form what is called an antagonistic pair, which means that they perform opposite actions. When one of the pair contracts, the other muscle relaxes. Each member of an antagonistic pair usually moves the bone in a direction that is opposite to the other member of the pair. Most skeletal muscles consist of antagonistic pairs. Complex motions require many sets of antagonistic pairs of muscles at each joint.

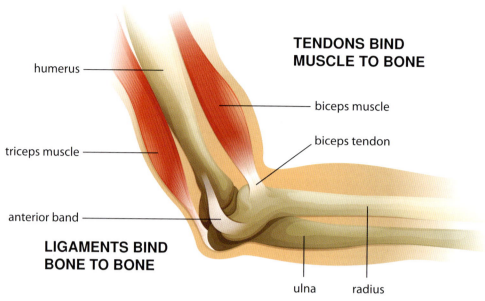

Figure 7-15. Tendons and ligaments.

Section Review — 7.3 A, B & C.

1. What are the three main types of muscles in the body?
2. What is the difference between a tendon and a ligament?
3. The _____ of a muscle is the place where the muscle attaches to the non-moving bone and the _____ of a muscle is the place where the muscle attaches to the moving bone.
4. The bending of a joint is called _____, and the straightening of a joint is called _____.
5. Identify striated muscle. What is its function?

CHAPTER 7: THE MUSCULOSKELETAL AND INTEGUMENTARY SYSTEMS

D. Cardiac Muscle and Smooth Muscle

Cardiac muscle refers to the type of muscle that makes up the heart. It shares some similarities with skeletal muscle. For instance, both types of muscle contain sarcomeres and mitochondria. The two types of muscle differ in that the cardiac muscle has shorter myofibrils and many more mitochondria.

Cardiac muscle, obviously, has a different function than skeletal muscle. Cardiac muscle is involuntary muscle, which means that it is under control of the automatic signals produced by the brain. A person cannot consciously control his heartbeat like he can control his muscles.

Additionally, cardiac muscle does not tire as easily as skeletal muscle. For example, a person could open and close his fist 80 times in one minute; however, the hand muscles would quickly tire out. On the other hand, cardiac muscle relaxes and contracts at this rate each and every minute, every day, for a person's entire life. In order for this work to be performed by the heart muscle, a constant supply of ATP is necessary. To meet these high-energy requirements, cardiac muscle contains many more mitochondria (the organelles considered the powerhouse of the cell) than skeletal muscle.

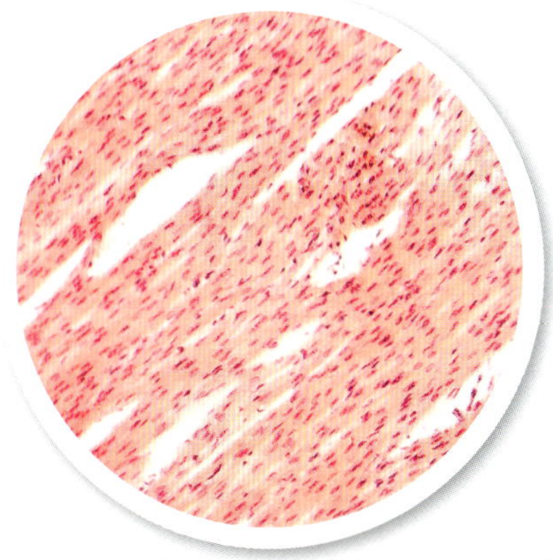

Figure 7-16. Cardiac muscle. It is created to be highly resistant to fatigue, having a large number of mitochondria (tube-like organelles considered the powerhouses of a cell), numerous myoglobins (oxygen-storing pigments), and a good blood supply, which provides nutrients and oxygen. Coordinated contraction of the cells of the cardiac muscle in the heart propels blood from the atria and ventricles of the heart to the blood vessels of the circulatory system.

Saints and the Heart

Saint Catherine of Siena loved God so much that she wanted to love Him perfectly and do only God's will, not her own. Saint Catherine prayed that Our Lord would take her heart and will from her that she might love and follow Him completely. She recited the words of Psalm 51:12 repeatedly, "Create a clean heart in me, O God." Our Lord answered her appeal and appeared to her holding a human heart and said, "Dearest daughter, as I took your heart away, now, you see, I am giving you Mine." As a sign of the miracle, a scar remained near St. Catherine's heart. May her prayer be our prayer, that Our Lord Jesus Christ would create a clean heart in us. May we repeat this petition throughout each day, so that one day our hearts will be pure for all eternity.

Saint Philip Neri, when he was twenty-nine, was praying with a great devotion for the gifts of the Holy Spirit in preparation for the feast of Pentecost. As he was praying, his chest became increasingly hot and swollen, and his heart began to beat very loudly. This strange phenomenon continued for the rest of his life. After his death, physicians and surgeons examined his chest and found that two ribs over his heart were broken and arched outwards to make room for his enlarged heart. Not only was his heart unusually large, but the main artery was twice the normal size, yet there was no sign of disease. It seems obvious that St. Philip Neri was so filled with love for God that his heart miraculously expanded as his love for God increased. May we try to follow his example and love God so much that it seems our heart must expand to contain our love for Him.

Biology For Life

Smooth muscle is the third type of muscle tissue. Each cell in the smooth muscle has its own nucleus and is long, thin, and tapered at both ends. Actin and myosin (the muscle proteins) fibers are present in the cytoplasm, but they are not arranged in repeating units, as they are in skeletal and cardiac muscle.

Smooth muscle is found mostly in internal organs, such as in the digestive system and blood vessel walls. Contractions of the smooth muscles of the digestive system move food through the digestive process, and contractions of the blood vessels regulate blood flow. Smooth muscle, like cardiac muscle, is involuntary, which means that the muscular contractions are not under conscious control.

E. Special Concerns Regarding the Muscles of the Young and Elderly

Muscular dystrophy refers to a group of genetic diseases that are marked by progressive weakness and degeneration of the skeletal muscles which control movement. The muscles of the heart and some other involuntary muscles are affected in some forms of muscular dystrophy, and a few forms involve other organs as well. There are nine different types of this illness. A few types are congenital, and the onset of the other types can begin between the ages of two to sixty. These illnesses are caused by defects in genes contained in a person's DNA. These genes are found in chromosomes in each body cell that codes for muscle proteins.

The goal of treatment for individuals with **muscular dystrophy** is to make the symptoms more manageable. Moderate exercise programs and physical therapy can minimize contractures, a condition in which shortened muscles around joints cause abnormal and sometimes painful positioning of the joints. The use of a cane or powered wheelchair may help maintain mobility and independence as long as possible.

Care is needed also for the respiratory and cardiac (heart) problems that are associated with muscular dystrophy. Because researchers now know that this disease is inherited due to a gene defect, gene therapies are being studied as possible ways to correct this disease in the future. Thank God for this scientific research which is aimed at understanding these painful and debilitating illnesses and finding a cure.

Cerebral palsy is a complete or partial paralysis of the muscles, primarily those of the arms and legs. Cerebral palsy may result from abnormal development of the brain or from brain damage that occurred before, during, or shortly after birth.

Physical therapy is needed to help children with **cerebral palsy** learn how to walk with the aid of braces and crutches. Special schools are available for those whose mental capabilities have been affected. Hearing aids and glasses are available to help with vision or hearing problems. Operations done by an orthopedic surgeon may alter the fixed stiffness in some deformed limbs and make movement easier.

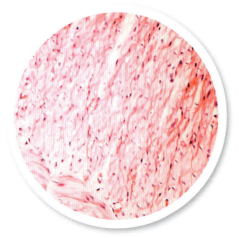

Figure 7-17. Smooth muscle. It is found within the layers of the large and small arteries and veins, the bladder, uterus, male and female reproductive tracts, gastrointestinal tract, respiratory tract, the ciliary muscle, and the iris of the eye.

> ### Section Review — 7.3 D & E.
>
> 1. Where is smooth muscle found?
> 2. What is muscular dystrophy?
> 3. What is cerebral palsy? What is the cause?
> 4. Identify cardiac muscle. How is it specialized?

7.4 The Human Skin

Section Objectives

- Know the two basic skin layers.
- Know the basic functions of the skin.

Skin is the largest organ of the body and is like a plastic container or large envelope that holds muscles and organs in place. It protects the body from germs and minor bumps and keeps body fluids in. Skin helps you breathe and controls the temperature of your body. It is an amazing, living, and important functioning part of the human body.

A. Structure and Function of Human Skin

Human skin is composed of two basic layers of tissue; the outermost layer, known as the **epidermis**, and the lower, thicker layer, called the **dermis**. There are many layers of **epithelial** [ĕp-ĭ-thē-lĭ-al] cells in the epidermis. The cells in the lowest layer are constantly undergoing mitosis. **Mitosis** is cell division that causes new cells to form. As new cells are produced, they are pushed toward the surface of the epidermal layer.

The outermost layer of the epidermis consists of dead cells. These outermost cells form the protective surface of the skin and serve as a protective barrier by preventing harmful substances and microorganisms from entering the body. The epidermis also protects the body from the harmful ultraviolet radiation of the sun.

Surprisingly, the skin, in the presence of ultraviolet light (sunlight), can produce more than enough Vitamin D that your body needs in as little as 15 minutes! By exposing your arms and legs to the midday sun for 10–15 minutes a day in the spring, summer, and fall, your skin can manufacture about 5,000 units of vitamin D, which is what is needed for good health. How fascinating it is that sunlight

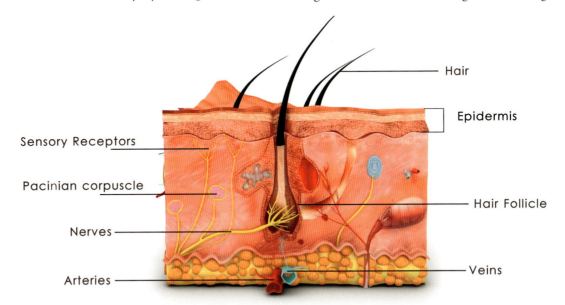

Figure 7-18. Structure of the skin. The skin plays a very important role in protecting the body against pathogens (agents which cause disease: bacteria, viruses, and fungi). Functions of the skin include insulation, temperature regulation, sensation, synthesis of vitamin D by sunlight, and the protection of vitamin B folates (a group of substances related to vitamin B). Severely damaged skin will heal by forming scar tissue. This is often discolored and depigmented (lacks color).

Biology For Life

or ultraviolet can be used to balance our nutritional requirements! Again, this shows that Our Creator God spared no detail when he made us.

The dermis contains adipose tissue, hair follicles, sweat glands, and nerve endings. The adipose tissue is made of fat cells. The body stores fat in these cells. Fat is typically used as an energy reserve but in the skin, it plays an important role against bumps and falls. In addition, it acts as an insulator by preventing the body from becoming to hot or too cold.

Hair follicles on the skin produce hair. Next to each hair follicle is a **sebaceous gland**. The sebaceous glands secrete oil, which helps to prevent the skin's outer layer from drying and cracking.

When the body becomes overheated, the **sweat glands** cause the skin to secrete water. When the sweat evaporates, the body carries heat away from the skin and cools the body. Thus, the skin helps maintain the body's internal temperature by regulating heat loss.

$$H2O\ (l) + heat \rightarrow H2O\ (g)$$

H2O is water. Here, liquid water in the body, H2O (l), is evaporated by the body heat and gives off water vapor, H2O (g), or water in the form of a gas.

The dermis contains many nerve endings that are sensitive to pressure, touch, heat, cold, and pain. The skin, the largest organ in the body, and the one we are most concerned about every day to keep clean and safe, is like a safety net or safety wrap to keep out germs and foreign substances. It helps regulate our body temperature when we are faced with extreme temperature changes. We need to take good care of our skin and protect it from too much sun, too much heat, or too much cold. We need to protect it from becoming too dry and cracked or too soft and thus losing its protective ability.

B. Special Concerns Regarding the Skin of the Young and Elderly

As we age, the skin becomes thinner. This is because the underlying fat, so abundant at infancy, is slowly lost. The skin's connective tissues, called collagen and elastin, undergo changes, causing the skin to lose firmness and become dry. The sweat- and oil-secreting glands in the skin also decrease, which makes it difficult for the skin to hold moisture. The blood vessels naturally become more fragile, so they are more likely to rupture and leak into the skin. This explains why some of the elderly may bruise easily.

There is little that we can do about the signs of aging. Prolonged exposure to the sun, especially during the peak hours of 10 a.m. to 2 p.m. apparently speeds up the aging process of the skin. However, we can reduce this effect by avoiding such over-exposure, or by wearing sunscreen creams with Sunscreen Protection Factors (SPF) of at least 15 when we are out in the sun. A hat and light, protective clothing which blocks the sun's rays also can help.

While too much sun can cause serious problems with sunburn now and skin problems in the future, some sunshine is necessary for a healthy lifestyle. Everyone, and not just the elderly should spend a little time in the sunlight during the week to help produce Vitamin D. Vitamin D is essential for maintaining healthy bones and for fighting many diseases. Vitamin D deficiency is a factor in various maladies, including multiple sclerosis, heart disease, osteoporosis, and some cancers. In the winter or on cloudy days, more time will be needed to produce an adequate supply of vitamin D and obtaining sunlight may not always be a possibility. However, in general, being outdoors will lift your spirits, put fresh oxygen in your system, aid in creating vitamin D, and give you an appreciation of the gifts of fresh air, sunshine, rain, and the view of the land that God created for us.

Section Review — 7.4 A & B.

1. Human skin is composed of two basic layers of tissue; the outermost layer known as the _____ and the lower layer known as the _____.

2. The epidermis consists of an outermost layer of dead cells that form a protective surface of the skin. What does it protect against?

3. What is contained within the dermis?

Chapter 7 Supplemental Questions

Answer the following questions.

1. List the type of tissue that makes up the skeletal and muscular systems.
2. Identify the tough membrane that covers the bones.
3. Know what type of cell secretes bone matrix.
4. Know what type of protein gives strength to bone.
5. Know the term for the lubricating fluid of the joints.
6. Give an example of a joint from each category: ball and socket, gliding, hinge, pivot, immovable.
7. List the major bones discussed in the chapter and be able to identify these bones on a drawing of the human skeleton. Know the diagrams on section 7.2 C.
8. Describe a congenital and age-related problem with the bones and muscles.
9. Know the two major proteins found in sarcomeres.
10. Know the purpose of tendons and ligaments.
11. Define antagonistic pair.
12. Know what substance is broken down to provide energy needed for ATP synthesis during vigorous exercise.
13. Know the scientific term for heart muscle.
14. How are cardiac muscles different from and similar to skeletal muscles?
15. Know the two basic skin layers.
16. Name the channels in bone through which blood vessels run.
17. Compare and contrast cardiac, skeletal, and smooth muscles.
18. Identify the following: myofibril, Z-band, and sarcomere.
19. Describe the energy source for striated muscular action.
20. Study all diagrams in this chapter. Be able to draw and label all diagrams.

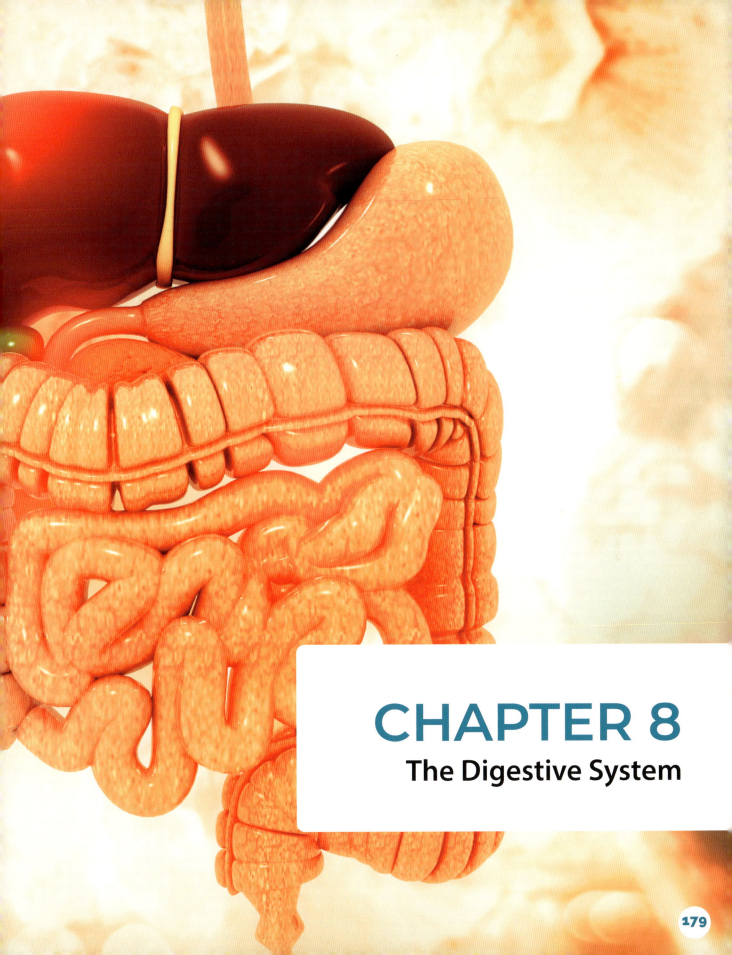

CHAPTER 8
The Digestive System

Chapter 8
The Digestive System

Chapter Outline

8.1 Introduction

8.2 The Mouth, Esophagus, and Stomach
- A. Mouth
- B. Esophagus
- C. Stomach

8.3 The Small and Large Intestines
- A. Pancreas and Liver
- B. Structure of the Small Intestine
- C. Function of the Small Intestine
- D. Large Intestine

8.4 Nutrition
- A. Carbohydrates and Fats
- B. Proteins
- C. Vitamins, Minerals, and Water
- D. Importance of a Balanced Diet
- E. Further Considerations: You Are What You Eat!

8.5 Special Concerns for the Newborn and Elderly
- A. The Newborn
- B. The Elderly

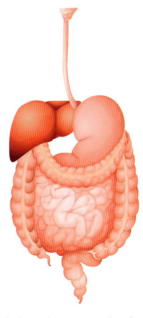

8.1 Introduction

When one thinks of the digestive system, one usually thinks of eating. For most healthy people, the consumption of food is a pleasurable activity. However, digestion involves far more than the process of eating. In the hidden recesses of our bodies, the very sustenance of human life takes place. God designed digestion to be a series of orderly physical and chemical steps by which the body obtains energy for life. For example, when you eat an apple, about 100 calories of energy are released as it is digested. Calories are units of measurement commonly associated with energy in foods. One Calorie is the amount of energy needed to raise one kilogram (2.2 Lb.) of water 1 °C.

The energy that we obtain from food is necessary to sustain life. When we begin depleting the energy stores in our bodies, we become hungry. Hunger is an essential part of our lives. Without it, we would not want to take food and would eventually die. The hunger we feel is insistent and sometimes painful. When we begin to eat, the feelings of hunger disappear and are replaced with an immediate sense of satiety.

People enjoy eating, and most of us have our own specific tastes in food. While the enjoyment of food is something good, the Church considers occasional voluntary fasting from food to be beneficial for our spiritual lives. All Fridays of the year are regarded as days of penance. Voluntarily fasting and abstaining from certain foods on Fridays are exemplary means to growth in holiness. Jesus provided our best example of fasting and went without food for forty days.

Throughout the centuries, the Church has required Catholics to abstain from meat. Why? Catholics offer their sacrifices in commemoration of Our Lord's most holy passion and death. During Lent, the Church obliges us to observe the rules of fasting and abstinence, most particularly on Ash Wednesday and Good Friday. Our Lord has made the most profound sacrifice for us, and thus faithful Catholics are pleased to offer Him, in return, due and reverent worship and sacrifice.

The digestive system may be thought of as a long tube with openings at each end. The passageway through the digestive system twists and turns. Various organs contribute to the process of food digestion as the food travels through this tube. The small intestine is the part of the digestive system where nutrients from food are absorbed into the blood stream. Inside the small intestine are hundreds of thousands of finger-like projections that create an enormous surface for absorption of food molecules. The total surface area of the small intestine is about 300 square meters, that is, the size of a tennis court. All of the nutrients needed for life's processes are obtained through this large surface of the small intestine. In this chapter, you will learn how all of the parts that make up the digestive system work together so that energy for our bodies can be obtained from the food we eat.

Biology For Life

CHAPTER 8: THE DIGESTIVE SYSTEM

8.2 The Mouth, Esophagus, and Stomach

The food that we eat is a complex mixture of proteins, carbohydrates, lipids, and other substances. Food must be broken down into small molecules in order to pass into the cells of our body. The cells will use these molecules for a variety of purposes, including energy use or storage, and cell growth and replication. The term **digestion** refers to the process of breaking down food into small molecules that can be absorbed by the body.

During digestion, food passes through a long tube-like structure known as the **alimentary canal**. This canal begins at the mouth and ends at the anus. Different parts of the alimentary canal are modified into specialized organs that have specific functions. Both chemical and mechanical processes are involved in digestion.

Mechanical digestion is the physical breakdown of food into smaller particles. Mechanical digestion occurs before chemical digestion. It is accomplished by chewing in the mouth and muscular churning of

Section Objectives

- Explain the difference between mechanical and chemical digestion
- Describe how the teeth and salivary glands begin initial digestion of food before the food is swallowed
- Label the parts of a tooth and know the purpose of each part
- Know the four different types of adult teeth
- State the function of peristalsis
- Explain the function of each organ making up the gastrointestinal (digestive) system
- List the three secretions of the lining of the stomach and state the functions of each
- Know the purpose of the sphincter muscles in the digestive system

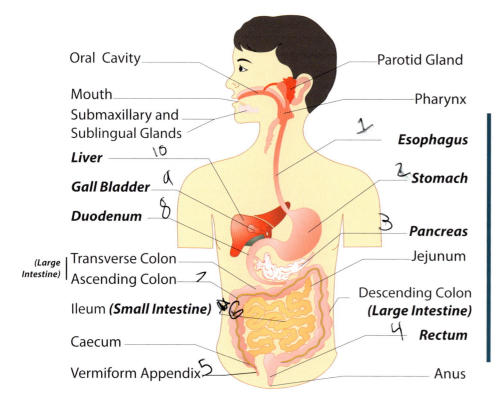

Figure 8-1. Human Digestive System. It serves two major functions: breaking down food into molecules the body can use and moving nutrient molecules into blood vessels or other vessels.

The major parts of the digestive system are shown in **bold italics**.

The large intestines are composed of the three parts of the colon.

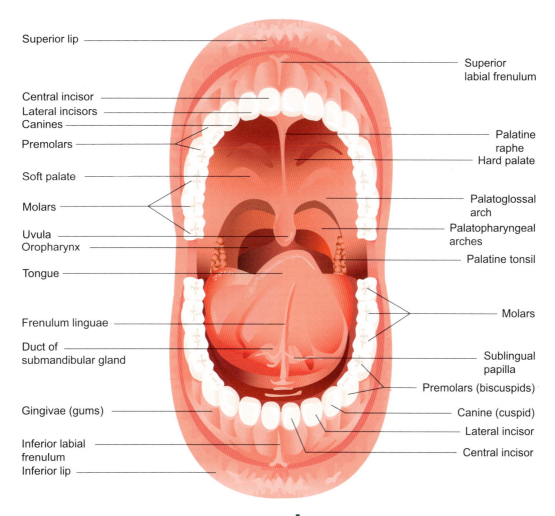

food in the stomach. Mechanical digestion speeds up chemical digestion, because it increases the surface area of the ingested food. The breaking up of food into smaller pieces allows the digestive enzymes to continue with chemical digestion. This is similar to the ease with which grains of sugar dissolve in water, much more easily than a cube of sugar.

Chemical digestion refers to the enzymatic breakdown of large molecules into smaller ones. This breakdown of large molecules is necessary, as the cells in the body more effectively use small molecules. In addition, the transport of large molecules is cumbersome and would require a great deal of costly energy. Certain large molecules are broken down into smaller molecules by combining with water, a process known as hydrolysis. Proteins are broken down into amino acids, and polysaccharides (known as complex carbohydrates) are broken down into simple sugars, mainly glucose. Triglycerides are hydrolyzed into fatty acids and glycerol.

Figure 8-2. Teeth in the mouth. Teeth are among the most distinctive (and long-lasting) features of the body. Teeth are often used to identify people after accidents when the rest of the body is unrecognizable.

A. Mouth

The passage of food through the alimentary canal starts at the mouth. Mechanical digestion begins in the mouth, with the teeth chewing and grinding food. Some chemical digestion begins as well, with the addition of saliva to the food; for example, the enzyme amylase is found in saliva and begins to breakdown starch before it enters the stomach.

Look at Figure 8-2. In the middle of the upper and lower jaws are four flat teeth called **incisors**. These teeth cut chunks of food from larger pieces. The pointed teeth are called **canines**, because they are most prominent in dogs, and are used for piercing and tearing food. The **premolars**, also known as bicuspids, are located next to the canine teeth. The **molars** follow the premolars. Both the premolars and molars have flat surfaces, and their function is to grind up food.

Figure 8-3, the structure of a tooth, shows that each tooth has roots that extend into the jawbones (maxilla = upper jaw and mandible = lower jaw). The **neck** of the tooth extends through the gums, which is the tissue that covers the jawbones. The chewing surface that we can see is known as the **crown** of the tooth.

A **pulp cavity** is located inside each tooth. This cavity contains nerves and blood vessels. A hard bone-like material called **dentin** surrounds the pulp cavity. A thin layer of harder material called **cementum** covers the dentin of the tooth's roots. **Enamel** is the hardest material in the body and covers the crown of each tooth. The **periodontal membrane**, or **periodontal ligament**, serves to anchor the roots of each tooth to the jawbones. This membrane is composed of fibrous connective tissue.

There are several secretions which are added to food in the mouth. Mucus comes from the cells lining the mouth. It moistens the food and makes it easier to swallow. **Saliva** is also a lubricant, which is added to food by three pairs of salivary glands. These three glands are known as the

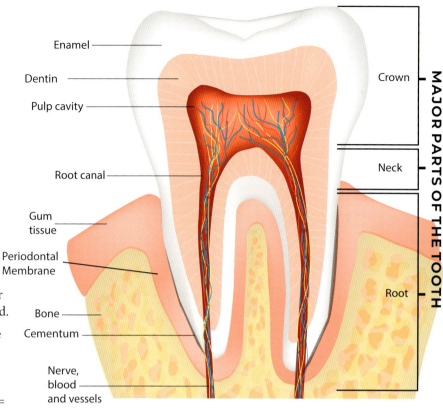

Figure 8-3. Structure of the Tooth.

parotid glands, the **submandibular glands**, and the **sublingual glands**. Saliva contains the enzyme **amylase**, which breaks down starch (a polysaccharide found in food) into **maltose**. You will recall that all enzymes end with "-ase".

You undoubtedly know that the tongue is involved in speech. It also has other important functions that are involved in the digestion of food. The tongue helps position food where it can be chewed by the teeth. In addition, the tongue pushes food to the back of the mouth during swallowing. The taste buds are also located on the tongue, as we learned in Chapter 6.

The process of swallowing requires complex muscular actions. Food is initially forced into the large area in the back of the mouth, called the **pharynx**. Muscles then push the food down the throat. The food then passes from the point where the air passage to the lungs branches off from the **esophagus** (ē-sŏf-a-gus), or food tube. Each time you swallow, a structure called the **epiglottis** (ĕp-ĭ-glŏ-tĭs) closes over the air passage. The purpose of the epiglottis is to prevent food and liquids from entering the lungs. When you "swallow down the wrong pipe," you swallow food or liquid too fast, before the epiglottis has a chance to cover the air passage. Your body responds by coughing violently to bring the material back up and into the esophagus.

> ### Section Review — 8.2 A
>
> 1. Distinguish what the difference is between chemical and mechanical digestion.
> 2. The upper jaw is the _____, and the lower jaw is the _____.
> 3. What are the three glands that produce saliva?
> 4. Name the nine major features of the human digestive system as identified in Figure 8-1.
> 5. The part of the tooth covered by enamel is the _____.
> 6. The _____ anchors the tooth's roots to the jawbone.
> 7. Name the three major features of the tooth as identified in Figure 8-3.
> 8. What is the purpose of the epiglottis?

B. Esophagus

As food is swallowed, it passes from the pharynx to the esophagus. The esophagus then carries food from the mouth to the stomach. The esophagus and the rest of the alimentary canal are lined with layers of cells known as the **mucosa**. These are cells that secrete mucus, which lubricate the lining of the tube, thus allowing for easier passage of food. The hollow interior space of the alimentary canal is known as the **lumen**. This includes the esophagus, the small intestine, and the large intestine.

Beneath the mucosa are two layers of muscle. In the innermost layer, muscle fibers wrap around the esophagus. The muscle fibers of the outermost layer run the length of the esophagus. These muscle layers alternately relax and contract. This causes the esophagus to squeeze together in places with a rhythmic pattern. These contractions move along the length of the esophagus in a wave-like pattern. The food is pushed ahead of these waves and through the digestive system. This rhythmic muscular action is known as **peristalsis** (pĕr-ĭ-stăl-sĭs).

C. Stomach

The **stomach** is a large, J-shaped organ at the end of the esophagus, on the left side of the body.

When food reaches the end of the esophagus, it passes a sphincter muscle, known as the **cardiac sphincter**. A **sphincter** is a circular muscle that closes a tube when it contracts. When the sphincter relaxes, food is allowed to enter the stomach. A second sphincter muscle, called the **pyloric sphincter**, is located at the end of the stomach. The two sphincters work together to keep food in the stomach. When the second sphincter (pyloric) relaxes, partially digested food moves out of the stomach and into the small intestine. One can think of the two sphincter muscles as two basketball coaches, getting their players in and out of the game but always keeping five players on the court.

There are three types of cells in the stomach mucosa (inner lining). One type of cell secretes mucus. Another type secretes enzymes, and the third type secretes hydrochloric acid and water. The enzymes, water, and hydrochloric acid combine to form **gastric juice.** Gastric juice is a very strong acid mixture having a pH of 2, which is similar to the pH of lemons and limes. The level of acid within gastric juice would burn your skin, but it does not harm the stomach, because the mucus layer provides a safe barrier. Gastric juice helps break up the connective tissues and cell membranes of the foods that we eat, thus releasing nutrients. It also kills many harmful bacteria and is one of the ways that we are protected from disease.

Biology For Life

CHAPTER 8: **THE DIGESTIVE SYSTEM**

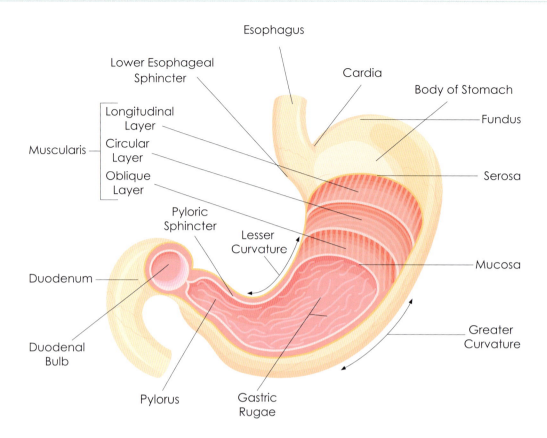

The digestion of all food occurs in sequence that begins in the mouth, then in the stomach, and then in the small intestines. All digestion is accomplished by mechanical and chemical means and we will use protein digestion as an example. Both types of digestion involve churning food for several hours. The final consistency of the stomach contents can be described as a thick soup and is called **chyme**. The enzymes produced by the stomach for chemical digestion are mostly **proteases**, which break down proteins. The primary stomach protease is called **pepsin**. It breaks down protein into polypeptides (long chains of amino acids). The polypeptides eventually are broken down into amino acids, which become the building blocks for new proteins in your body. It is at this point that chyme travels to the next digestive organ, the small intestine.

Figure 8-4. The stomach. It has three main functions: (1) breaking down food into smaller pieces, thus creating a larger surface area for easier digestion, and (2) holding food and releasing it at a constant rate, and (3) killing most harmful bacteria that are ingested.

Section Review — 8.2 B & C.

1. The hollow interior space of the alimentary canal is known as the _____.

2. What is the purpose of sphincter muscle in the digestive system? Name the sphincter muscles in the digestive system.

3. What is the function of peristalsis?

4. List the three secretions of the stomach lining, and state the functions of each.

Biology For Life

CHAPTER 8: **THE DIGESTIVE SYSTEM**

8.3 The Small and Large Intestines

Section Objectives

- Explain the roles of the pancreas and liver in digestion.
- Understand why the structure of the small intestine makes it well-suited for the process of absorption.
- Identify the absorption of water as the main function of the large intestine.
- Match digestive enzymes to the organ where they originate.
- Know the functions of the digestive enzymes.

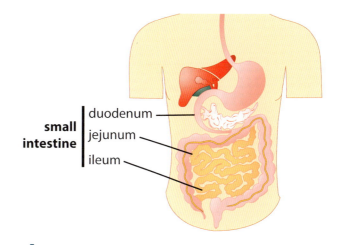

Figure 8-5. The small and large intestines. The small intestine is composed of the duodenum, jejunum, and ileum, and it is followed by the large intestine. Within the small intestine, nutrients diffuse through the villi (projections sticking out of the walls of the small intestine) into blood.

After the food in the stomach has reached the proper consistency, the food passes through the pyloric sphincter at the end of the stomach. Food then begins to move through the **small intestine**. If the small intestine were straightened out instead of coiled up inside the body, the length would be about six meters, or about twenty feet long. Most chemical digestion and absorption of food molecules occurs in the small intestine. Two digestive organs associated with the small intestine are the **pancreas** and **liver**.

The first section of the small intestine is known as the **duodenum**, and its length is about 20 centimeters (8 inches). The next section, about 2.5 meters long, is known as the **jejunum**. The last half of the small intestine is called the **ileum**. The final stages of digestion occur in the **large intestine**, where water and salts are absorbed and undigested wastes are expelled from the body.

A. Pancreas and Liver

Both the pancreas and liver aid the process of digestion. However, these two organs are not a part of the alimentary canal.

The pancreas has two main functions. First, it produces hormones that regulate the amount of glucose (sugar) that is carried in the blood. Second, the pancreas produces **pancreatic juice**, which passes into the duodenum through the pancreatic duct. Pancreatic juice neutralizes the acidity of the chyme (food material) from the stomach before the contents move into the small intestine. To neutralize

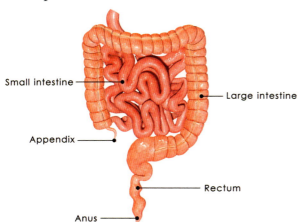

Figure 8-6. The large intestine. It is about 1.5 meters (4.9 ft) long, which is about one-fifth of the whole length of the intestinal canal.

Biology For Life

means to remove the acidity from the chyme. If you have ever taken an antacid tablet, such as Rolaids, you experienced the neutralization of the extra acid in your alimentary canal, which was giving you "heartburn" or making you feel uneasy.

Pancreatic juice contains a number of digestive enzymes. A few of these enzymes are proteases, which continue to break down proteins and turn them into amino acids. Recall from the above section that the stomach produces a protease known as pepsin. However, the pancreas produces several more proteases. Another enzyme that the pancreas secretes is **lipase**. It breaks down fat molecules into fatty acids and glycerol. Pancreatic juice also contains amylase, which is similar in function to the amylase secreted by the salivary glands of the mouth. That is, it helps break down starch and glycogen into simple sugars.

The liver has many functions. It converts excess glucose (sugar) in the blood into glycogen (a polysaccharide), which is the body's storage form of glucose. Glycogen is stored in the liver until the body needs it. The liver also produces **bile**, which is used in digestion. Bile does not contain any enzymes; however, bile contains substances that aid in the digestion of fats, or **lipids**. Lipids do not dissolve in water. Just as oil does not dissolve in vinegar in salad dressing, lipids in the diet do not dissolve in the watery gastric juice. The bile contains chemicals known as **emulsifiers**, which break apart fats and oils in the small intestine. The fats and oils are turned into tiny droplets, making the digestion of lipids by pancreatic lipase easier.

The **gall bladder** is the organ that stores bile. Bile passes through the **common bile duct** and into the duodenum when it is needed for digestion. A person can live without their gall bladder, and sometimes it is necessary to remove it from a person's body through surgery. People who have had their gall bladders removed must be careful to limit their intake of fat.

> ### Section Review — 8.3 A.
> 1. What are the three sections of the small intestine?
> 2. What are the two main functions of the pancreas?
> 3. What are the functions of the liver?
> 4. The liver produces a fluid called _____, which aids digestion but does not contain enzymes.
> 5. What is the purpose of the gall bladder?

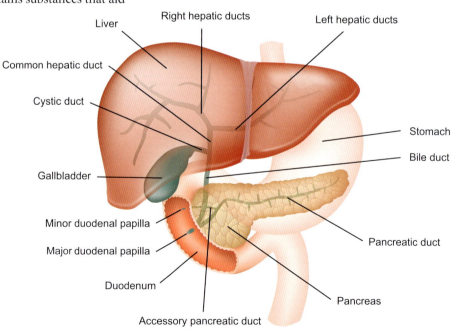

Figure 8-7. The pancreas and liver.

Biology For Life

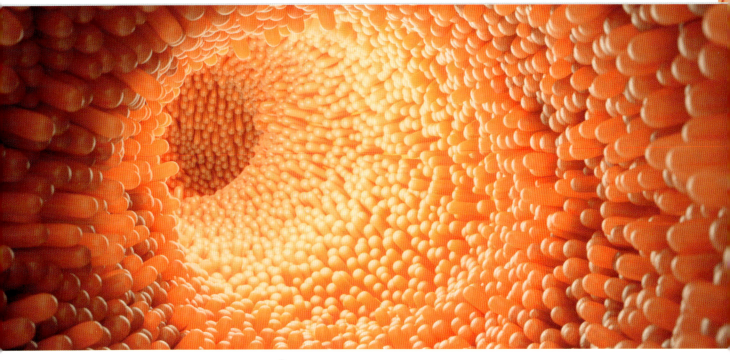

Figure 8-8. An illustration of the intestine lining, showing the villi.

B. Structure of the Small Intestine

The interior of the small intestine has a very large surface area, which maximizes the absorption of a very large number of food molecules. This large surface area is made possible by its many folds, which make the surface area twice as large as it would otherwise be. These tiny folds contain many finger-like projections called **villi**. These further increase the surface area of the small intestine by a factor of ten. In summary, it can be seen that the small intestine is not simply a hollow tube with minimal surface area. Villi within folds further increase the surface area and its digestive capabilities.

Many blood vessels are associated with each villus (singular form of villi). In addition, within each villus, there is a **lacteal**, or central tube, that is part of the lymphatic system. A single layer of epithelial cells covers each villus. Food molecules or the products of digestion must pass through these epithelial cells before reaching the blood vessels and lacteals within the villi. The plasma membrane of each epithelial cell is folded into many tiny projections called **microvilli**. These microvilli, along with the folds of the mucosa and the projections of the villi, further increase the surface area of the small intestine. The surface area of the small intestine can be likened to a cotton face cloth. The face cloth has many small threads and fibers, which greatly increase its surface area. Accordingly, it holds a great deal more soap and water than a flat piece of satin.

C. Function of the Small Intestine

In the duodenum, the first section of the small intestine, chyme is mixed with bile and pancreatic juice. This soup-like mixture then moves into the second section of the small intestine, the jejunum, and then into the third section, the ileum. Figure 8-5 shows this section.

Cells of the intestinal **epithelium** produce digestive enzymes. These enzymes are not released into the lumen of the digestive tract, but rather stay embedded in the cell membranes of the epithelial cells. One of these enzymes is **peptidase**. It breaks apart short polypeptides into individual amino acids. Other epithelial enzymes break disaccharides (two sugars chemically attached) into monosaccharides. For example, the enzyme **maltase** breaks apart maltose into two molecules of glucose. Many enzymes are produced throughout the digestive process.

The primary function of the jejunum is chemical digestion. Small molecules are absorbed through the villi and then into the blood vessels. Most chemical digestion is complete once the chyme reaches the ileum. It is at this point that the food has been completely broken down. The products of digestion have been absorbed into the bloodstream. Polysaccharides (these are listed as carbohydrates on food labels) have been broken down into simple sugars; proteins have been split into amino acids; and fats (lipids) have been broken down into fatty acids and glycerol. You should re-read this last section and make a special note of it, because it contains very important facts about foods and what happens to them in your body.

After sugars and amino acids cross the epithelial cell layer of the small intestine, they move into the bloodstream and are carried directly to the liver. Glucose is returned to the liver via the bloodstream to meet the demands of the body.

Fatty acids and glycerol follow a different path. These molecules pass into the intestinal epithelial cells and are immediately converted into triglycerides (lipids) and other important cellular building blocks before they move into the lacteals, which are connected with the lymphatic system. The lymphatic vessels bypass the liver and empty their triglycerides directly into the bloodstream.

D. Large Intestine

The digestive structures below the small intestine are primarily concerned with **excretion**. When the chyme passes from the small intestine to the large intestine, it is composed mostly of indigestible material that the body considers waste.

The **large intestine** has a wider diameter than the small intestine, and it is much shorter in length. The large intestine is only about 2 meters long (6-1/2 ft). Another name for the large intestine is the **colon**. A small projection called the **appendix** is located near the point where the small and large intestines meet. At one time it was believed that the appendix was a useless, or "vestigial," organ. Recent research, however, has shown that the appendix contains many infection-fighting cells and that it helps regulate intestinal bacteria. During times of illness, bacteria are flushed from the intestines. It has been suggested that the appendix serves as a safe haven for these

ORIGINS AND FUNCTIONS OF SOME DIGESTIVE ENZYMES

Enzyme	Origin	Function
Amylase	Salivary glands Pancreas	Starch breakdown
Pepsin (a Protease)	Stomach	Protein breakdown
Trypsin (a Protease)	Pancreas	Further breakdown of protein
Lipase	Pancreas	Fat breakdown
Peptidase	Intestinal epithelium (small)	Breakdown of short polypeptides
Maltase	Intestinal epithelium (small)	Maltose breakdown

bacteria and allows for the re-colonization of bacteria once the illness has passed.

The primary function of the large intestine is to absorb water and salts from the lumen. As water is absorbed from the indigestible wastes, these become more solid and are known as **feces**. These pass down the colon to the final 20 or 30 centimeters (8 to 12 inches) of the large intestine, called the **rectum**. The feces are stored in the rectum until they are eliminated from the body through the **anus**.

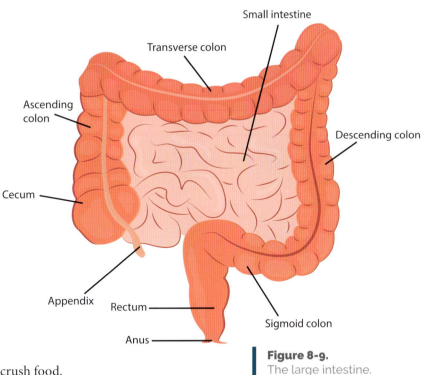

Figure 8-9.
The large intestine.

Summary of Events That Take Place in the Digestive System

1. **Mouth**: Teeth are used to crush food.
2. **Salivary glands**: Enzymes are secreted which begin starch digestion.
3. **Esophagus**: Peristalsis moves food to the stomach.
4. **Stomach**: Churning of food and the actions of gastric juice begin protein digestion.
5. **Liver**: Bile produced here; extra glucose converted to glycogen.
6. **Gallbladder**: Storage organ for bile until it is released into the small intestine to emulsify fats.
7. **Pancreas**: Pancreatic juice is produced to neutralize acidic stomach contents, further digest protein, and break down triglycerides into fatty acids and glycerol.
8. **Small intestine**: Intestinal enzymes digest carbohydrates, fats, and proteins; digested food is absorbed into the bloodstream and carried to the liver.
9. **Large intestine**: The main task of the large intestine is to absorb water from the lumen. The now drier waste, called feces, is stored in the colon, then moves in to the rectum, eventually leaving the body from the anus, the final part of the alimentary canal.

Section Review — 8.3 B, C & D

1. Why is the large surface area of the small intestine important?
2. Why does the structure of the small intestine make it well suited for the process of absorption?
3. List the digestive enzymes and the organs where they originate.
4. What are the functions of the digestive enzymes?
5. What is the main function of the large intestine?

CHAPTER 8: THE DIGESTIVE SYSTEM

8.4 Nutrition

> **Section Objectives**
>
> - Explain how the energy content of food is measured.
> - Explain the role of carbohydrates, proteins, and fats in proper nutrition.
> - List the functions of some of the important vitamins and minerals.
> - List the vitamins that are fat soluble and the vitamins that are water-soluble.
> - Know what is meant by a balanced diet.
> - List the major food groups.

We have reviewed the importance of food; our bodies need it to live. In the "Our Father," the prayer spoken by Our Lord Himself, we recite the words, "Give us this day our daily bread." Pope St. Pius X told us that the first meaning of "bread" in the "Our Father" is the Bread of Life in Holy Communion. Our Lord told us, "Truly, truly, I say to you, unless you eat the flesh of the Son of Man and drink His blood, you will not have life in you; he who eats my flesh and drinks my blood has eternal life, and I will raise him up on the last day" (Jn. 6:53-54). Bread and wine are the sensible elements of the sacrament of Holy Eucharist. Because we have bodies, God uses ordinary bread and wine to be the conduits for Himself in the Bread of Life. The second meaning of the word bread in the "Our Father" is that we might have food to sustain our lives. We ask God for our daily food while recognizing that many people in the world go to bed hungry.

Our bodies use food for energy. In addition, the food that we eat provides the materials for the growth, maintenance, and replacement of tissues of the body. The process by which organisms obtain and use food is called **nutrition**. The term **nutrients** applies to all the chemical substances that are eaten in order to grow and remain healthy.

Nutrients are divided into two groups: (1) those that can provide energy, which includes carbohydrates, fats, and proteins, and (2) those that do not provide energy but are essential for life, which includes vitamins, minerals, and water.

Food calories, or **kilocalories**, are the units used to measure the energy content of food. A calorie (with a lower case "c") is a unit of energy and is defined as the amount of energy required to raise

the temperature of one gram of water by one degree Celsius. One thousand calories is equivalent to one kilocalorie, or one Calorie (with an upper case "C"), or one food calorie. The Calorie, or kilocalorie, is the amount of energy needed to raise the temperature of one kilogram of water by one degree Celsius. When nutritionists list the number of calories in foods, they are referring to kilocalories. All food molecules contain chemical bonds that release energy when these bonds are broken. The Calorie content of a food is an indication of how much energy will be released after the food is completely broken down. The unit by which food energy is measured is a metric system unit called the Joule. One food calorie, or Calorie, is equal to 4,184 Joules. You will learn more about **Joules** when you take chemistry or physics.

A. Carbohydrates and Fats

Carbohydrates are primarily used to supply our bodies with energy. Most carbohydrates are broken down to the simple sugar glucose. In the process of respiration, glucose combines with oxygen, and the compound ATP is produced. You will recall that ATP provides the cells with usable energy and is known as the "universal currency."

Sugars, starches, and cellulose are the most common carbohydrates found in foods. Some single molecule sugars are the monosaccharides; for example, fructose, glucose, and galactose. **Fructose** is the sugar found naturally in fruits. **Glucose** is the sugar found naturally in plants and leaves. Galactose is found in dairy products and sugar beets.

Other sugars are disaccharides, which are composed of two monosaccharides linked together to form a larger sugar. Some examples of disaccharides are **sucrose** (table sugar), **maltose**, and lactose (sugar found in milk).

Sugar molecules can join to form long chains called **polysaccharides**, known as starches. Starches take longer to digest than the smaller monosaccharides and disaccharides. Grains and grain products, such as breads and rice, contain large amounts of starch. Certain vegetables, such as potatoes, also contain a great deal of starch. Persons who are diabetic need to limit their intake of the simple sugars glucose, sucrose, and fructose,

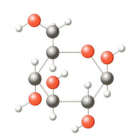
Glucose

Fructose

Figure 8-10. A ring diagram of the monosaccharides glucose and fructose. These carbohydrates are isomers, having the same atoms but in different arrangements.

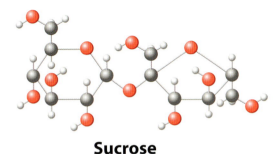
Sucrose

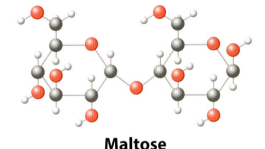
Maltose

Figure 8-11. Ring diagrams of the disaccharides maltose and sucrose. Maltose is made of two condensed glucose molecules; sucrose is made of glucose joined to fructose.

since they will more quickly enter the blood stream than the polysaccharides, such as bread, potatoes etc.

Cellulose is another polysaccharide. Cellulose forms the cell walls of plant cells. Humans are unable to digest cellulose; however, it is still important in our diet. It provides fiber, or roughage, against which the muscles of the intestine push. This pushing motion causes food to move more quickly through the alimentary canal.

When digestible carbohydrates are eaten, the body uses these first, to satisfy its immediate energy needs and to maintain the standing supply of glycogen in the muscles and the liver. Any excess carbohydrates, or those that will not be used, may be converted to fat. The undigested cellulose, or fiber, passes through the digestive tract and is eventually expelled from the body with feces.

Most of the fats that we consume are in the form of triglycerides. Triglycerides provide energy for the body and contain twice as many food calories per gram than do carbohydrates. Fats are stored in special cells called adipose cells. The body can mobilize these fats when it needs energy. The body also uses fat for purposes other than energy. Fat is used to make cell membranes. It surrounds the nerves, which aids in the conduction of nerve impulses. In addition, fat is a good insulator and helps retain warmth in the body. It is not surprising, then, that many mammals and birds (whales, walruses, penguins) that spend time in cold water have a great deal of fat. Non-migrating birds that spend winter in the northern half of the United States also "put on a layer of fat" before winter. Finally, certain vitamins are found only in fatty foods, for example, vitamins D and E. Foods that have a high fat content are margarine, butter, salad oils, cheeses, nuts, and some meats.

B. Proteins

Proteins supply the body with energy, though that is not their primary function. Proteins make up much of the structure of the cells and are needed for cell growth and repair. Enzymes are complex proteins that play a major role in chemical reactions that take place inside the body. Accordingly, many of the proteins that we digest are eventually used to make enzymes.

Like fats and carbohydrates, proteins contain the elements carbon, hydrogen, and oxygen. Proteins differ from fats and carbohydrates in that they also contain the element nitrogen. You will recall that proteins are nothing more than long chains of folded amino acids. During the digestive process, proteins are broken down into amino acids. These are transported to the cells of the body via the bloodstream. They are then reassembled to

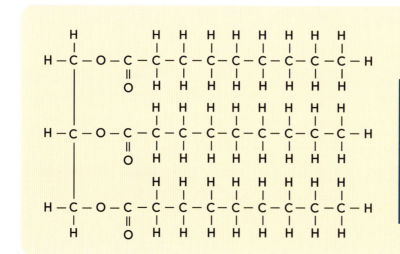

Figure 8-12. A triglyceride molecule. Three fatty acid molecules have bonded to a glycerol molecule. Triglycerides are the main constituents of vegetable oils and animal fats.

make new proteins. Some good sources of protein are meat, milk, fish, eggs, dried beans, seeds, and cheese.

The human body makes use of all 20 amino acids found in nature. Eight of these twenty must be obtained from food and are known as the essential amino acids: phenylalanine, valine, threonine, tryptophan, isoleucine, methionine, leucine, and lysine. The remaining twelve amino acids must either be synthesized by the body or obtained from food. It is worth noting that the body cannot make protein unless it has all of the essential amino acids present at the same time. In addition, the body is not capable of storing amino acids, as it does with fat.

There are certain foods known as complete protein foods, which contain all of the essential amino acids. If you eat one of these foods, you will have all of the essential amino acids in your body at one time. Meat, fish, poultry, milk, cheese, and eggs are all complete protein foods. Some foods, such as soybeans, do not contain all of the essential amino acids. Soybeans lack the amino acids valine and methionine. Rice does not contain the amino acids lysine or leucine. However, if you eat both rice and soybeans together, you obtain all of the eight essential amino acids.

Section Review — 8.4 A & B

1. List the nutrients that provide energy and those that do not.
2. The energy content of food is measured in _____.
3. What is a monosaccharide?
4. Which nutrients provide the most energy?
5. What is the primary function of proteins?
6. When proteins are digested, they are broken down into ____ ____.

C. Vitamins, Minerals, and Water

Vitamins, minerals, and water do not supply the body with energy, but they do perform other functions. **Vitamins** function as coenzymes, which are a necessary part of many chemical reactions that take place in the body. You may recall from Chapter 2.7.C that enzymes are proteins that act as catalysts. **Coenzymes** are smaller molecules, not proteins, that are not directly catalysts, but that do act to enhance the catalytic action of the enzymes involved in those reactions. The food sources of many vitamins and minerals, as well as their functions, are described in the tables at the end of this section.

Conditions known as **deficiency diseases** can result if vitamins are missing from the diet. For example, people may develop scurvy if they do not get enough vitamin C in their diets. Their gums may swell and bleed; their teeth may fall out; and their joints can become sore. A condition known as rickets occurs with a deficiency of vitamin D. Rickets causes the bones to become soft and can result in deformities.

Vitamins are divided into two groups based on the substance in which the vitamin will dissolve. The water-soluble vitamins will dissolve in water. These vitamins include vitamin C and the B complex vitamins. Water-soluble vitamins can be lost from foods if the food is cooked in water for too long. The second group of vitamins includes those that dissolve in fat and are not soluble in water. This group includes the fat-soluble vitamins A, D, E, and K. Nutritionists believe that it is healthier to get your supply of vitamins from eating the proper foods, rather than relying on vitamin supplements, due to the fact that supplements do not contain other important nutrients that are found in foods.

A few decades ago, some believed that raw vegetables were better for you than those that were cooked. It was thought that cooking removed vitamins and minerals from vegetables. However, researchers found that people who ate cooked vegetables obtained more nutrients than those who ate raw ones. The reason for this is that cooking helps break down the cell walls of vegetables, making the vitamins more accessible to the digestive system. However, it is important that foods are not overcooked or reheated several times.

Minerals are inorganic substances, which means they do not contain the element carbon. The minerals that we obtain from our diet have a variety of functions. They maintain the acid-base balance in the body, regulate the amount of water in the blood, and are an important part of many body structures, such as bone and red blood cells.

Finally, the last and most important of nutrients is water. Life cannot exist without it! It is possible to go without food for some time or thrive with a malnourished diet, but the body will not function properly for long without water. Most people do not drink enough water, and the average person requires about 8 glasses a day. You can obtain water by drinking it alone or by drinking fruit juices and milk. Water is also present in many foods. Water has many functions in the body, such as dissolving food materials so that they can pass into the bloodstream. Also, many of the chemical reactions that occur in the body take place in water or aqueous solutions.

The functions of the vitamins in our diet are listed in the table below.

DIETARY VITAMINS

Vitamin	Source	Function	Symptoms of Deficiency
A	Egg yolks, butter, green and yellow vegetables, organ meats, fish liver oils	Growth, healthy skin and eyes	Night blindness, epithelial changes in cells, retarded growth
B1 (Thiamin)	Seafood, poultry, meats, whole or enriched grains, green vegetables, milk, soybeans	Carbohydrate metabolism; growth; heart, muscle, and nerve function	Retarded growth, beriberi, nerve disorders, fatigue
B2 (Riboflavin)	Milk, eggs, poultry, yeast, meats, soybeans, green vegetables	Carbohydrate metabolism, growth	Retarded growth, premature aging
Niacin	Leafy vegetables, peanut butter, potatoes, whole or enriched grains, fish, poultry, meats, tomatoes	Growth, carbohydrate metabolism, growth	Digestive and nervous disturbances
B12	Liver	Growth, carbohydrate metabolism, digestion, nerve function	Anemia

DIETARY VITAMINS CONTINUED

Vitamin	Source	Function	Symptoms of Deficiency
C	Citrus fruit, tomatoes, leafy vegetables	Growth, healthy gums	Sore gums, susceptibility to bruising
D	Milk, liver, eggs, fish liver oils, sunlight	Growth, calcium and phosphate metabolism	Poor tooth development, rickets
E (tocopherol)	Vegetable oils, butter, milk, leafy vegetables	Protects cell membranes, reproductive function	Peripheral nerve damage, breakdown of red blood cells
K	Green vegetables, tomatoes, soybean oil, most made by intestinal bacteria	Blood clotting, liver function	Hemorrhaging

The functions of the minerals in our diet are listed in the table below.

DIETARY MINERALS

Mineral	Source	Function	Symptoms of Deficiency
Calcium	Milk and other dairy products, bean curd, dark green vegetables	Tooth & bone formation, nerve transmission, muscle contraction	Rickets, osteoporosis
Phosphorus	Most foods	Bone development, transfer of energy in cells	Fragile bones
Sodium	Meats, dairy products, salt	Nerve transmission, muscle contraction	Dehydration, shock
Chlorine	Salt	Formation of HCl used in stomach acid	Abnormal contraction of muscles
Potassium	Fruit	Regulation of heart beat, maintenance of water balance, nerve transmission	Heart dysfunction
Magnesium	Nuts, grains, dark green vegetables, seafood, chocolate	Catalyst for ATP formation	Weakness, mental confusion

CHAPTER 8: THE DIGESTIVE SYSTEM

DIETARY MINERALS CONTINUED

Mineral	Source	Function	Symptoms of Deficiency
Iodine	Seafood, iodized salt	Thyroid activity	Goiter
Iron	Meats, dark green vegetables, dried fruits	Hemoglobin formation	Anemia

D. Importance of a Balanced Diet

Twice in the New Testament, we read about the taking of food to be proof of life. During Jesus' visible stay on earth, He performed many great miracles, none more dramatic than raising the dead. On one occasion, Jesus came upon people who were weeping, for a little girl had died. "He said to them, 'Why do you make a tumult and weep? The child is not dead but sleeping.'…Taking her by the hand he said to her, "Talitha cumi"; which means, 'Little girl, I say to you, arise.' And immediately the girl got up and walked…And immediately they were overcome with amazement. And he strictly charged them that no one should know this, and told them to give her something to eat." (Mk. 5:39-43). The dead cannot eat; perhaps Jesus wanted to prove to the witnesses that the girl was indeed alive again, giving glory to His heavenly Father. The fact that He told the witnesses to give her something to eat shows the importance of food.

Later, after Jesus' resurrection from the dead, He appeared to His disciples one morning at daybreak. They had been fishing but had not caught any fish. Jesus told them to cast their nets back into the water, whereupon they caught 153 fish. He then invited them to eat breakfast with Him, proving that He had really risen from the dead and was not just a figment of their imaginations (Jn. 21:4-14). The incredible events of those several days came together for the apostles, who were now eyewitnesses to proof of Christ's resurrection.

Nutritionists divide the sources of nutrients into five main groups, and they recommend eating servings of each group every day. These are the meat group, the milk and dairy group, the fruit group, the vegetable group, and the cereal-grain group.

- The **protein group** or **meat group** includes meat, poultry, fish, eggs, and legumes. The legumes, which can serve as meat substitutes, include beans, lentils, peas, and peanuts.
- The **dairy group** includes foods such as milk, cheese, and yogurt.
- The **fruit group** contains all of the fruits.

CHAPTER 8: THE DIGESTIVE SYSTEM

- The **vegetable group** contains all of the vegetables. The fruits and vegetables may be fresh, frozen, canned, or dried.
- The **grain group** includes breads, cereal, rice, pasta, and tortillas.

Servings from the main food groups should be eaten daily. Sweets and desserts are listed as the sixth group, the minor group, but, while these foods are delicious to eat, they have high levels of fat and sugar and should be eaten only sparingly. By consuming foods from each of the main groups each day, you get a well-balanced diet of a variety of the necessary vitamins, minerals, proteins, carbohydrates, and fats needed for life.

Malnutrition is a serious health problem that may develop if a person does not eat a balanced diet. It is worth keeping in mind that a person can eat more calories than necessary, or be overweight, and still suffer from malnutrition. Malnutrition affects about 1/6th of the world's population. Millions of people are actually starving to death because they are not getting adequate amounts of the right kinds of foods. People who are undernourished are more susceptible to disease. Poor diets can harm the growth and development of children. For example, brain damage can occur if babies do not get enough protein.

Back in the days when sea travel was the main method to cross the oceans a disease known as scurvy began to appear in sailors. This hideous

Figure 8-13.
Food Group Pie Chart

Biology For Life

199

disease killed an estimated 2 million people between the 15th and 18th centuries. A breakthrough came during a voyage to the West Indies, in 1779, when a Scottish physician, Sir Gilbert Blane, enforced the use of lemon juice and other fresh foods in the diet. During Britain's Napoleonic wars, he organized the distribution of limes to seamen that reduced scurvy among sailors. Because of this important use of limes, British sailors were referred to as "limeys."

Obesity is a condition characterized by an excessive amount of body fat. This condition is not as serious as undernourishment, but obesity is also unhealthy. People who are overweight frequently have health problems, such as diabetes and heart disease. In our present day you are probably aware of the many different types and varieties of "junk food" which exist for you to consume. Many of these foods contain too much sugar and fat and lack a balance of essential vitamins and minerals. While these foods are tempting, the desire to eat them has to be calmed, and they should be eaten as little as possible. Otherwise one risks becoming overweight and possibly obese.

People often go on diets to lose weight; however, the diet must be well-planned and nutritionally balanced. An exercise regimen approved for the person by a physician will also be helpful, because exercise helps maintain muscle and lean body weight. People need to avoid going on fad diets, which promise rapid weight loss but often lack important nutrients. It is dangerous for the body to lose too much weight too quickly. Your body cannot function properly if your calorie intake is too low. Also, if large amounts of weight are lost quickly, the pounds usually accumulate again once the person goes off of the fad diet. Very often the person gains back all of the weight that was lost on the fad diet and even gains extra weight, especially if exercise is absent. During the fad diet, muscle and lean body mass, more than stored fat, are lost. When the diet ends and the person goes back to "normal" eating habits, the body reacts to the weight loss stress and adds fat to prevent that from happening again. When exercise is present the body does not lose the lean mass in the first place, but it reduces stored fat. If exercise is maintained when the "normal" eating habits are resumed, the fat will not be able to re-accumulate as quickly. Proper nutrition is extremely important, and a balanced diet of moderation in all food groups is necessary for optimum health.

E. Further Considerations: You are What You Eat!

In addition to a balanced diet, it would be wise to consider the importance of natural, fresh foods and ingredients. As modern living continues to accelerate in pace, concerns about convenience and expedience conflict with those of wholesomeness and nourishment. Highly processed or prepackaged foods tend to include many artificial ingredients such as monosodium glutamate (MSG), high fructose corn syrup, hydrolyzed proteins, modified cornstarch, and hydrogenated oils. These foods may be convenient, but a steady diet of them can harm a person's health. If we "are what we eat," then we need to pay attention to what kinds of things make up our diet.

Examining just one aspect of the many food processing methods, Dr. Michael T. Murray, a physician and author of more than twenty books, recommends, "Eliminate the intake of margarine and foods containing trans-fatty acids and partially hydrogenated oils." Dr. Murray continues, "During the process of margarine and shortening manufacture, vegetable oils are hydrogenated. This means that a hydrogen molecule is added to the natural unsaturated fatty acid molecules of the vegetable oil to make it more saturated. This hydrogenation changes the structure of the natural fatty acid to many 'unnatural' fatty acid forms which interfere with the body's ability to utilize essential fatty acids."

Dr. Murray then warns, "Trans-fatty acids and hydrogenated oils have been implicated as contributing to … many health … disorders." Refined white sugar and table salt are also related to serious conditions like diabetes and high blood pressure. So it is wise to minimize our intake of these ingredients.

A rule of thumb when purchasing food is to buy fresh food, and it does not necessarily have to be labeled "organic." If it is possible, always buy fresh food and prepare it fresh, so that it retains most if not all of its nutrients. Freshly prepared foods that are cooked and eaten immediately are a good choice. It is important to recognize that the closer food is to fresh food, the better it is for you. Many people plant vegetable gardens. What they do not eat in the summer can be given away to friends or family or canned for use in the winter. Why not consider growing a vegetable garden of your own? It is work, but is a good way to learn about plants. In addition, you will be assured of fresh food. Finally, it is a good way to learn about God's Creation, obtain exercise and spend time together as a family.

Section Review — 8.4 C, D & E

1. _____ are usually coenzymes, and they are needed for chemical reactions to occur in the body.
2. The fat-soluble vitamins are ____, ____, ____, and ____.
3. The water-soluble vitamins are ____, and ____.
4. List the major food groups.
5. What is meant by a balanced diet?

Biology For Life

8.5 Special Concerns for the Newborn and Elderly

> **Section Objectives**
> - Explain some of the differences between the digestive systems of the newborn and elderly.

Following birth, newborn babies are not capable of producing all of the enzymes necessary to digest all types of food. Some of the digestive organs need more time to mature. However, they can process their mothers' milk, the food God provides them naturally, while their other organs and glands develop.

In the elderly, there are different problems. While the function of some digestive organs may decline over time, the main problem with older people tends to be a loss of function in the muscles involved with digestion.

A. The Newborn

The ability of the newborn to digest, absorb, and metabolize foods is adequate, but it is limited in certain functions. Enzymes are available to break down proteins and carbohydrates (mono- and disaccharides), but, due to a deficiency in the production of pancreatic amylase, complex carbohydrates (polysaccharides, namely starch) cannot be broken down. Also, a deficiency of pancreatic lipase will limit the absorption of fats, especially foods with a high saturated fat content, such as cow's milk. Babies under one year of age should not be given cow's milk, but, rather, they should be given breast milk or formula specially prepared for infants.

The liver of the infant is the most immature of the digestive organs. The activity of the enzyme **glucuronyl transferase** is reduced, which contributes to a temporary condition known as **jaundice** (yellowing of the infant's skin) that is often seen in newborns until about the eighth day after birth. The liver also cannot form plasma proteins well. A decrease in the amount of plasma proteins plays a role in **edema** (swelling due to an excess of tissue fluid) often seen at birth.

Some of the blood's clotting factors are also low in the newborn. The liver stores less glycogen (the storage form of glucose) at birth compared than later in life. Consequently, the newborn is more prone to **hypoglycemia** (low amounts of sugar in the blood), which may be prevented by early and effective feeding, especially breast-feeding.

Some salivary glands are functioning at birth; however, the majority of these glands do not secrete saliva until about age 2 to 3 months, when drooling is common. The infant requires frequent small feedings, because the stomach capacity is limited to about 90 ml (less than 1 cup of volume). Infants who are breast-fed usually have more frequent feedings and more frequent stools than infants who receive formula.

The infant's intestine is longer when compared to its body size than the intestine of an adult. Therefore, there are a larger number of secretory glands and a larger surface area for absorption of nutrients as compared to the intestine of an adult. There are rapid peristaltic waves (peristalsis) and simultaneous nonperistaltic waves, which occur along the entire esophagus. These waves, combined with an immature, relaxed cardiac sphincter (the muscular valve at the top of the stomach), cause the infant to frequently spit up small amounts of formula or breast milk. The infant's pattern of stools will change progressively during the first week of life. These changes indicate a properly functioning digestive tract.

B. The Elderly

The gastrointestinal (digestive) tract remains resilient throughout life and does not lose its ability to function. However, there can be some loss in the capacity of the muscles to coordinate their movements, which adversely affects the regularity of peristalsis. Consequently, constipation is a common problem in seniors. The contractions that take place

in the upper gastrointestinal tract during swallowing can also weaken and become less well-coordinated. This lack of coordination in this area of the digestive system can lead to **gastroesophageal reflux** (food which has mixed with acid in the stomach backing up into the esophagus), which is more commonly called **heartburn**. You probably have not experienced heartburn, but your parents and grandparents will likely tell you that they have. Once again, proper diet can help avoid some or possibly all of this uncomfortable heartburn.

The production of the gastric juices pepsin and hydrochloric acid, as well as saliva, tends to decrease as a person ages. The rate of absorption of vitamins and minerals can change over time. Because the ratio of body fat to body water increases as we age, compounds that are soluble in fat, such as vitamin A, may have more rapid absorption and metabolize more easily than compounds that are soluble in water. Compounds that are water-soluble may not be used as efficiently. Changes may take place in the way the body absorbs minerals, particularly calcium and iron.

Another common characteristic of the bowels of aging people is the formation of diverticula in the large intestine (diverticulosis). **Diverticula** are small, pouch-like formations, which can result from increased pressure on the walls of the bowel and decreased resilience of the muscles. This disorder can be present in as many as one-third of seniors. The problem can be harmless, unless bits of stool become lodged in the diverticula, causing inflammation and infection. When diverticula become infected, the condition is known as **diverticulitis**.

CHAPTER 8: THE DIGESTIVE SYSTEM

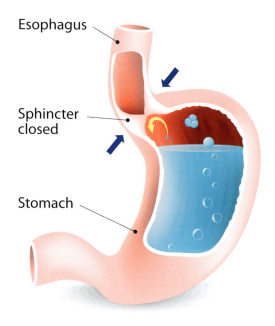

Healthy

Another negative result of aging involves the liver, one of the most important organs in the body. The liver becomes smaller over time and loses some of its blood supply. The liver is also less capable of repairing damaged tissue, and is less efficient in its filtration of toxins from the body. However, the liver should function well throughout life. The pancreas also should function well throughout life. However, the position of the pancreas can shift slightly, and it can accumulate fat and scar tissue with age. By thoroughly chewing and swallowing food and maintaining a well-balanced diet, you can enable your digestive tract to work well throughout life.

Section Review — 8.5

1. _____ and _____ are two conditions of the newborn that can be attributed to immature development.

2. Early and proper feeding of the newborn can prevent _____, a condition of low blood sugar.

3. _____ are small pouch-like formations in the large intestine that can form from high pressure on the walls of the bowel. If these become infected a condition called _____ occurs.

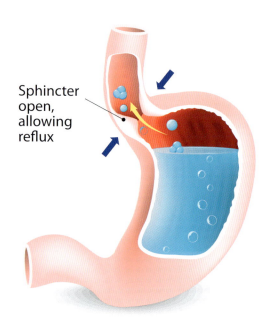

GERD

Figure 8-14. Gastroesophageal Reflux, more commonly called heartburn, is when food and stomach acid come back up the esophagus.

Biology For Life

Chapter 8 Supplemental Questions

Answer the following questions.

1. Explain the difference between mechanical and chemical digestion.
2. Describe how the teeth and salivary glands begin initial digestion of food before the food is swallowed.
3. List the parts of the alimentary canal.
4. Label the parts of a tooth and know the purpose of each part.
5. Know the four different types of adult teeth.
6. State the function of peristalsis.
7. Explain the function of each organ making up the gastrointestinal (digestive) system.
8. List the three secretions of the lining of the stomach and state the functions of each.
9. Know the purpose of the sphincter muscles in the digestive system.
10. Explain the roles of the pancreas and liver in digestion.
11. Understand why the structure of the small intestine makes it well-suited for the process of absorption.
12. Identify the main function of the large intestine.
13. Know the digestive enzymes and the organs where they originate.
14. Know the functions of the digestive enzymes.
15. Explain how the energy content of food is measured.
16. Explain the role of carbohydrates, proteins, and fats in proper nutrition.
17. List the functions of some of the important vitamins and minerals.
18. What are the vitamins that are fat soluble, and what are the vitamins that are water-soluble?
19. Know what is meant by a balanced diet.
20. List the major food groups.
21. Explain some of the differences in the digestive systems of the newborn and elderly.
22. Know the diagrams in section 8.2.
23. Study all diagrams in this chapter. Be able to draw and label all diagrams.

Biology For Life

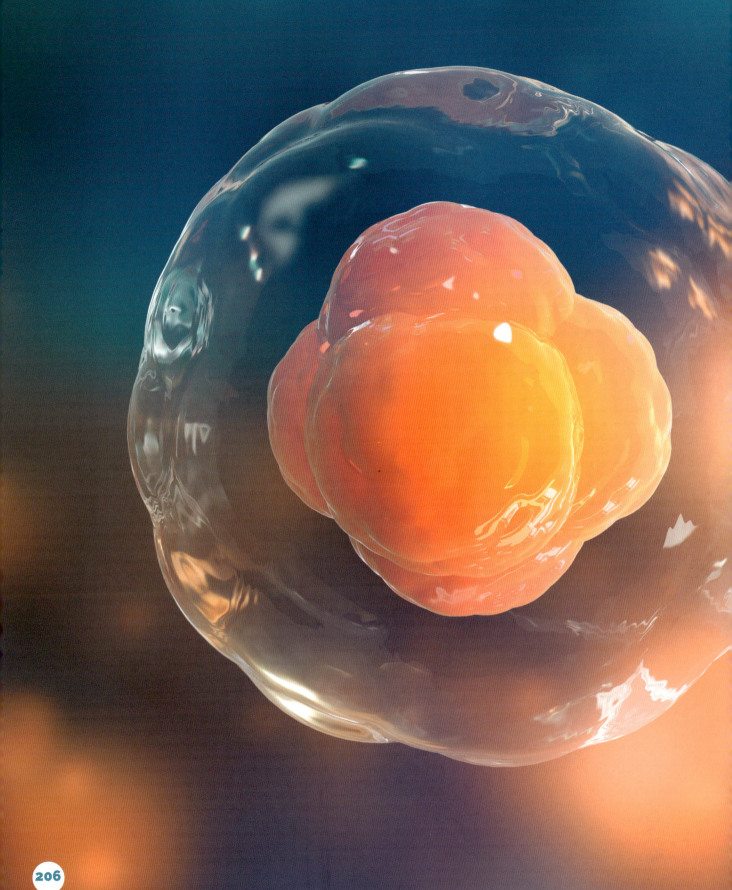

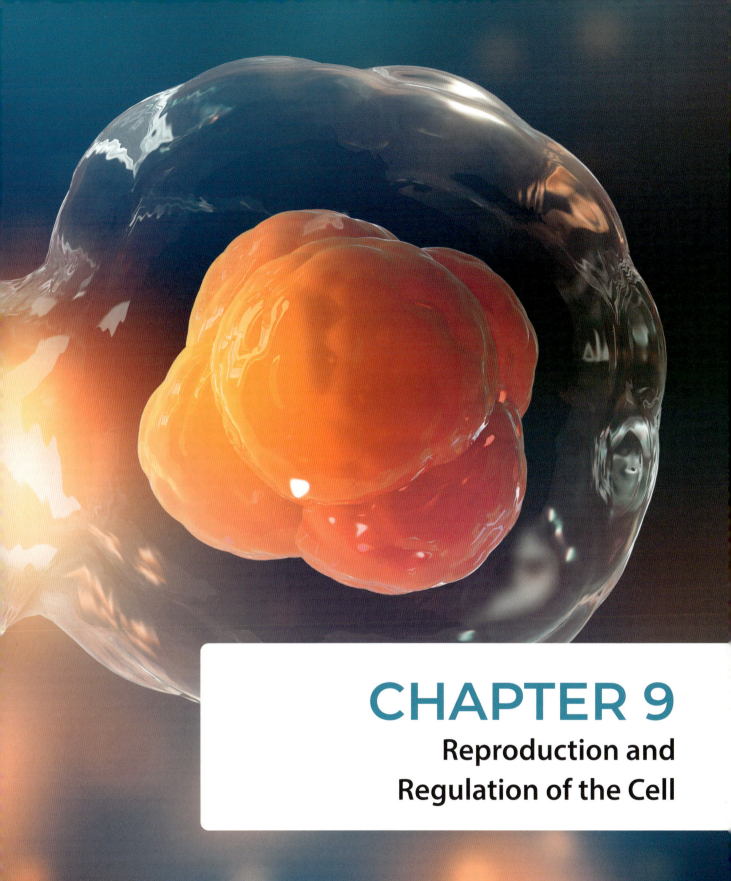

CHAPTER 9
Reproduction and Regulation of the Cell

Chapter 9

Reproduction and Regulation of the Cell

Chapter Outline

9.1 Introduction

9.2 DNA, RNA, and Protein Synthesis
    A. DNA: The Double Helix
    B. The DNA Code
    C. RNA Structure
    D. Protein Synthesis

9.3 Cell Reproduction
    A. The Reproduction of Cells
    B. Mitosis
    C. Cell Growth and Differentiation

9.4 Unlocking the Mystery of DNA and Protein Synthesis
    A. Why are Proteins Important to Us?
    B. What is DNA?
    C. The Structure of DNA
    D. The Process of Transcription
    E. Translation: Changing from RNA to Protein

9.5 Genetics
    A. Mendel's Laws
    B. Punnett Squares
    C. Meiosis and Mitosis

9.1 Introduction

To understand what is happening to an organism as it grows and changes, one must consider what is taking place at the cellular level. In this chapter, we will study how cells reproduce and are regulated. We will explore the mysteries of what goes on in the cell as a whole.

Because even the whole cell is so tiny and cannot be seen with the unaided eye, its even tinier parts are difficult to imagine. Yet, as we will see, even the parts have parts in many cases. This incredible complexity is necessary because each cell, as long as it lives, does a great deal of work for the body. Considering that our bodies are composed of trillions of these cells, imagine how much activity must be going on inside a large living body, such as that of an oak tree or an elephant!

Everything that is alive—from bacteria, to fungi, to plants, to animals—is composed of cells. While cells all serve the same basic purpose as the building blocks of the living

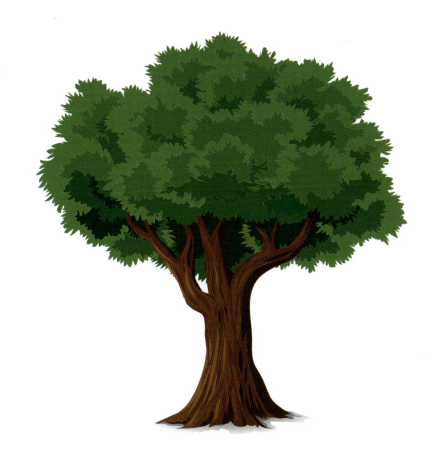

Biology For Life

body, there are many differences among the cells of different creatures. The composition and functions of different kinds of cells vary according to the organism's needs.

The cells of the human body, however, are unique in more than their physical composition. Unlike those of any other bodily creature, these cells, so long as they are part of the living body, are united to a spiritual soul destined to live forever. When that soul is in the state of grace, its body is a dwelling place of God, a temple of the Holy Spirit (cf. 1 Corinthians 6:19). In the Incarnation, God Himself took on a human body, with all of these cellular processes, sanctifying them by sharing in them. And someday, at the end of time, our bodies will be raised to a new life free of sickness, injury, and death. Never will the cells of the resurrected body wear out and die or be corrupted by viruses. Thus, even our cells share in the tremendous blessings that God has given to us as human beings.

In this life, of course, our cells don't always function perfectly, but they nonetheless carry out a variety of important tasks for our bodies. We will now explore what these are and how the cell performs them.

9.2 DNA, RNA, and Protein Synthesis

Section Objectives

- Recognize the similarities and difference between the structures of DNA and RNA.
- List the three main parts of a nucleotide unit of DNA and of RNA.
- Understand how the codes produced by DNA regulate protein synthesis.
- List the functions of messenger RNA, transfer RNA, and ribosomal RNA.
- Understand the functions of ribosomes and polysomes.
- Use the base-pairing rule to match up nucleotides to a section of DNA or RNA.
- Describe what takes place in the cell during the processes of transcription and translation.
- Draw a model section of DNA and describe what is happening during DNA replication.

God gave us life on earth. The proteins made by our bodies are essential to life and are important parts of our body's cells. Proteins are made from amino acids and 20 of these are found in nature. Since the human body can only manufacture twelve amino acids, we must daily ingest the eight essential amino acids in order to sustain our bodies and remain healthy.

Proteins make up parts of the plasma membrane and cytoplasm of the cell, and the organelles within it that are needed for its functions. Enzymes are necessary proteins that catalyze chemical reactions occurring inside the cell. All of our cells are equipped to make all of the different proteins our bodies need. How do our cells accomplish this task of making proteins? After studying this chapter, you will appreciate the beauty and complexity of Our Lord's wonderful design of the cells of life. In fact, the work that each cell performs is so deeply complex, intricate, and amazing that it is difficult to comprehend how anyone could not believe in a Creator and Intelligent Being who we know as God. As St. Paul says in Romans 1:20: "For the invisible things of him, from the creation of the world, are clearly seen, being understood by the things that are made; his eternal power also, and divinity: so that they are inexcusable."

A. DNA: The Double Helix

The letters **DNA** are an acronym, which stands for deoxyribonucleic (dē-äk-si-rī-bō-nu-klā-ik) acid. The DNA not only controls the production of proteins in the cell—it is directly responsible for everything that our body makes! The cell relies on its DNA to control the making of proteins that are essential to life. The cell uses the instructions contained in its DNA to repeatedly make cellular proteins.

DNA is composed of nucleotides. A **nucleotide** is the basic unit of structure of DNA. It is composed of three parts: a five-carbon sugar called deoxyribose [dē-äk-si-rī-bōz], a phosphate group, and a nitrogenous base that is either a purine or pyrimidine.

Purines and pyrmidines are nitrogen-containing bases. Purines have a more complex structure than pyrmidines, although both bases start with a six-sided hexagon ring of carbon and nitrogen, and sometimes hydrogen. Pyrimidines have other elements bonded at some of the joints, but the basic shape stops at the hexagon. Purines have a five-sided pentagon-shaped ring attached to the hexagon ring. When purines are oxidized, they tend to form uric acid. Pyrimidines do not form uric acid.

There are four types of nitrogenous bases: the purines, **adenine** (ăd-ĕ-nēn), **guanine** (gwä-nēn); the pyrimidines, **thymine** (thī-mēn), and **cytosine** (sī-tō-sēn). These bases are often abbreviated by the first letter of the word: A, G, T, C, respectively. The bases are pictured in Figure 9.1.

Study the pictures for a moment. Do you see any patterns among the four kinds of bases? In 1953, two scientists, James Watson, an American biologist, and Francis Crick, a British physicist, discovered the three-dimensional structure of DNA. Watson and Crick concluded that DNA is composed of two long nucleotide chains.

The sides of this ladder (the backbone or rails), are composed of alternating deoxyribose (sugar) and phosphate molecules. These molecules, both

Figure 9-1. The four bases of DNA

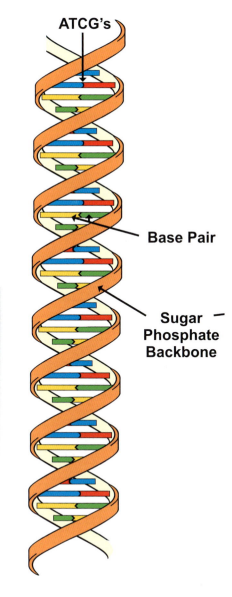

PURINES

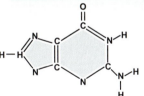

PYRIMIDINES

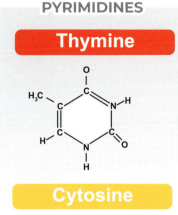

Biology For Life

sugars and phosphates, bind to each other to form linkages or the rails. Restated, the phosphate group of one nucleotide form a covalent bond with the sugar of another nucleotide.

The rungs of this ladder are composed of the four nitrogenous bases; each rung of the DNA ladder is made up of a pair of nitrogenous bases. The bases pair up according to a specific rule: adenine pairs with thymine, and thymine pairs with adenine (A:T or T:A). Likewise, cytosine pairs with guanine, and guanine pairs with cytosine (C:G or G:C). These nucleotides have been designed to readily form pairs, and this process is referred to as complementary base pairing. The reason base pairs form easily is because of hydrogen bonds.

Unlike a real ladder, the actual arrangement of DNA is twisted into a spiral shape, more like a spiral staircase. This shape is known as a double helix. The rails are linked with covalent bonds and the rungs are held together by hydrogen bonds. Look at the picture of the DNA molecule.

Amazingly, one DNA molecule can contain millions of nitrogenous bases. The order in which the bases are arranged will determine the types of proteins that are eventually synthesized, or produced, for the cell.

B. DNA Code

DNA is simply coded information that is carried by the sequence of nitrogenous bases. Recall from section 1.2.B that the sum total of all DNA within a organism is known as the genetic code or genome.

If you consider the nitrogen bases as letters of the alphabet, then groups of three bases can be compared to words. Each group of three nucleotides, or "word" in a strand of DNA, is called a **codon** (kō-dän), which corresponds to the production of one amino acid. For example, the codon consisting of adenine, cytosine, and cytosine (ACC) codes for the amino acid glycine. All other amino acids can be produced by two or more codons. That is, one codon will produce one amino acid, but that amino acid may be produced by several different codons. As an example,

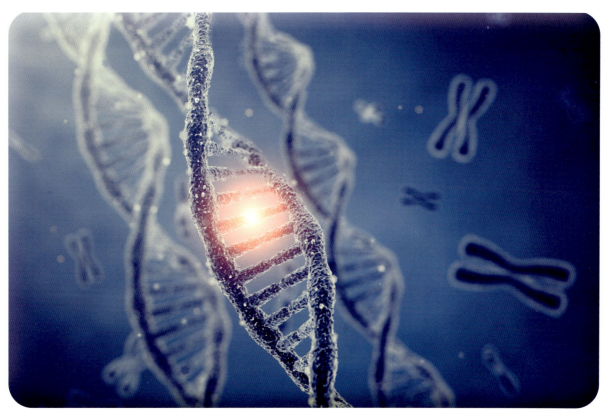

alanine can be produced by the codons CGA, CGG, CGT, and CGC. Each of these four codons will produce alanine on its own. (It is like a baseball team that has three catchers, but only one can play at a time.) We can combine the "words," or codons, into "sentences" to make amino acids and then put the "sentences," amino acids, together as "paragraphs," which would be the proteins.

Molecular biologists use the term **redundancy** to describe more than one way to produce the same amino acid. Although there is redundancy in the production of amino acids, there is **no ambiguity**. That is, no single codon can specify another type of amino acid. This is what we might expect from a Creator God. He built a system of protein production based on three building blocks (codons). In turn, these three building blocks are used to construct each and every living organism on the planet.

Some codons are not associated with any amino acid; these are known as start and stop codons. The start signal is the codon AUG, which specifies the production of the amino acid methionine. All growing proteins begin with this amino acid, even those of the prokaryotes. Conversely, there are three stop or terminator codons (ATT, ATC, ACT) that cause the termination of protein synthesis.

The section of DNA that codes the production of a protein is called a **gene**. All cells in an organism have exactly the same set of genes, except the egg and sperm cells. Accordingly, each cell has the instructions to produce each and every protein that the organism needs. Nevertheless, they do not. That is, each cell within an organism is able to draw upon the information in its genetic code (genes) to produce the proteins that it requires at the moment. The remainder goes unused, but may be called upon at a later time as the organism changes, matures, or faces an emergency. This may be part of God's plan to protect the organism during emergencies.

The genes determine all of the physical traits for an organism. You may recall that we used the term **genome** to describe all of the hereditary information possessed by an individual. How this genetic information is outwardly expressed in an organism is referred to as phenotype; for example, hair and eye color, skin tone, height, etc. This information is unique to that person and is contained in the DNA. During criminal investigations, law enforcement officials routinely obtain biological materials from a suspect. Typically these are hairs of the head and cells that have been collected from the inside lining of the mouth. These are then used to construct the DNA profile of the suspect in an attempt to associate him with a crime. This technique is referred to as **DNA fingerprinting**.

Section Review — 9.2 A & B

1. What controls the production of proteins in the cells?
2. What are the four nitrogenous bases?
3. What does DNA provide to a living organism?
4. The sugar that makes up part of the backbone of DNA is called _____.
5. The nucleotide that links to adenine is _____.
6. DNA is wound into a spiral shape known as a _____.
7. If a particular protein is made up of 225 amino acids, how many nucleotides are needed to code for these amino acids?
8. ____ are linked in a chain, and form the building blocks of substances called proteins.

C. RNA Structure

Our wonderful body cells also contain smaller nucleic acids called **RNA**. The letters RNA stand for ribonucleic (rī-bō-nu-klā-ik) acid. RNA is similar to DNA but differs in a few important structural details. In the cell, RNA is single-stranded, while

DNA is double-stranded. RNA contains **ribose**, a five-carbon sugar, and DNA contains **deoxyribose**, a different five-carbon sugar. The difference between ribose and deoxyribose is that ribose has one more oxygen atom. Finally, in RNA the nucleotide uracil is substituted for thymine, which is only present in DNA. The table above lists these differences.

COMPARISON BETWEEN DNA AND RNA

DNA Structure	RNA Structure
contains the five-carbon sugar deoxyribose	contains the five-carbon sugar ribose
contains the nitrogenous base thymine (T)	contains the nitrogenous base uracil (U) instead of thymine
is double-stranded and forms a spiral called a double helix	is single-stranded and does not form a helix
length of the chain is longer than RNA	is shorter than DNA

When you look at the picture of DNA and RNA in Figure 9-2, do you see what is similar about the two structures?

There are three types of RNA within all of the cells in our body. Each type of RNA has a unique structure and function. These include messenger RNA (**m-RNA**), transfer RNA (**t-RNA**), and ribosomal RNA (**r-RNA**). Each of these types is necessary for the synthesis of proteins.

D. Protein Synthesis

Transcription and Translation: DNA to RNA to Protein

You learned in Section 1.2.A that DNA contains all of the instructions and acts as a "blueprint" for constructing an organism. The DNA is located in the nucleus of all cells and its information must pass to the cytoplasm, where protein synthesis occurs. How do these instructions find their way into the cytoplasm, if the DNA never leaves the nucleus?

RNA is the chemical messenger that codes and decodes the information stored in the DNA molecules in the nucleus of a cell. These messengers transport the genetic code to the ribosomes in the cytoplasm where proteins are made. This process is called transcription and contains three steps: initiation, synthesis of the RNA, and termination. The first step in transcription, initiation, involves the unwinding of the double helix DNA molecule. This can be likened to the partial opening of a zipper. Initiation does not alter the DNA molecule or the order of the nucleotides. The unzipping begins at an area of the DNA molecule known as the promoter region, by a type of enzyme known as a helicase.

Once the DNA is unwound, step two, synthesis of the RNA, begins. In this process, an enzyme called RNA polymerase binds to the open DNA molecule. The RNA polymerase then "slides" along the single strand of DNA. As the enzyme slides along the DNA, it facilitates the binding of complementary RNA bases, according to the DNA/m-RNA base pair rule, to the existing DNA strand; for example, adenine binds with uracil and cytosine binds with guanine. As a result, an RNA molecule is formed that carries the exact genetic code stored in the DNA molecule. Eventually, the RNA polymerase enzyme reaches a portion of the DNA strand known as the terminator. As the name implies, this is where the transcription process stops; this is the third and final stage of transcription. Once a segment of RNA is formed and the RNA polymerase passes by, the original complementary DNA strands reconnect to form the original, unaltered double-stranded helix.

The new RNA molecule dissociates from the DNA strand. Because enzymes are neither altered nor consumed, the RNA polymerase also dissociates from the DNA strand and is free to be used again in another transcription reaction. The new strand of RNA is called messenger RNA, or m-RNA, because it now serves as a messenger to carry the genetic code taken from the DNA molecule to the cytoplasm.

All m-RNA is synthesized during a specific point in cell division and then transported to the cytoplasm until it is needed. When the cell has a demand for protein synthesis, m-RNA will be found in association with the ribosomes.

Recall that ribosomes are cytoplasmic organelles that are made of proteins and contain specific types of RNA molecules, called ribosomal RNA (r-RNA). The process of protein synthesis, which occurs in association with ribosomes, is called translation. It is easy to remember because the m-RNA "translates" the genetic code into proteins.

In order for the genetic code to be translated, it is decoded in units of three nucleotide bases along the m-RNA molecule. A group of three bases on the m-RNA molecule is called a codon. At this point, it might be helpful to revisit section 9.2.B on DNA, to help illustrate that a codon of the m-RNA originates from its mirror image codon of the DNA template. Each codon is like a specific abbreviation. Most codons specify an amino acid, but some signal start and stop. A start codon signals the start of the translation process, and the stop codon signals the

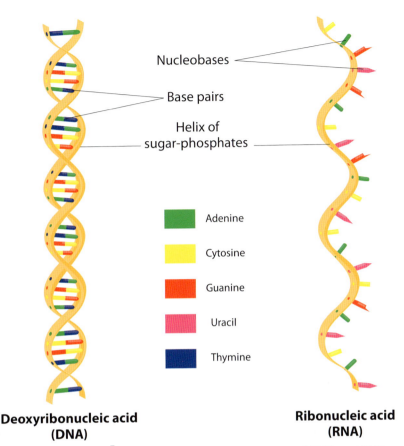

Figure 9-2. The 3-D structures of DNA and RNA

end of the translation process. Stop codons are also known as termination codons.

The coding sequence is the region between the start and stop codons that specifies the sequence of amino acids of the protein. There are 64 codons that code for 20 amino acids, plus the starts and stops. Each codon is specific for one amino acid, but each amino acid has several different codons that code for it. During the process of translation, amino acids, specified by triplet codons, are linked to form polypeptide chains and eventually proteins.

Once the m-RNA molecule is in the cytoplasm and is associated with a ribosome (ribosomal subunit), translation begins. A third type of RNA molecule called transfer RNA (t-RNA) transfers the amino acid specified by the codon on the m-RNA to the m-RNA-ribosome complex. Just like there are 64 codons, there are 64 t-RNAs — one t-RNA for each codon.

Figure 9.3 demonstrates how t-RNA is a single-stranded RNA molecule folded onto itself to form a cross. The bottom loop is the area of the molecule that recognizes the m-RNA codon. This area is made up of a triplet of bases called an **anticodon** that are complementary to the codon of the m-RNA to allow the t-RNA to "recognize" the m-RNA and to allow placement of the proper amino acid. The anticodon pairs bases with the appropriate codon of the m-RNA, thus delivering the appropriate amino acid to the growing peptide chain. Notice also on this figure the uppermost portion of the t-RNA molecule; this is the attachment site for the amino acid.

In the t-RNA represented in Figure 9-3, the anticodon sequence is CGA. This would bind with a GCU on the m-RNA strand. GCU is the codon for the amino acid alanine. Therefore, the t-RNA above would transfer an alanine to the growing polypeptide chain. The process then continues until a stop codon is reached; the protein is then complete.

The t-RNAs are not altered during the process of translation. As a result, the t-RNA molecules are available to transfer additional amino acids to other growing peptide chains. How efficient is God's work! Accordingly, once the beginning of the code has been translated by a ribosome and a portion of the polypeptide is formed, another similar ribosome can start translating the same m-RNA again. That is, the second ribosome can make another copy of the same protein before the first ribosome has finished the first protein. Groups of ribosomes attached to the same strand of m-RNA and translating the m-RNA into proteins are known as **polysomes**, or polyribosomes. Multiple copies of the same protein can be made very quickly by this polysomal mechanism — again, the wonder of God's wisdom!

Section Review — 9.2 C & D

1. The three bases on the bottom of a transfer RNA molecule are known as _____.
2. Compare and contrast the structure of DNA to the structure of RNA.
3. What is a codon?
4. List the three types of RNA.

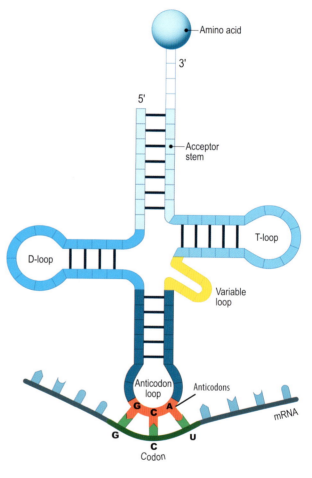

Figure 9-3. Structure of the t-RNA.

9.3 Cell Reproduction

Section Objectives

- Match the cellular events during mitosis with the particular mitotic stage (i.e., prophase, metaphase, etc.).
- Define the terms "cytokinesis" and "differentiation."

A. The Reproduction of Cells

All of the cells in the body continually divide and form new cells. The body's cells continue to reproduce so that bodies can grow and change as a person grows to become an adult. There are three steps that must occur in order for cells to divide and reproduce. First, the DNA code must be copied so that the new cell has a complete set of instructions in order to carry out its functions. Next, the division of the nucleus must occur, and, finally, the cytoplasm must divide so that two new cells can be formed.

1. Replication of DNA

During cell division, each cell must not only duplicate itself, but also duplicate the DNA within it. This process, where an exact copy of the DNA is produced, is called **replication**. During replication, the left and right sides of the DNA ladder are "unzipped" by an enzyme known as helicase. This process results in two strands; the nucleotides of the left strand will pair up with the nucleotides of a newly synthesized right strand, following the base-pairing rule. The identical process occurs for the right strand. Take a moment to study the diagram of DNA replication. Notice that there are two identical copies of the DNA strand.

2. DNA Is Stored In The Cell

The DNA of eukaryotic cells is tightly bound together in a network of proteins within the nucleus of the cell. Certain proteins called histones act as spools that wind up small

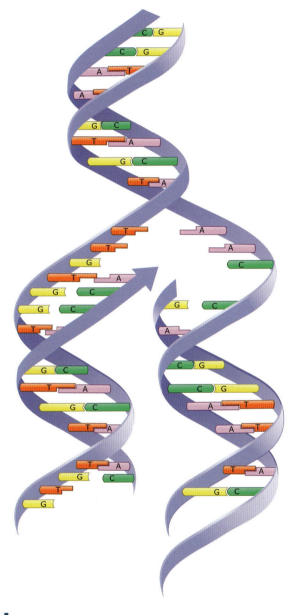

Figure 9-4. DNA duplication. Each strand of the original double-stranded DNA molecule serves as a template for the reproduction of the complementary strand. Hence, following DNA replication, two identical DNA molecules have been produced from a single double-stranded DNA molecule.

Biology For Life

CHAPTER 9: REPRODUCTION AND REGULATION OF THE CELL

stretches of DNA. Other histones stabilize and support these spools by producing a complex network of protein coils. This entire structure is called a **chromosome**.

During most of their life, chromosomes are not visible with a microscope. However, before cell division, each chromosome is duplicated and condenses into short (~ 5 μm) structures. When a cell biologist wishes to observe the chromosomes of a cell, a dye or stain is added to a group of cells or very thin tissues. The dye binds to the chromosomes and makes apparent those chromosomes that are in the process of **replication**. These are visible with the aid of a microscope. Duplicated chromosomes are called **dyads** and are shown in figure 9.6.

All of the genes involved in the production of a living organism are carried in its chromosomes. Accordingly, there are many unique chromosomes within the nucleus of each cell. The number of chromosomes is dependent on the organism – humans have 46 chromosomes in the nuclei of each cell; cats have 38; horses have 64; but crayfish have 200! It is worth keeping in mind that one half of an individual's chromosomes come from the father and the other half comes from the mother.

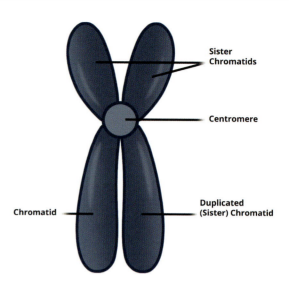

Figure 9-6. A dyad is a duplicated chromosome having two sister chromatids joined at the centromere.

3. Cellular Division

Cell division is a process by which a cell, called the parent cell, divides into two or more cells, called daughter cells. Cell biologists typically recognize cell division as one part of a larger cell cycle. Within eukaryotes, this process is known as **mitosis** [mī-tō-sĭs], and the daughter cells are capable of further division. The corresponding sort of cell division in prokaryotes is known as prokaryotic fission. Earlier terminology referred to this process as **binary fission**, but this term is currently used for describing reproduction among flatworms and some other organisms. **Meiosis** is a special type of cell division found in eukaryotes. It refers to germ cells (male and female reproductive cells) that are permanently transformed into their final products, either sperm or eggs.

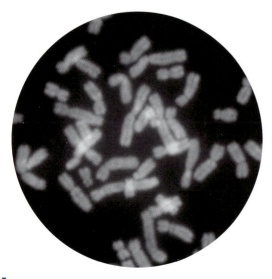

Figure 9-5. Human chromosomes during metaphase magnified 440 times.

> **Section Review — 9.3 A**
>
> 1. The process in which a cell divides into two or more cells is called _____.
>
> 2. What is meiosis?

Biology For Life

B. Mitosis

There are two parts to cell reproduction, **mitosis** and **cytokinesis**. Mitosis is a process in the cell cycle where the genome duplicates. That is, during mitosis the chromosomes in the cell nucleus prepare for duplication and distribution of chromosomes. There are five distinct stages to the process of mitosis: **interphase**, **prophase**, **metaphase**, **anaphase**, and **telophase**.

Mitosis is followed by cytokinesis, a division of the cytoplasm. The end result is the formation of two identical daughter cells, each having the same number of chromosomes, organelles and cell membrane. These daughter cells are exact replicas of the parent cell.

1. Process of Mitosis

a. Interphase

The first of the five stages in mitosis is known as interphase. For many years, interphase was believed to be the resting stage of the cell. We now know that interphase is a period of growth for the cell. During interphase, the cell is not yet dividing, but growing and replicating its DNA in preparation for cell division. At this point in the mitotic cycle the chromosomes are not visible as long spaghetti-like strands but are bunched together and called chromatin.

After each chromosome has been copied, the identical pairs of chromosomes are known as **chromatids** [krō-ma-tĭds] or sister chromatids.

b. Prophase

During this stage, the nucleolus and nuclear membrane dissolve. Structures called **centrioles** [sĕn-trē-ōlz] begin to appear on the outside of the nuclear membrane. These are made up of microtubules and move to the opposite ends of the nucleus. The centrioles form what

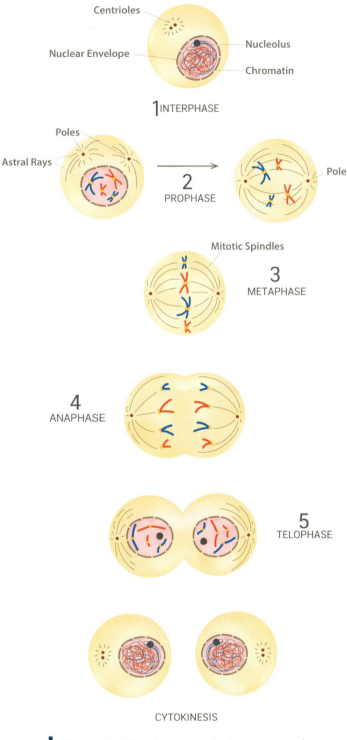

Figure 9-7. The five stages in the process of mitosis, the division of a cellular nucleus.

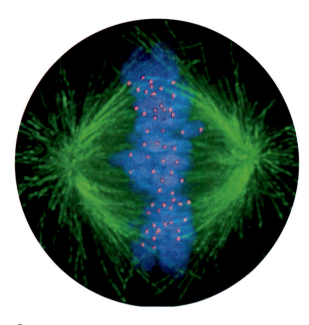

Figure 9-8. Image of a human cell showing microtubules in green, chromosomes (DNA) in blue, and kinetochores in pink.

are known as the **poles** of the cell. **Astral rays** form around each centriole. These rays may be compared to the crepuscular rays, which streak through the openings in the clouds on cloudy days, or those rays emanating from the heart of the risen Jesus. Each centriole, with its astral rays, forms what is known as the **mitotic spindle**. The purpose of the mitotic spindle is to help the chromatids (chromosome pairs) line up in the middle of the nucleus between the two poles. Identical chromatids are attached by a kinetochore [kĭ-ne-tō-kor].

Kinetochores form late in prophase. Kinetochores develop in the centromere area of each chromatid and are attached by microtubules on opposite sides of the spindle. Kinetochores are so small that they can only be seen through an extremely powerful microscope, known as an electron microscope. Kinetochores are complex, and are made up of a minimum of 45 proteins in 3 layers.

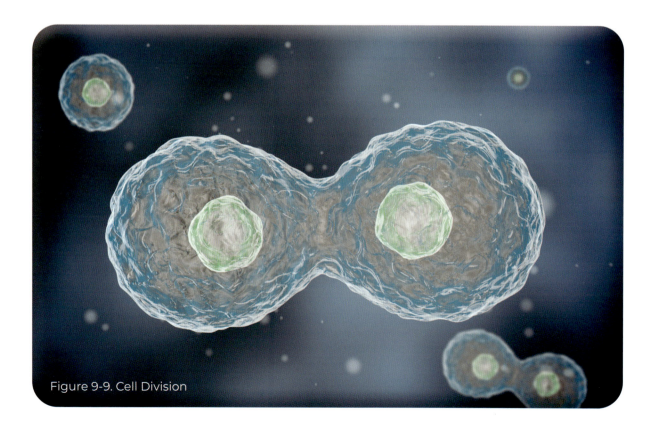

Figure 9-9. Cell Division

c. Metaphase

The chromatid pairs arrive at the middle of the nucleus, and the kinetochores split apart. The chromatid pairs begin to unwind and separate, and, once again, the term "chromosomes" is used to describe these structures. The spindle attaches to each chromosome's kinetochore, and the chromosomes line up at the equator.

d. Anaphase

The two chromosomes that made up each chromatid pair begin to move away from each other and go to the opposite poles of the cell. One part of the pair moves to one pole of the cell, and the other part moves to the opposite pole.

e. Telophase

Telophase begins when the chromosomes have finished moving to the opposite poles of the nucleus. The chromosomes bunch up on each end of the nucleus, and two new nuclear membranes and nucleoli form. The spindle disappears. The interphase stage can begin again, and the entire process of mitosis can repeat itself.

In order for distinct cells to be formed, the entire cell must divide, and this involves **cytokinesis** (sī-tō-kĭ-nē-sĭs) or division of the cytoplasm. Although cytokinesis usually initiates during the late stages of anaphase, it is completed in telophase. When the nucleus is dividing, the cytoplasm of the cell surrounding the nucleus is also undergoing division. The organelles within the cytoplasm must be transported to each incipient cell. Cytokinesis helps ensure that the chromosome number is maintained from one generation to the next.

C. Cell Growth and Differentiation

The growth of an organism occurs because existing cells grow, and new cells are produced. New cells may become specialized for specific functions as the organism develops. For example, in humans, some cells develop into skin cells, some develop into nerve cells, and others develop into bone cells. The process by which cells become specialized to perform one type of function for the body is known as **differentiation**.

Section Review — 9.3 B & C

1. Name the 5 stages of cell division by mitosis in order.
2. When the nucleus is dividing, the cytoplasm of the cell surrounding the nucleus also divides in a process called _____.
3. What are chromatids?

9.4 Unlocking the Mystery of DNA and Protein Synthesis - Summary

In an effort to reinforce our knowledge of how DNA controls the process of protein synthesis, let us consider the following points, questions, and items for reflection.

Why are Proteins Important to Us?

Proteins are important for these three reasons:

1. Proteins are components of every cell in the body.
2. Proper functioning of cells requires enzymes. Enzymes are a special type of protein that catalyze (increase the rates of) chemical reactions.
3. Proteins make up the structure of the plasma membrane, cytoplasm in the cell, and organelles within each cell.

What is DNA?

DNA is a large molecule contained in the cell's chromosomes. Key points to remember:

1. Chromosomes are located in the nucleus of each cell.

2. DNA can be compared to a library that stores information for the cell.
3. Like a library book, DNA's instructions can be used over and over again by the cell.
4. The gene is a segment of DNA, and a pair of genes specifies a trait for an organism.
5. The total complement of genes in an organism or cell is known as its genome, which is stored on one or more chromosomes.

The Structure of DNA

DNA is made up of nucleotides linked together. The nucleotide has three parts: a sugar with five carbon atoms called deoxyribose, a phosphate group, and a nitrogen base. DNA has four different nitrogen bases:

1. adenine (abbreviated with a capital **A**)
2. cytosine (abbreviated with a capital **C**)
3. guanine (abbreviated with a capital **G**)
4. thymine (abbreviated with a capital **T**)

These nitrogen bases can link together in only one way:

- **A** can pair only with **T**.
- **C** can pair only with **G**.

When each nucleotide is linked together and all the bases are paired, a DNA molecule looks like a spiral staircase or a double helix. Sugar and phosphate groups make up the sides of the ladder. Paired nitrogen bases are the rungs of the ladder.

How Does the Structure of DNA Carry the Information the Cell Needs to Make a Protein?

Consider these points:

1. The code of DNA presents a specific order of nitrogen bases.
2. The arrangement of the bases determines which amino acid will be made.
3. Many amino acids link together and create a specific type of protein.
4. The order in which the amino acids are linked determines the particular function of the protein.

Groups of DNA Bases Represent Specific Amino Acids:

1. A group of 3 bases represents one amino acid.
2. Each group of 3 bases is called a codon.
3. Codons are like three-letter words. For example, **CAG** represents the amino acid **valine**. Several different codons can stand for the same amino acid.
4. One codon is needed to code for each amino acid, which contains 3 nucleotides. So 3 nucleotides are necessary to code for one amino acid.
5. Some codons do not stand for any amino acid; they are stop codons that tell RNA to stop synthesizing proteins. The codon AUG, which specifies the amino acid methionine, is the start signal, so all protein synthesis begins at an AUG codon.

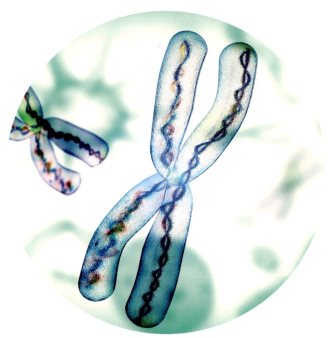

Figure 9-10. A Chromosome

COMPARISON OF RNA AND DNA STRUCTURE

RNA	DNA
nucleotides contain a ribose sugar	nucleotides contain a deoxyribose sugar
single-stranded molecule	double-stranded molecule
bases - A, C, G, U (the base uracil replaces the base thymine for RNA)	bases - A, C, G, T
A pairs with U and U pairs with A	A pairs with T and T pairs with A
C pairs with G and G pairs with C	C pairs with G and G pairs with C

The Process of Transcription: Copying from DNA to RNA

Before a protein can be made, the cell must copy its DNA. This copying process is called **transcription**. To understand the process of transcription we need to learn about another complex molecule called RNA (ribonucleic acid).

There are three types of RNA: m-RNA, t-RNA, and r-RNA. Protein is synthesized in ribosomes that are found in the cytoplasm of the cell.

Translation: Changing from RNA to Protein

Transfer RNA (t-RNA) is used for this process. Amino acids attach to the t-RNA according to the anticodon of three letters at the bottom of each t-RNA. Anticodons on t-RNA are the same groups of letters that were originally on the DNA codon, before these were transcribed to and became the codons on the m-RNA template. The t-RNA – amino acid complexes move toward the m-RNA on the ribosome. Next, the t-RNA anticodons are paired with the m-RNA codons, resulting in a linkage of the amino acids to form a protein. This pairing, linking, and synthesizing is made possible by the r-RNA of the ribosomes.

Section Review — 9.4

1. Why are proteins important to us?
2. Before a protein can be made, the cell must copy its DNA. This copying process is called _____.
3. The groups of three bases on the messenger RNA are known as _____.

CHAPTER 9: REPRODUCTION AND REGULATION OF THE CELL

9.5 Genetics

Section Objectives

- Describe the results of Mendel's laws.
- Use a Punnett square to describe the characteristics of the offspring of a parent population.
- Describe the similarities and differences between mitosis and meiosis.

All the characteristics of a particular organism are contained in the DNA of its chromosomes. The scientific study of how parents pass on their characteristics (hereditary traits) to their offspring is called **genetics**.

A. Mendel's Laws

Beginning in 1856, Fr. Gregor Mendel, an amateur scientist and Augustinian monk at the monastery in Brno, located in what is now called the Czech Republic, experimented with pea plants that grew in his garden. He crossed long-stemmed plants with short-stemmed plants, plants that bore round peas with those that were wrinkled, and plants with yellow peas with those that were green. By crossing subsequent generations of offspring, he was able to observe what happened to certain physical traits over time. These crosses indicated the inheritance patterns of many traits, which were explained by simple rules and ratios.

Figure 9-11. Fr. Gregor Mendel, the founder of genetics.

Mendel crossed plants that had produced green peas with those that produced yellow peas. His crosses resulted in offspring that were either green or yellow. None of his crosses produced a blend, which, at the time, was expected by the scientific community. He also found that traits appeared with a predictable 3-to-1 frequency. Mendel called the more common trait "**dominant**" and the less common one "**recessive**."

Because traits did not blend, Mendel concluded that if either parent contributed one dominant part of a gene pair, the offspring's physical appearance (**phenotype**) would reflect that of the dominant trait. This would occur even if its genetic makeup (**genotype**) contained the **recessive trait**. Thus, tall peas would appear with greater frequency than small ones, because tall was the **dominant trait** for pea height. Notice that the words "physical" and "phenotype" both begin with "ph," while "genetic" and "genotype" both begin with "ge."

Biology For Life

The initial letters of each of these words may help you remember the difference between the two. Restated, the phenotype of an organism refers to its expressed physical traits, and the genotype refers to its genetic makeup.

Interestingly, Mendel's work demonstrated that even if a trait was not evident in an offspring's phenotype, it could still be passed on through subsequent generations. When a hybrid parent – one that has a genotype consisting of one dominant and one recessive trait (we will call this Tt) – crosses with another hybrid parent (also Tt), the chances are three in four that an offspring will display the dominant trait. This is because both TT and Tt genotypes exhibit the dominant trait. But there is also a one-in-four chance that an offspring will be tt and instead display the recessive trait.

The results of his research are summarized as follows:

- Organisms inherit traits (**alleles**) in pairs. One trait from each parent is present in the gene.
- Some traits are dominant, and some traits are recessive.
- Dominant traits express themselves when present in the gene, and the recessive traits stay hidden.
- Recessive traits only express themselves when the dominant trait is absent from the gene.

B. Punnett Squares

A **Punnett square** is a chart which shows/predicts all possible gene combinations in a cross of parents of which the genes are known. In addition, it allows us to see the simple ratios that follow. Punnett squares are named for an English geneticist, Reginald Punnett, who developed them around 1910.

Even though Fr. Mendel's research was conducted more than 150 years ago, we can use his experimental framework to investigate the inheritance of traits. In crosses between a homozygous (purebred) tall pea plant and a homozygous short pea plant, we can predict both the genetic makeup (genotype) and physical appearance (phenotype) of the offspring.

1. We designate letters that represent the genes, or traits, of the peas. Capital letters represent dominant traits, and lowercase letters represent recessive traits. **T** = tall pea and **t** = short pea.

2. We write down the genotypes (genes) of each purebred parent; that is, tall = **TT** and short = **tt**.

3. We list the alleles that each parent can contribute.

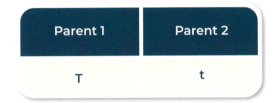

4. We draw a Punnett square – 4 small squares in the shape of a window. Write the possible alleles of one parent across the top and the alleles of the other parent along the side of the Punnett square. Fill in each box of the Punnett square by transferring the letter above and in front of each box into each appropriate box. As a general rule, the capital letter goes first, and a lowercase letter follows.

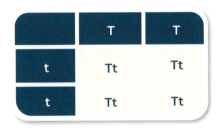

The letters inside the boxes indicate possible **genotypes** (genetic makeup) of offspring resulting from the cross of these particular parents. There are 4 boxes, and the genotypic results can be written either as fractions or percents. In this case, all 4 boxes show the Tt genotype. Therefore, each of the offspring has a **4/4 or 100%** chance of showing the Tt genotype.

5. Since capital letters indicate **dominant** genes, T (tall) is dominant over t (short). The phenotype (physical appearance) in each offspring can then be written in the box:

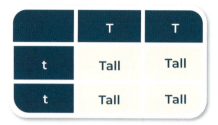

Therefore, each of the offspring has a **4/4 or 100%** chance of being tall.

If we use a Punnett square to predict the genotypic and phenotypic outcome (offspring) of a cross between two heterozygous/hybrid tall (Tt) pea plants, then we have

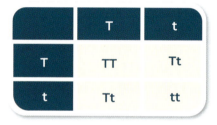

In this case there is

- 1 box out of 4 that show the TT genotype
- 2 boxes out of 4 that show the Tt genotype
- 1 box out of 4 that show the tt genotype.

So an offspring has a 25% chance of showing either of the TT or tt genotypes and a 50% chance of showing the Tt genotype. Since a capital letter indicates a dominant gene, the T (tall) phenotype is dominant over the t (short) phenotype and can be written as:

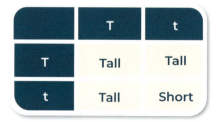

Therefore, each of the offspring has a **3/4 or 75%** chance of being tall and a 25% chance of being short.

C. Meiosis and Mitosis

On the cellular level, Mendel's laws take into account the fact that the sex cells from the parents (sperm and egg) contain only half the number of the chromosomes of the organism. So, when an egg is fertilized, it contains a complete set of chromosomes associated with the organism. Each parent has contributed one gene in the gene pair that constructs the genotype of the offspring. This part of the gene pair may be either the dominant (T) or recessive (t) part of the trait, and the two together give rise to the possible TT, Tt, and tt traits.

Meiosis is the process by which sperm and eggs are created. It uses many of the same mechanisms employed during mitosis to accomplish the distribution of chromosomes. The most important difference is the pairing and genetic recombination between **homologous chromosomes**. What are homologous chromosomes? During meiosis, a chromosome pairs with another similar chromosome. These are similar in that they have the same structure, length, and genes at the same location, or **locus** (plural *loci*). The primary difference between each member of a pair is that they may have different alleles. An allele is a member of a gene pair, on the same chromosome, and in the same loci, that controls the expression of a trait. Since most genes exist in pairs, it is the dominant allele which will express itself in the organism. For example, in humans, eye color is controlled by a dominant allele of a gene pair. Similarly, height is also controlled by a dominant allele.

An easy way to remember the differences between mitosis and meiosis is as follows. Mitosis involves one replication of the cell's chromosomes and is followed by one cell division. Meiosis involves one replication of the cell's chromosomes and is followed by two cell divisions. Mitosis results in two daughter cells having the same number of chromosomes, or the diploid number, as the parent cell. Conversely, meiosis results in the daughter cells having half as many chromosomes as the parent

CHAPTER 9: **REPRODUCTION AND REGULATION OF THE CELL**

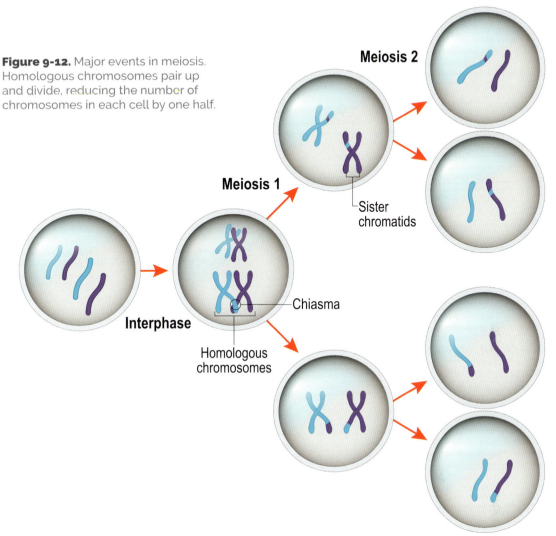

Figure 9-12. Major events in meiosis. Homologous chromosomes pair up and divide, reducing the number of chromosomes in each cell by one half.

cell. In summary, mitosis involves one replication, one division, and the full number of chromosomes. Meiosis involves one replication, two divisions, but only half the number of chromosomes.

While the other cells in your body contain 46 chromosomes, the sex cells (sperm or egg) only contain 23. You no doubt have seen a person with Down Syndrome. This is a condition in which the cells of a person's body do not contain the normal 46 chromosomes, but instead contain 47 chromosomes. These individuals are said to have an extra chromosome. They are special people and are characterized by their unique simplicity and affability – they are God's "special children" and just as "priceless" as you and I.

Section Review — 9.5

1. Who was Fr. Gregor Mendel?
2. What are the findings of Mendel's research?
3. What are alleles?
4. What is a Punnett square?
5. What are homologous chromosomes?

Biology For Life

Chapter 9 Supplemental Questions

Answer the following questions.

1. Recognize the similarities and differences between the structures of DNA and RNA.

2. List the three main parts of a nucleotide unit of DNA and of RNA.

3. Understand how DNA codes regulate protein synthesis.

4. What is DNA fingerprinting and why is it possible?

5. List the functions of messenger RNA and transfer RNA.

6. Understand the functions of ribosomes and polysomes.

7. Protein synthesis takes place in ribosomes. Explain the roles of amino acids, codons and anticodons in this process.

8. Describe what takes place in the cell during the processes of transcription and translation.

9. Match the cellular events during mitosis with the particular mitotic stage (that is, prophase, metaphase, etc.).

10. Define the terms "cytokinesis," "differentiation," "chromatids," "centriole," and "replication."

11. Briefly explain "mitosis" and "meiosis."

12. Define the terms "genetics," "gene," "chromosome," and "allele." What are the relationships among these?

13. Identify "homozygous," "heterozygous," "dominant," and "recessive" in genetic science.

14. Identify Fr. Gregor Mendel. Describe his major scientific discovery.

15. What are the possible genotype(s) of a tall plant?

    What are the possible genotype(s) of a short plant?

    What would be the phenotype of TT?

    What would be the phenotype of tt?

    Why is the phenotype of Tt tall and not medium/average?

Chapter 9 Supplemental Questions

16. In pea plants, yellow peas are dominant over green peas. Use a Punnett square to predict the phenotypic and genotypic outcome (offspring) of a cross between a plant heterozygous/hybrid for yellow (Yy) peas and a plant homozygous/purebred for green (yy) peas.

17. In pea plants, yellow peas are dominant over green peas. Use a Punnett square to predict the phenotypic and genotypic outcome (offspring) of a cross between two plants heterozygous for yellow peas.

18. In pea plants, round peas are dominant over wrinkled peas. Use a Punnett square to predict the phenotypic and genotypic outcome (offspring) of a cross between a plant homozygous for round peas (RR) and a plant homozygous for wrinkled peas (rr).

19. In pea plants, round peas are dominant over wrinkled peas. Use a Punnett square to predict the phenotypic and genotypic outcome (offspring) of a cross between two plants heterozygous for round peas.

20. Study all diagrams in this chapter. Be able to draw and label all diagrams.

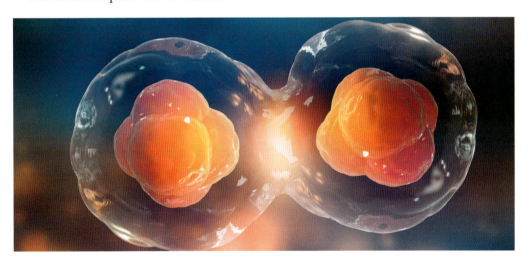

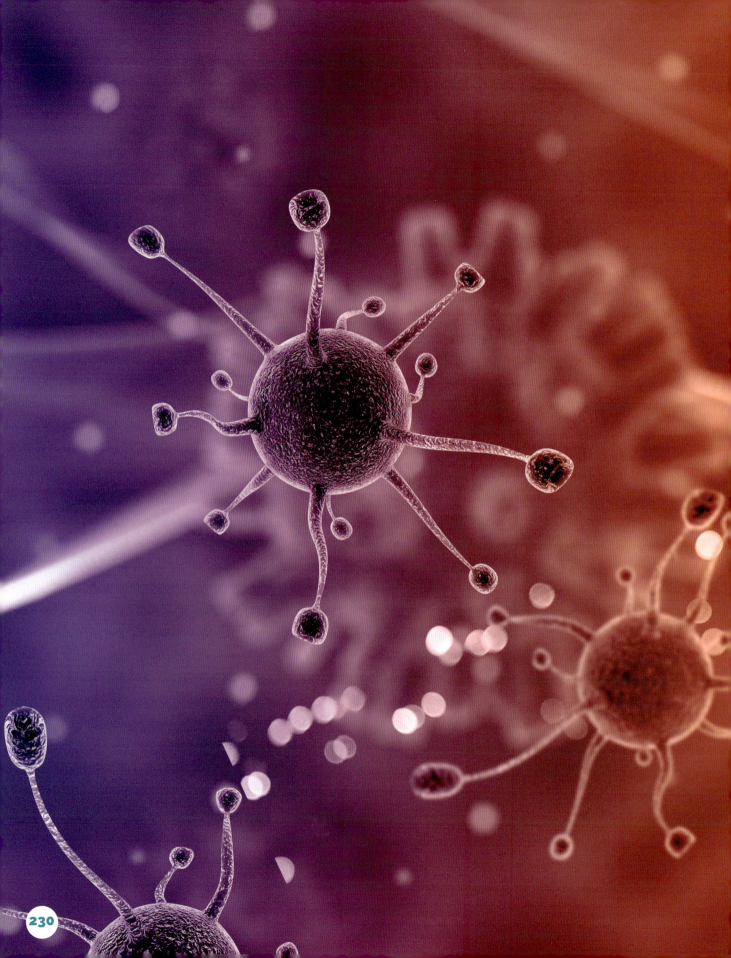

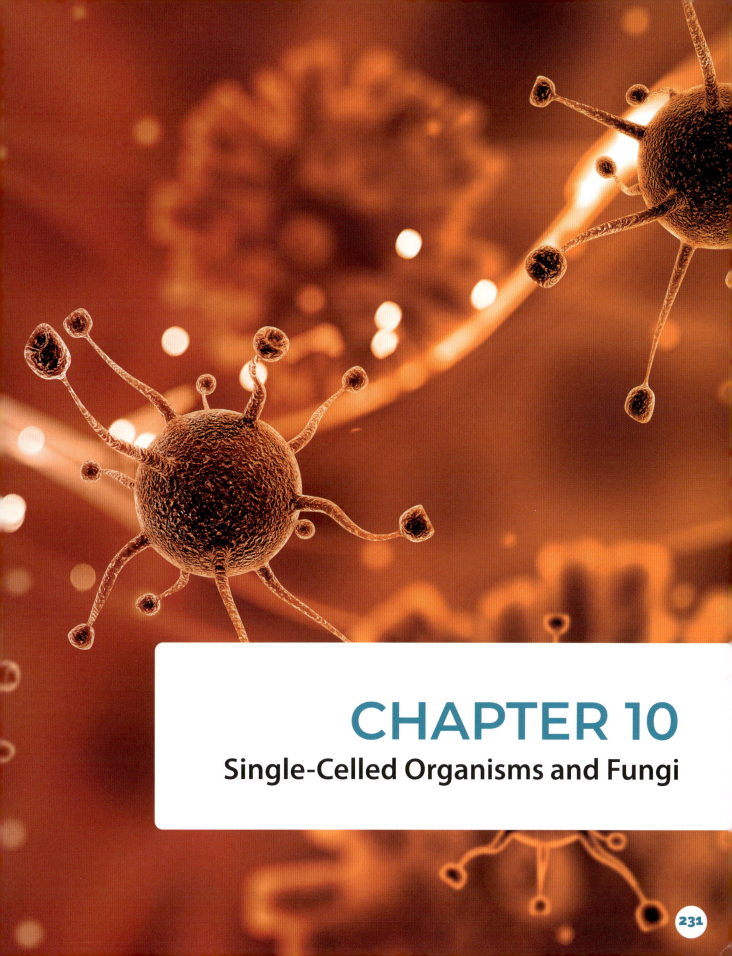

CHAPTER 10
Single-Celled Organisms and Fungi

Chapter 10

Single-celled Organisms and Fungi

Chapter Outline

10.1 The Classifications of Living Things

10.2 Kingdom Archaea and Bacteria
  A. Characteristics of the Monera
  B. Nutrition in the Bacteria
  C. The Nitrogen-fixing Bacteria
  D. Growing Bacteria in Laboratories
  E. Reproduction in the Prokaryotes
  F. Exchange of Genetic Material
  G. Bacteria and Disease

10.3 Diversity of the Bacteria
  A. Division Schizophyta
  B. Division Cyanophyta

10.4 Viruses
  A. Structure of Viruses
  B. Viral Reproduction
  C. Viroids

10.5 Kingdom Protista
  A. Plant-like Protists
    1. Phylum Euglenophyta
    2. Phylum Pyrrophyta
    3. Phylum Chrysophyta
  B. Animal-like Protists
    1. Phylum Mastigophora: the Flagellates
    2. Phylum Sarcodina
    3. Phylum Ciliophora: the Ciliates
    4. Phylum Sporozoa

10.6 Kingdom Fungi
  A. Characteristics of Fungi
  B. Fungi Structure
  C. Fungi Nutrition
  D. Reproduction in Fungi
  E. Variety of Fungi
    1. Subkingdom Gymnomycota
    2. Subkingdom Dimastigomycota
  F. True Fungi
    1. Phylum Zygomycetes
    2. Phylum Ascomycetes
    3. Phylum Basidiomycetes
    4. Phylum Deuteromycetes: Imperfect Fungi
  G. Lichens

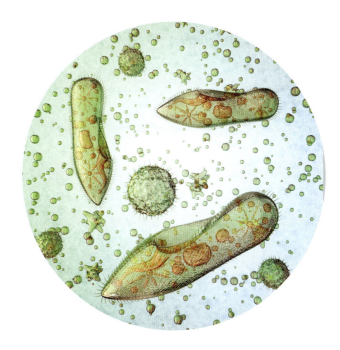

10.1 The Classifications of Living Things

Since the beginning of time, people have tried to understand the world around them. Philosophers and scientists divided the living things that they observed into two commonly understood categories, animals and plants. The great philosopher and naturalist, Aristotle, tried to become more specific with his "Ladder of Life," which eventually gave rise to the Systema Naturae that was produced by Carl Linne or Carolus Linnaeus. Written in 1735, this publication became known as the Linnean System of binomial nomenclature. It remains the standard that all scientists follow today when classifying organisms. Modern science refers to the classification of living organisms as taxonomy or biosystematics. Accordingly, the scientists who work in this field are called taxonomists or biosystematists.

The earliest naturalists or taxonomists examined and compared anatomical and physiological attributes when deciding how to classify organisms. For example, living things that could move on their own were

Life		
Domain Eukarya	Domain Archaea	Domain Bacteria
Kingdom Protista Kingdom Plantae Kingdom Fungi Kingdom Animalia	Kingdom Archaea	Kingdom Bacteria

Figure 10-1. Three Domain Biological Classification System: Domains and Kingdoms.

Biology For Life

CHAPTER 10: SINGLE-CELLED ORGANISMS AND FUNGI

classified as animals, while things "planted" in the ground and which could not move on their own were called plants. These two kinds of organism, plants and animals, became known as Kingdoms. Animals were further classified by whether they ate plants or other animals, by whether they had horns, hoofs, or fur, and so on.

Within the past decade, the five-Kingdom system has been replaced by six Kingdoms, and is the current means by which taxonomists classify living organisms. The names of these Kingdoms are Bacteria, Archaebacteria, Protista, Fungi, Plantae, and Animalia. For many years, only the Plant and Animal Kingdoms were recognized; Monera, Protista, and Fungi were added due to knowledge gained by scientific methods and microscopy. Recent changes by microbiologists changed the name of the Kingdom Monera into the Kingdom Bacteria. The bacterial Kingdom includes the bacteria and the cyanobacteria or blue-green algae. A specific type of bacteria was removed from the Monera and given its own Kingdom, the Kingdom Arachaea. You will read about this Kingdom later in this chapter.

Additionallly, further advances in scientific understanding, particularly the knowledge and examination of cellular organelles and DNA, have added an entire new level to the classification system, that of the "Domain," which is considered to be a level higher than the kingdom level. Under this system,

which takes DNA and specific attributes of rRNA in ribosomes into account, there are three domains with six kingdoms under them.

A system of just two domains, one for creatures with eukaryotic cells and one for those with prokaryotic cells, might be suggested. However, the differences between the prokaryotic Archaea and the prokaryotic Bacteria seemed so great that they were separated not only as different kingdoms, but also as different domains.

In biology, the taxonomic classification system has been, and continues to be, a convenient way to classify living organisms. The Domain Eukarya is made up of creatures that have eukaryotic cells.

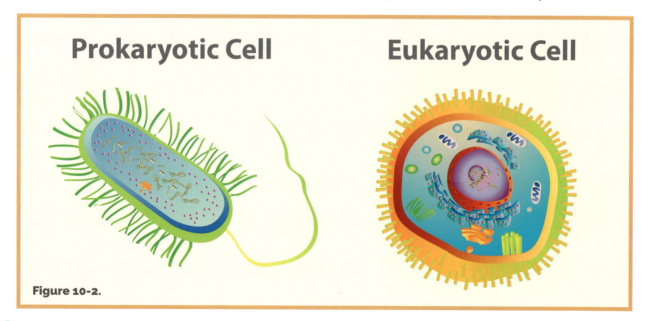

Figure 10-2.

Biology For Life

CHAPTER 10: SINGLE-CELLED ORGANISMS AND FUNGI

Eukaryotic cells are cells that have a complete nucleus, within a nuclear membrane. The other two domains, Archaea and Bacteria, are comprised of single-celled creatures that have no nucleus, and usually do not have membrane-bound organelles of any type. Archaea and Bacteria differ in their cell wall construction, RNA makeup, and lipid compositions. Archaea have the ability to live in some of the harshest conditions on Earth and seem to be able to metabolize non-organic compounds for energy, for example hydrogen, ammonia, and even metal ions. The six Kingdoms fit into the Domain system as follows: Bacteria, in the Domain Bacteria; Archaea, in Domain Archaea; and Protista, (prō-tis-ta), Fungi (fun-gĪ), Plantae, and Animalia, in Domain Eukarya. The three-domain system was developed by Carl Woese, an American microbiologist and physicist from the University of Illinois at Urbana-Champaign.

It is worth noting that the taxonomic nomenclature (naming) of many organisms frequently changes. Scientists make their reputations and careers from grouping, ungrouping, and then regrouping organisms. Those who remove organisms from groups are called "splitters"; those who add them are called "lumpers." Sometimes an organism will be removed from a group, and the scientific community welcomes this change. Several years later, perhaps sooner, new evidence appears which suggests the organism should have remained in its former classification. This is usually followed by debates that take place in scientific journals. At some point, the organism will, once again, be listed in its former classification. As simple as this process sounds, the debates often become heated and the losers often endure some ridicule, especially if they did not follow proper procedures. In the meantime, the organism itself has not changed!

The classification scheme is based on similar characteristics:

Kingdom Archaea is composed of organisms composed of prokaryotic cells (cells without a membrane-bound nucleus). Archaea have significantly different cell wall construction that permits them to live in extremely harsh conditions, and they can survive on certain non-organic substances.

The Six Kingdoms

Archaea - Halobacteria

Bacteria - Klebsiella Pneumoniae

Protista - Paramecium

Fungi - Mushroom

Plantae - Rose

Animalia - Koala

Figure 10-3. The Six Kingdoms: Archaea, Bacteria, Protista, Fungi, Plantae, and Animalia

Biology For Life

CHAPTER 10: SINGLE-CELLED ORGANISMS AND FUNGI

Kingdom Bacteria contains organisms composed of prokaryotic cells. Bacteria make up the largest total biomass on Earth. Typically, there are more live bacteria in a human body than there are human cells. Many of these are helpful; however, some cause serious diseases, such as bacterial meningitis, cholera, diphtheria, and pneumonia. "Good" bacteria are useful for fermenting wine, cheese, and yogurt.

Kingdom Protista contains those organisms composed of only one eukaryotic cell, as well as algae, because they do not have specialized components, such as roots and leaves.

Kingdom Fungi contains mostly organisms that feed on dead organisms. Yeast is a common example of a single-celled member of Kingdom Fungi.

Kingdom Plantae is composed of organisms that are made of many eukaryotic cells and produce their own food. Trees, shrubs, forbs, and grasses are members of the Plantae.

Kingdom Animalia is composed of multicellular organisms that eat other organisms (usually living) rather than making their own food. Insects, birds, cats, fish, and snakes are members of the Animalia.

Kingdoms are further divided into **phyla** (singular – **phylum**), which classifies organisms based on their general body plan. Phyla are divided into classes, which are determined by one or more particular attributes. For example, all creatures with mammary glands are called mammals. Classes are divided into orders, which are also determined by one or more different attributes; for example, mammals that eat meat are called carnivores. Orders are divided into families; families are divided into genera (singular – genus); and genera are divided into species.

So the total classification scheme is: **Domain – Kingdom – Phylum – Class – Order – Family – Genus – Species**. An easy way to remember this classification hierarchy is by using a mnemonic (memory) device:

"**D**ear **K**ing **P**hilip **C**ried **O**ut **F**or **G**oodness **S**ake."

Each and every living organism that has been described by biologists has a unique name consisting of two parts, referred to as a **binomen**. Thus, the categorization of living organisms is typically referred to as the Linnean system of binomial nomenclature. Organisms have a place in this system based on their apparent features, or taxonomic characteristics. For example, a tiger is a member of the Kingdom Animalia, or the Animal Kingdom. It is also a member of the Phylum Chordata, a group consisting of animals with backbones. It is placed in the Class Mammalia because it has hair and nurses its young with milk. It is a member of the Order Carnivora, since it is a meat eater with enlarged canine teeth. Tigers and other cats are members of the Family Felidae. Tigers, lions, and old world panthers are members of the genus *Panthera*; tigers are put in the species *tigris*. An African or Asian lion belongs to all the same higher taxonomic categories as a tiger, but

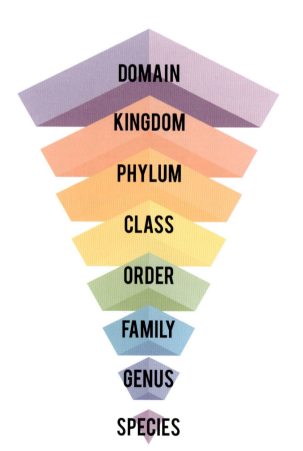

Figure 10-4. Hierarchy of Biological Classification

Biology For Life

CHAPTER 10: SINGLE-CELLED ORGANISMS AND FUNGI

Figure 10-5. Classification of a tiger: Kingdom - Animalia; Phylum - Chordata; Class - Mammalia; Order - Carnivora; Family - Felidae; Genus - *Panthera*; Species - *Tigris*

it is a different species, *leo*. Genus and species names are always italicized or underlined.

Humans share some characteristics with tigers and lions. We are in the same Kingdom, Phylum, and Class. However, humans are in the Order Primates and the Family Hominidae. More commonly, we are known by the binomen *Homo sapiens* (genus *Homo*, species *sapiens*).

You may have noted in the first paragraph of this chapter that the six-kingdom system of binomial nomenclature had its roots in the mid-eighteenth century and is credited to Carl von Linnaeus, a Swedish scientist. Linnaeus was a man of great piety, devoted to Sacred Scripture, and he loved nature deeply. He believed that God created the world. Linnaeus' desire was to equate the Biblical term "kind" with species. Furthermore, he believed that species were fixed, and that the study of nature would reveal the divine order of God's creation. He believed that by studying nature we can appreciate God's wisdom, which would reveal His order in the universe.

Note that viruses are not classified in any of the six Kingdoms, because they are not considered to be living; that is, they cannot reproduce on their own. In this chapter, you will study the single-celled organisms of the Archaea, Bacteria, Protista, and the viruses. It is truly amazing that God gave these small creatures the intricate structures and numerous functions to survive in our world. In addition, you will study Fungi, which are multi-cellular structures. You will learn how these organisms obtain nutrition and reproduce, and you will also learn about their diversity and ecology.

> **Section Review — 10.1**
>
> 1. What are the 3 domains?
> 2. Name the 6 kingdoms.
> 3. List the classification scheme in order.

Figure 10-6. Carl von Linnaeus

Biology For Life

CHAPTER 10: SINGLE-CELLED ORGANISMS AND FUNGI

10.2 Kingdom Archaea and Bacteria

Section Objectives

- List differences between prokaryotic and eukaryotic cells.
- Describe the ways in which the Archaea and Bacteria obtain energy.
- Give an explanation of the process of nitrogen fixation.
- Describe conditions needed to grow bacteria in the laboratory.
- Discuss the reproduction of bacteria.
- Explain the processes of conjugation and transformation.
- List examples of how humans are affected by the Monera.

A. Characteristics of the Monera

You read in the earlier paragraph about the splitting of the Kingdom Monera into the Kingdoms Archaea and Bacteria. For the purposes of this course we will recognize the taxonomic differences between them but refer to the Archaea and Bacteria as **prokaryotes**. So when you see the term prokaryote, think of Archaea or Bacteria.

The primary attribute of prokaryotes or prokaryotic organisms has always been that they lacked internal membranes. In addition, prokaryotes do not have a true nucleus, while eukaryotic cells, eukaryotes, possess a true nucleus. These are the main differences between these two groups. The name prokaryote means "pre-nucleus," and the name eukaryote means "true-nucleus."

CHART 10.1 - COMPARISON OF PROKARYOTES AND EUKARYOTES

Prokaryotes	Eukaryotes
0.5 - 2 microns	50 - 200 microns
no nuclear membranes	have nuclei covered with nuclear membranes
no mitochondria	have mitochondria
none have chloroplasts	some have chloroplasts
cell walls contain either murein or pseudomurein	cell walls, if present, lack murein
some are motionless, others move by gliding or use of flagella	move by means of flagella, cilia, contractile fibrils, and amoeboid movement; some are motionless
most are unicellular	many are multicellular

Biology For Life

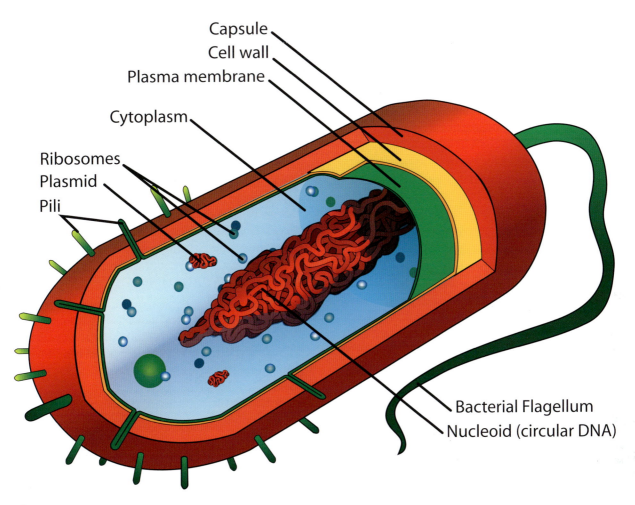

Figure 10-7. Cell structure of a bacterium, one of the most numerous members of the Kingdom Bacteria. Note the pili (singular - pilus), which are cellular cylindrical appendages that allow the transfer of genetic material between bacterial cells in a process called conjugation. Flagella are rigid protein structures, about 20 micrometers in length, which are used for movement.

Prokaryotes are much smaller than eukaryotic organisms. For example, a typical bacterium is about two micrometers (microns) long (1 micrometer = 1/1000 millimeter), whereas the average eukaryotic cell is about 50 to 200 micrometers long. As a point of comparison, a human hair is said to be about 50 micrometers wide.

Prokaryotes are very small, and 5,000 to 10,000 of these could lie end-to-end across your thumbnail. Though prokaryotes are tiny, they are complete organisms. They metabolize food, grow, produce and excrete wastes, and reproduce. All of the necessary nucleic acids, enzymes, and other substances are present in the prokaryotes, enabling them to carry out their life processes. In studying these smallest of God's creatures, one may appreciate the infinite mind of God.

Biology For Life

CHAPTER 10: SINGLE-CELLED ORGANISMS AND FUNGI

B. Nutrition in the Bacteria

There are two major groups of Bacteria, the bacteria and the **cyanophytes**. Most bacteria are **heterotrophs** that feed by secreting enzymes into their environment. A heterotroph is an organism that cannot make its own food and must take in nourishment from its environment. The enzymes secreted by the bacteria break down large molecules of food. The smaller food molecules that result are absorbed through the cell membrane.

All of the cyanophytes are **autotrophic**, which means that they produce their own food from simple substances. Cyanophytes contain **chlorophyll** that is located on cytoplasmic membranes; however, they lack **chloroplasts**. The photosynthetic process in cyanophytes is similar to that of eukaryotes, since the chlorophyll traps light from the sun and converts the energy to transform carbon dioxide and water into glucose, oxygen, and water.

Saprobes are heterotrophic organisms that feed on dead plants or animals. These types of bacteria and fungi live in soil or water. The saprobes grow and multiply by converting the dead material into simple chemicals. Some unused compounds, or byproducts, will be released back into the environment and will eventually be taken up by plants as natural fertilizer. Saprobes are also called **decomposers**, because they break down decaying material and are part of God's design for keeping the world clean. Imagine what the world would be like if, for example, all of the animals that were hit by cars did not break down by decomposing organisms. In a short period of time, our highways would be littered with animal carcasses.

Some bacteria live on or in the bodies of other living organisms. These bacteria depend on their host for food. In some cases, the bacteria are beneficial to their living hosts. For example, bacteria live in the digestive tracts of many large grazing animals. Sheep and cows would not be able to digest the cellulose of grass without these bacteria. The bacteria are provided with a warm, moist environment while absorbing food from the animal's intestine. This association is mutually beneficial for both the cattle and the bacteria. We humans have bacteria in our digestive tract (mostly in our large intestines) to help in the breakdown of food – these bacteria have a different DNA structure than we do! Organisms that live in close association with one another for the benefit of both are said to be **symbiotic**. Sometimes, on the other hand, certain species of bacteria enter and harm their hosts by causing disease. These bacteria are known as **parasites**.

Many heterotrophic bacteria are **anaerobes**. This means that they break down food without using oxygen. These bacteria survive in areas with no oxygen, such as in mud at the bottom of still ponds. Some bacteria are poisoned by oxygen. These bacteria are known as **obligate anaerobes**. The organisms that do not use oxygen but are not harmed when oxygen is present are called **facultative**

Figure 10-8. Curvularia. A common species of saprobe found in soils, plants, cereals, and cellulosic materials such as paper. Some species are plant pathogens. Some can be found indoors and are allergenic.

$$6CO_2 + 12H_2O \xrightarrow{chlorophyll} C_6H_{12}O_6 + 12H_2O$$

Figure 10-9. The Balanced Chemical Equation for Photosynthesis.

anaerobes. Anaerobic bacteria are classified according to the metabolic wastes they produce. For example, some anaerobes produce acetic acid, and others produce lactic acid.

Aerobic bacteria require oxygen for cellular respiration. Recall that prokaryotes do not possess mitochondria. Therefore, cellular respiration is conducted by electron transport proteins that are built into the cell membrane. Recall from Chapter 5.3.A that the term aerobic mean the presence of oxygen is required and the term anaerobic means that oxygen is not required.

Some bacteria are autotrophic and do not contain chlorophyll. The pigment responsible for photosynthesis in these autotrophs is either purple or green and is very different from chlorophyll. Whereas water is the source of electrons for plants during photosynthesis, the compound hydrogen sulfide (H2S) provides the source for these bacteria. Sulfur, rather than oxygen, is produced as a reaction byproduct. These types of reactions are called REDOX reactions, which you will learn about in chemistry.

C. The Nitrogen-fixing Bacteria

All living things need nitrogen for the synthesis of proteins and nucleic acids. The air is about 78% nitrogen gas (N2); however, plants cannot use nitrogen in the gaseous form. Some plants, especially legumes (those in the bean family) possess nodules in their roots that contain nitrogen-fixing bacteria. The bacteria in the roots capture the nitrogen in the atmosphere by the process of **nitrogen fixation**. It is an enzyme that converts nitrogen gas into nitrates (chemicals that contain nitrogen). The nitrates are a usable form of nitrogen by these bacteria. If you have ever applied fertilizer to a plant, you were mostly giving nitrates to the plant.

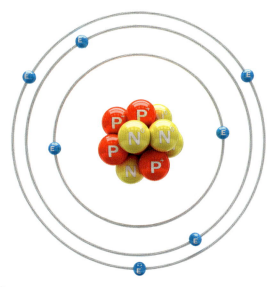

Figure 10-10. A Nitrogen Atom

The nitrogen-fixing bacteria and cyanophytes supply nitrates to eukaryotic organisms. Legumes such as peanuts, clover, beans, and alfalfa have large populations of bacteria in the nodules in their roots. These bacteria receive food and support from the plant and permit the plant to grow in nitrogen-poor soils. This type of mutual benefit to both organisms is referred to as symbiosis.

D. Growing Bacteria in Laboratories

Scientists are able to cultivate bacterial organisms in the laboratory because they understand the nutritional requirements of bacteria. A **growth medium** is a substance that will allow the growth and reproduction of bacteria. For heterotrophic bacteria, the growth medium must contain water, food (such as sugars and amino acids), and minerals.

Bacteria may be grown in flasks or on a plastic dish with a lid (called a Petri dish), either of which

$$12H_2S + 6CO_2 \rightarrow C_6H_{12}O_6 + 12S + 6H_2O$$

Figure 10-11. The Balanced Chemical Equation for bacterial photosynthesis involving hydrogen sulphide found in purple sulfur bacteria and green sulfur bacteria.

CHAPTER 10: SINGLE-CELLED ORGANISMS AND FUNGI

must contain a growth medium. Typically the medium contains **agar**. Agar is a powder that is dissolved in hot water and then becomes a gel when it cools (similar to Jell-O). Bacteria absorb food molecules from the medium by diffusion and are able to grow on the gel surface. A single bacterium that settles on an agar dish will grow and reproduce until millions of descendants of the original cell form a visible colony on the agar surface.

Some bacteria require a simple growth medium, while others require more complex media. Microbiology laboratories in hospitals and clinics use agar plates in the diagnosis of bacterial infection. If you have ever had a throat infection, a doctor or nurse may have streaked a cotton swab near the back of your throat. The swab would then be placed in a sterile package and delivered to a laboratory. A laboratory technician would rub the swab on an agar plate and allow the bacteria to incubate and multiply for a specific time. At some point, tests would be conducted to determine the type of throat infection. This process helps the physician with his diagnosis as well as with treating with antibiotics.

E. Reproduction in the Prokaryotes

The Prokaryotes do not divide by the process of mitosis. However, they do have several methods of duplicating and passing copies of their genetic material on to subsequent generations. Simple cell division or **prokaryotic fission** is the most common method of reproduction in prokaryotes. This process is different from mitosis in eukaryotic cells. It differs in that there is no condensing of chromosomes and no formation of spindle fibers.

Prokaryotic cells have only one chromosome, consisting of a single, circular molecule of DNA. During prokaryotic fission, the molecule of DNA copies itself, and one copy moves to each end of the cell. The cell membrane then invaginates (grows inwards) and splits the parent cell into two daughter cells that are separated by a newly developed cell plate. See Figure 10-12, below.

Many cyanophytes and some bacteria are able to form spores. These are reproductive structures that originate from asexual division such as prokaryotic fission. Spores enable dispersal, as well as the survival for extended periods of time in unfavorable conditions. Even after many years of surviving inhospitable conditions, the spores quickly come to life when conditions change or they end up in a suitable environment.

F. Exchange of Genetic Material

It is important and necessary for the survival of all populations of living organisms to produce new combinations of genes. Most organisms do this through sexual reproduction and the process of meiosis. However, prokaryotes do not

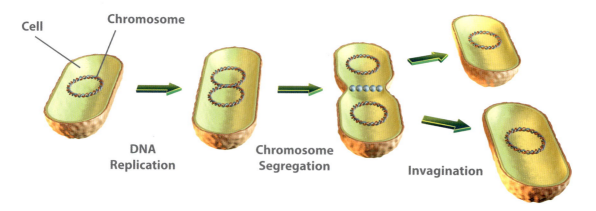

Figure 10-12. The process of prokaryotic fission is different from mitosis in that there is no condensing and formation of spindle fibers.

Biology For Life

reproduce sexually, so they neither undergo meiosis, nor do they form gametes (eggs or sperm). So how do bacteria reproduce to produce new combinations of genes? They do so through methods called conjugation and transformation.

The process of **conjugation** (see Figure 10-13, steps 1-4) involves an extra circular molecule of DNA, which is known as a **plasmid**. The plasmids of some bacteria carry a sex factor. Bacteria that carry the sex factor are able to conjugate with those bacteria that do not. In the conjugation process, a bridge of cytoplasm forms between two bacterial cells by means of a pilus (plural, pili). The cell that carries the sex factor (donor cell) makes a copy of its chromosome. This DNA copy moves across the cytoplasmic bridge and into the second cell or recipient. The bridge usually breaks before the entire chromosome is transferred; therefore, the cell receiving the new chromosomal material receives only part of its genes from the donor cell. The new genes are incorporated into the recipient cell's chromosome. Conjugation results in bacterial cells with new combinations of genes. This process permits some bacteria to overcome resistance to antibiotics. Occasionally you might read about these bacteria in newspapers or magazines, where they are referred to as "superbugs." A bacterium may already be genetically resistant to an antibiotic agent, so it survives. It then breeds new copies of itself that are also genetically resistant. The superbug is formed.

Transformation is another means by which bacteria may obtain new genetic material. In this process, bacteria absorb DNA molecules from their surrounding medium. For example, most

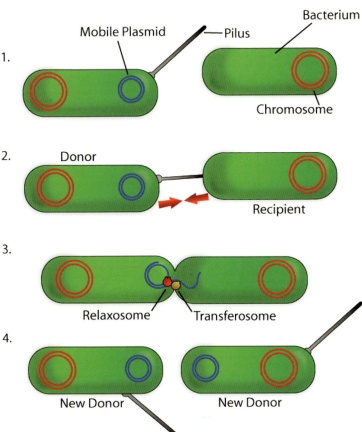

Figure 10-13. Process of conjugation, the transfer of genetic material between bacteria through direct cell-to-cell contact.
1- Donor cell produces pilus.
2- Pilus attaches to recipient cell, brings the two cells together.
3- The mobile plasmid is nicked and a single strand of DNA is then transferred to the recipient cell.
4- Both cells re-circularize their plasmids, synthesize second strands, and reproduce pili; both cells are now viable donors.

Escherichia coli bacterial cells make all of the necessary amino acids from simple chemicals in an agar medium. Some mutant *E. coli* cells lack the ability to make certain amino acids, such as lysine. This mutant form will not grow if lysine is not present in the agar. However, if a strain of *E. coli* that can produce lysine is added to a culture of mutant cells, which cannot make lysine, some of the mutant cells will be transformed. The lysine-making genes are incorporated into these new cells by recombination. The newly transformed cells are now able to grow on a medium that does not contain lysine.

G. Bacteria and Disease

Parasitic bacteria cause many human diseases. Food poisoning, tuberculosis, and strep throat are a few examples of bacterial illnesses. Farm animals and plants are also susceptible to a number of bacterial diseases. People spend billions of dollars each year on preservation techniques, such as heating, salting, refrigerating, adding chemicals to food, and even using ionizing radiation in order to slow the growth of bacteria that cause food to spoil. Vigorous hand washing with hot soapy water is a simple method for preventing many bacterial diseases; the bacteria are killed, either by the soap or by the mechanical action of hand rubbing.

Yet even some disease-causing bacteria can be helpful to people. For example, some bacteria cause disease in pest organisms. One such bacteria is *Bacillus thuringiensis*, which is lethal to all caterpillars. It kills caterpillars by causing holes to form in its stomach, or crop. If these bacteria are sprayed on crops, they kill the caterpillars and leave no residues that could occur with traditional chemical pesticides.

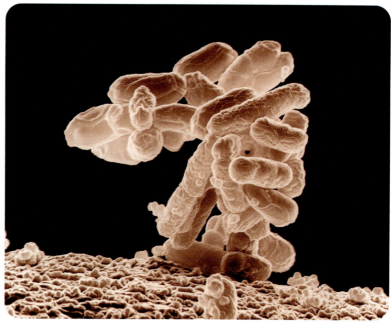

Figure 10-14. *Escherichia coli* bacteria that is commonly found in the lower intestine of warm-blooded animals.

Section Review — 10.2

1. Compare and contrast Prokaryotes and Eukaryotes.
2. What is a heterotroph?
3. What are Saprobes?
4. Some bacteria are poisoned by oxygen; what are these bacteria called?
5. _____ is a _____ that is dissolved in hot water and then becomes a gel when it cools.
6. What is prokaryotic fission?
7. Describe the process of conjugation.

CHAPTER 10: SINGLE-CELLED ORGANISMS AND FUNGI

10.3 Diversity of the Bacteria

> **Section Objectives**
>
> - List the characteristics used for separating the Bacteria into two divisions: schizophytes and cyanophytes.
> - Describe the characteristics of schizophytes and cyanophytes.

The Bacteria live in a wide range of locations, from the hottest deserts to the coldest, driest parts of the Antarctic. Bacteria are classified into two divisions, which are the **Schizophyta** and **Cyanophyta**.

A. Division Schizophyta

The Schizophyta contain about 15,000 different species of bacteria and related forms. The classification of the **schizophytes** is based on features such as cell shape, energy source, and the chemistry of their cell walls.

Phylum Firmicutes - This is the phylum of the germs. These bacteria, in their hundreds of trillions, were the reason that you were not allowed to pick up the candy you dropped on the floor or eat that egg salad that looked so good a week ago.

There are several shapes of Firmicutes, including the spherical **cocci**, the rod-shaped **bacilli** (ba-sĭl-ī), and the spiral-shaped **spirilla** (spĭ-rĭl-a). Some of the Firmicutes have distinctly colored pigments. Other types have **flagella**, which are whip-like structures that enable these to move through fluids. Many bacteria in this phylum have the ability to form thick-walled cells called **endospores**. An endospore is in a dormant stage, which allows the bacteria to withstand extreme conditions of temperature and drought.

If the environment becomes too dry or too hot, a thick wall forms inside the cell. The wall encloses the nuclear material and some cytoplasm, allowing the endospore to withstand boiling water and to survive extremely dry conditions. Some endospores have been known to survive dry conditions for as long as fifty years. Bacteria live in a wide variety of habitats, including some environments that are uninhabitable to most organisms. A few bacteria live in the Great Salt Lake of Utah, which has almost four times the salt concentration of the ocean. Other species of bacteria can live in hot springs, where the water is strongly acidic and can reach temperatures as high as 900° C (water boils at 100° C).

Phylum Proteobacteria – The **myxobacteria** are the closest thing to multicellular organisms in the Kingdom Bacteria. The cells within a colony of myxobacteria glide along the forest floor and absorb food from decaying leaves or wood. If conditions become unfavorable for feeding and movement, the colony stops moving, and cells pile on top of one another, forming a stalk. A swollen structure called a **fruiting body** forms at the top of the stalk. Individual cells within the fruiting body then form resistant spores.

Wind blows the spores to new locations, and upon the return of favorable conditions, each spore can reproduce and form a new colony. Cell specialization and chemical communication are

Figure 10-15. Myxobacterial Fruiting Bodies.

Biology For Life

CHAPTER 10: SINGLE-CELLED ORGANISMS AND FUNGI

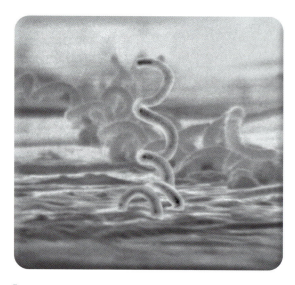

Figure 10-16. Treponema Pallidum Spirochetes. Spirochetes are chemo-heterotrophic in nature, about 5 and 250 μm long and around 0.1-0.6 μm in diameter. Note that a human hair is about 50 μm in diameter.

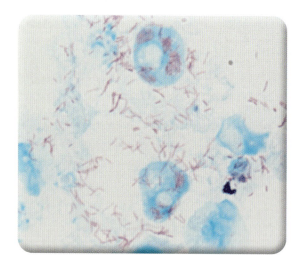

Figure 10-17. Gimenez stain of tick hemolymph cells infected with Rickettsia rickettsii, the causative agent of Rocky Mountain spotted fever. R. rickettsii is a small bacterium that grows inside the cells of its hosts. These bacteria range in size from 0.2 x 0.5 micrometers to 0.3 x 2.0 micrometers. They are difficult to see in tissues and require the use of special staining methods.

required for movement to occur and in order to form fruiting bodies. These traits are not found in the other schizophytes.

The Phylum Proteobacteria also contain the rickettsiae (rĭk-ĕt-sĭ-ē), which are very small, parasitic bacteria. These bacteria live and reproduce inside eukaryotic host cells. Rickettsiae are found in ticks, mites, lice, and fleas. When these arthropods are infected with rickettsia, they are capable of transmitting diseases such as Rocky Mountain spotted fever, scrub typhus, epidemic typhus, and murine typhus. Some of these diseases are potentially deadly, requiring quick diagnosis and treatment with antibiotics.

Phylum Actinobacteria - Most **actinobacteria** are soil organisms, and most are harmless. However, two species in the genus *Mycobacterium* cause tuberculosis and leprosy in people. The genus *Streptomyces* contains species that are the source of some very effective **antibiotics**. Antibiotics kill or slow the growth of other bacteria. It is interesting how two different genera in the same phylum, that is, the *mycobacterium* and *streptomyces*, lead to either terrible illnesses or to healing.

Phylum Spirochetes – Many **spirochetes** (spī-ro-kēt) live in mud or water and are anaerobic. A few species are parasitic, such as *Treponema pallidum*, which causes syphilis in people. Lyme disease is transmitted by the black-legged tick. The name for the disease-causing spirochete is *Borrelia burgdorferi*.

Spirochetes have a spiral shape, much like the spirilla; however, unlike spirilla, the spirochetes have an **axial** (ăk-sĭ-al) **filament**. This filament consists of a series of fibers between the cell wall and the cell membrane. The axial filament helps the organism move by rapidly twisting and coiling.

Phylum Tenericutes – **Mycoplasmas** are parasites that live inside cells. The mycoplasmas are even smaller than the rickettsiae and they have no cell walls. Despite this, the cells often present a certain shape that gives them the ability to thrive in different environments. Some mycoplasmas cause disease in animals and people. One type causes a mild form of pneumonia in children and young adults.

Biology For Life

B. Division Cyanophyta

There are about 1,500 species in this division. The cyanophyta are photosynthetic and are more commonly referred to as blue-green algae. These were once classified as plants; however, their prokaryotic cell structure indicates that they are more closely related to the bacteria than to plants or protists.

Cyanophytes are found in a wide variety of locations. Many live in fresh water or moist soils, and a few species live in oceanic environments. Some cyanophytes are even able to live in the acidic water of hot springs. Others are occasional pests of tropical fish aquariums.

Cyanophytes contain two photosynthetic pigments that are not found in eukaryotic plants. These pigments are called **phycocyanin**, which is a blue pigment, and **phycoerythrin**, which is a red one. Many cyanophytes are multicellular and have cells that are joined end-to-end in long green filaments – see Figure 10-19. The chains of cells break into pieces in order to reproduce. The cells at the broken ends divide, so that the length of the new filament is increased.

Some cyanophytes, such as *Nostoc*, contain thick-walled cells called **heterocysts**. A heterocyst contains enzymes for nitrogen fixation, but it does not contain chlorophyll. The heterocyst supplies the other cells of the filament with nitrogen compounds. In turn, the other cells supply the heterocysts with food produced from photosynthesis.

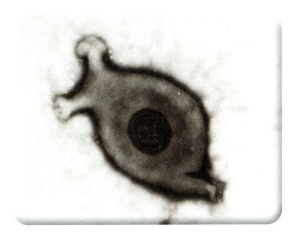

Figure 10-18. Mycoplasma gallisepticum is a bacterium belonging to the Phylum Tenericutes and the Family Mycoplasmataceae. It is the causative agent of chronic respiratory disease in chickens and infectious sinusitis in turkeys. It is the smallest cell known to man.

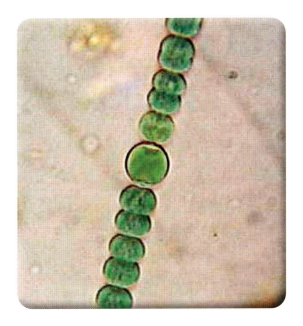

Figure 10-19. Anabaena sphaerica, a type of cyanobacteria, or blue-green algae, that is known as plankton. Because of their nitrogen-fixing abilities, they form symbiotic relationships with certain plants, such as the mosquito fern. Note the size of the bar is 10 micrometers = 0.000393701 inches.

Section Review — 10.3

1. Bacteria are classified into two divisions. What are they?
2. What are the different features of schizophytes?
3. How many different phyla are in Schizophyta and what are they?
4. The cyanophyta are photosynthetic and are more commonly referred to as _____.

10.4 Viruses

Section Objectives

- Describe the structure of a typical virus and discuss the two different reproductive cycles of a virus.

In 1892, Dmitri Iwanowski, a Russian biologist, attempted to isolate an organism that caused "mosaic disease" in tobacco plants. He assumed that the organism was a bacterium. He collected samples of the diseased areas, placed these in solutions, and attempted to collect these using a special filter. In an attempt to cause mosaic disease, he placed the filtered material on uninfected tobacco plants. Nothing happened. Next, he collected the solution, which passed through the filter, and placed it on the tobacco plants. This solution caused disease in the tobacco plants. Because the material easily passed through the bacteria filter, this was proof that the organism which caused the disease was smaller than the bacteria. However, the new organism was too small to be seen by the light microscope. It was not until almost 50 years later when these organisms were observed, after the development of the electron microscope. These organisms were called **viruses**.

Viruses are considered to be alive by some scientists and belong to their own Virus Kingdom. However, as is typical with "academic debates," there are many virologists who do not consider viruses alive and do not place these in a kingdom. When you "catch a virus" and come down with a fever and accompanying aches and pains, you will certainly believe that they are alive.

A. Structure of Viruses

Viruses are incredibly small, ranging in diameter from 0.03 to 0.30 micrometers (a human hair is about 50 micrometers in diameter). The largest virus is about one-tenth of the volume of the smallest bacterium. A virus consists of a single nucleic acid molecule that is surrounded by a layer of protein molecules, or protein coat. The

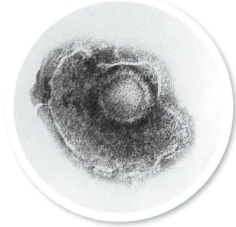

Figure 10-20.
Electron micrograph of negatively-stained herpes zoster virus - note the envelope around the virus particle.

protein coat is made up of several hundred protein molecules packed together in a geometric pattern. In some larger viruses, such as the human influenza virus, there is a complex capsule surrounding the protein layer.

The nucleic acid in a virus may be a double or single-stranded molecule of DNA, and some viruses have a single molecule of RNA. The amount of nucleic acid material in a virus is much smaller than that found in any type of cell. The tiniest bacterium has enough DNA for about 2,000 genes, whereas the largest virus has enough for about 100 genes. Many viruses have only enough DNA for ten genes.

The genes of a virus carry instructions for the production of new virus particles; however, there are no cytoplasmic structures such as ribosomes that use these instructions to make new viruses. How, then, are new viruses made? Viruses cannot live by themselves. They must live as a parasite inside a living host cell, where they use the energy of the host cell and its protein-producing organelles in order to create new viral particles. You have likely experienced the parasitic nature of viruses whenever you fell victim to a nasty cold virus or stomach virus.

CHAPTER 10: **SINGLE-CELLED ORGANISMS AND FUNGI**

Each type of virus infects only a specific host cell. For example, the tobacco mosaic virus infects only the cells of the tobacco plant. The rhinovirus infects the cells lining the human respiratory tract and is responsible for causing the common cold. Some viruses infect bacteria and are known as **bacteriophages** (bak-tir-ē-ō-fājs), or phages. Some bacteriophages have an unusual appearance, such as the phage that infects the *E. coli* bacterium. This phage contains a tail, which is attached to the protein coat, along with several long fibers at the base of the tail, which resemble spider legs.

B. Viral Reproduction

To understand how viruses reproduce, we will use the example of the bacteriophage that infects the *E. coli* bacterium. The tail fibers of the phage are made in the shape of a protein that matches the molecules in the cell membrane of the *E. coli* bacterium. The phage and the bacterium fit together much like a key in a lock. This "viral matching" explains why viruses can infect only a specific type of host cell.

After the virus has attached to its host, it injects its DNA into the host. The viral DNA then takes over the organelles in the cytoplasm of the host cell that become responsible for cell reproduction. The nucleotides of the host cell make many duplicate copies of the phage DNA. These DNA molecules then synthesize phage nucleic acids and proteins that form as many as several hundred new phage organisms. Finally, the phage DNA must instruct the host cell to self-destruct, so that these new phages can be released. The host cell then produces an enzyme, which **lyses**, or digests, the bacterial cell wall, allowing the new phage particles to be released. Each new phage repeats this reproductive cycle on another uninfected *E. coli* bacterial cell. A reproductive cycle like that just described is termed a **lytic** (lī-tĭk) **cycle**.

Some viruses carry out modifications of the lytic cycle. For example, the human influenza virus does not lyse the cells of the host, but, rather, these push out through the cell membranes to exit the cells. These types of viruses have capsules that consist partly of host cell membranes and viral proteins that are embedded in the membranes.

A second type of reproductive cycle in viruses is known as the **lysogenic** (lī-so-jĕ-nik) cycle. In this process, the virus inserts itself into the host cell by recombination. The term recombination means that the viral DNA becomes part of the host cell DNA. The host cell then reproduces its own genetic material along with the genetic material provided by the instructions from the viral DNA. For many generations of the host cells, there may be no production of active viral particles. However, the viral DNA may occasionally become active in the host cell and trigger reproduction.

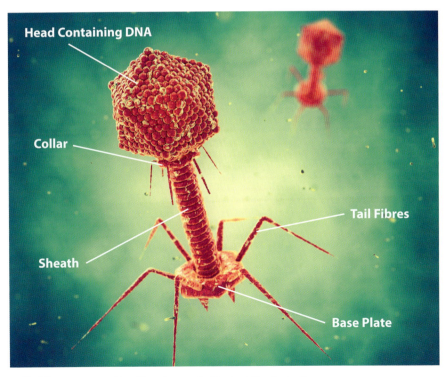

Figure 10-21. The Structure of a Typical Bacteriophage.

Biology For Life

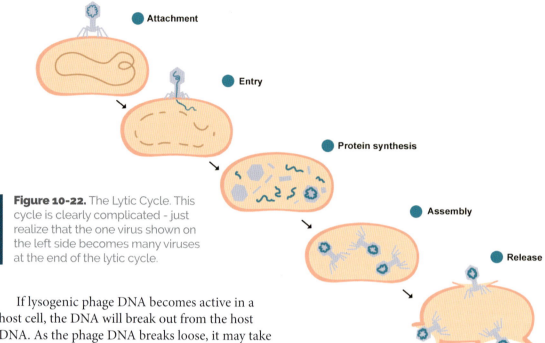

Figure 10-22. The Lytic Cycle. This cycle is clearly complicated - just realize that the one virus shown on the left side becomes many viruses at the end of the lytic cycle.

If lysogenic phage DNA becomes active in a host cell, the DNA will break out from the host DNA. As the phage DNA breaks loose, it may take several bits of host cell DNA with it. In the next cycle of infection, the phage carries the bacterial genes (DNA) from the original host cell to a new host cell. The process by which a virus carries bacterial genes from one bacterium to another in a lysogenic phage is known as **transduction**.

Lysogenic phages are used in recombinant DNA research, because the phages are able to transfer a desired gene into the chromosomes of a bacterium. Lysogenic phages are known to cause cold sores, and some types play a role in human cancers. Scientists have known that viruses cause some animal cancers; however, proof that viruses cause human cancers has been elusive. Strong evidence exists that lysogenic viruses may transform human DNA. These transformations may result in the host cell becoming cancerous.

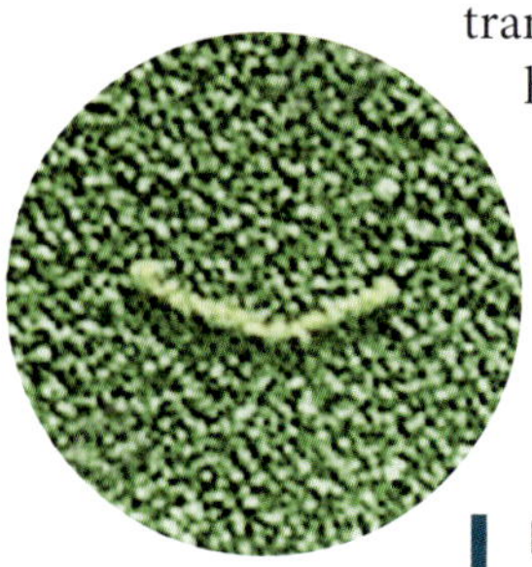

C. Viroids

The first **viroids** (vī-roid) were discovered in the late 1960s and early 1970s. Very little is known about **viroids**, except that they are known to cause diseases in plants and have a similar reproductive cycle to the viruses. Viroids lack protein coats and are about one-fourteenth the size of a virus particle. They contain short pieces of RNA that consist of about 350 nucleotide bases. In comparison, the nucleic acids of viruses contain about 5,000 nucleotide bases.

Figure 10-23. Viroid Magnified 350,000X

Section Review — 10.4

1. Viruses are incredibly small, ranging in diameter from _____ to _____ micrometers.

2. Since viruses cannot live by themselves, how do they survive?

3. What is a lytic cycle?

4. A second type of reproductive cycle in viruses is known as _____.

CHAPTER 10: **SINGLE-CELLED ORGANISMS AND FUNGI**

10.5 Kingdom Protista

Section Objectives

- List the characteristics of the protists.
- Give two differences between euglenophytes and other plant-like protists, like pyrrophytes and chrysophytes.
- List the characteristics of dinoflagellates.
- List the major characteristics of the most important chrysophytes.
- List an example of a mastigophoran.
- Describe the major characteristics of the Phylum Sarcodina.
- Identify the major structures found in a Paramecium.
- Describe the life cycle of Plasmodium vivax.

Figure 10-24. Euglena

propel them through water. If you have the chance to view living *Euglena* under a microscope, you will not be able to view them for very long, as they rapidly move in and out of the field of view.

At one time, biologists had difficulty classifying the *Euglena* as either a plant or an animal. The *Euglena's* chloroplasts identified it as a plant, but its ability to swim around made it seem more like an animal. Taxonomists created the Kingdom Protista to resolve this problem.

1. Phylum Euglenophyta

Most members of this phylum have chloroplasts and are autotrophic. Some are heterotrophic and do not have chloroplasts, and others are parasitic. The discussion of this phylum will center upon the *Euglena*, which is autotrophic.

Refer to the diagram below for the internal structure of the *Euglena*.

The protists are microscopic unicellular organisms that are divided into two major types, the plant-like and the animal-like protists.

Protists are organisms that do not strictly fit into any of the other Kingdoms. Protists are eukaryotic, while bacteria and archaebaceria are prokaryotic. Protists have nuclei with a nuclear membrane, mitochondria, and various other organelles. Most protists are single-celled organisms, though all reproduce by mitosis and cell division. Many forms of protists also reproduce sexually.

The protists are a very diverse group. There are about 10,000 species of protists that are in the plant-like category.

A. Plant-like Protists

Euglena is a single-celled protist that is common in our ponds, lakes, rivers, and roadside ditches. It is bright green, which is due to the photosynthetic pigment chlorophyll that is contained in its chloroplasts. *Euglena* possess whip-like tails that

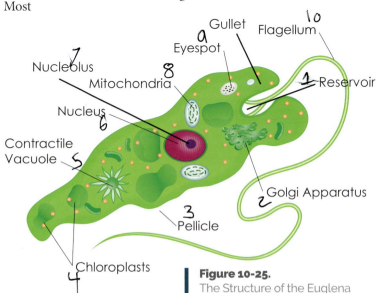

Figure 10-25. The Structure of the Euglena

Biology For Life 251

Each cell contains a large nucleus and a nucleolus. Bright green chloroplasts are distributed throughout the cytoplasm. The chloroplasts will disappear if the organism is placed in the dark for a few days. In this case, the cell survives by absorbing nutrients through the cell membrane. The chloroplasts reappear when the organism is, once again, exposed to light.

The *Euglena* moves by means of **flagella** (fla-jĕ-la), a whip-like tail structure. Each **flagellum** (pl. flagella) is an extension of the cell membrane and is composed of eleven microtubules, which are arranged with 9 surrounding the other 2 (9 + 2). Each *Euglena* has a long flagellum and a short flagellum. The long one propels the organism through water. Scientists remain uncertain about the role of the smaller flagellum. Both flagella are attached at a point on the organism called the **reservoir**, which is a depression at the front end of the cell. The long flagellum emerges through an area called the **gullet**, a canal that opens into the reservoir. There is a red-pigmented area known as the **eyespot**, which is light sensitive. When the eyespot senses light, the flagellum propels the *Euglena* toward the light source. Accordingly, this structure plays an indirect, albeit important, role in photosynthesis.

Euglena lack cell walls but have spiral strips of protein on the outer surface of their cells. The protein forms a flexible **pellicle**, which gives the cell its shape. The *Euglena* can crawl along by changing its shape, in a form of motion known as **Euglenoid movement**.

Euglenas live in fresh water, and the cytoplasm of the organism is always more concentrated (contains more dissolved ions) than fresh water. Because of this difference in concentration, water tends to diffuse into the cell by osmosis; that is, the net movement of water from a higher concentration to a lower one. Euglenas and other protists have a **contractile vacuole**, which allows the organisms to control the water balance of the cell. The water flows into this vacuole, causing expansion of the vacuole. When the vacuole has expanded completely, it contracts, releasing the excess water out through the gullet. *Euglena* reproduces asexually. The nucleus undergoes mitosis, and the cell splits lengthwise, forming two daughter cells.

2. Phylum Pyrrophyta

Most members of this phylum are marine, which means that they live in ocean waters. **Dinoflagellates** are the most common pyrrophytes. The small organisms that dwell near the surface of the ocean, the **plankton**, are dinoflagellates. There is often an assumption that most of the oxygen that we breathe in originates in forests. While forests supply a great deal of oxygen, plankton produce most of the earth's oxygen.

The chloroplasts of dinoflagellates contain chlorophyll; however, they also have red and yellow pigments. These other pigments give these organisms their characteristic red or brown color. Most dinoflagellates have a cellulose cell wall made of many segments that fit together like the pieces of a suit of armor. If you have the chance to view these organisms with a microscope, you will be struck with their intricate designs and beauty.

Most dinoflagellates have two flagella. One flagellum propels the organism forward, while the other vibrates and allows the organism to spin as it moves. The nucleus of dinoflagellates is

Figure 10-26. The dinoflagellate *Ceratium hirundinella*.

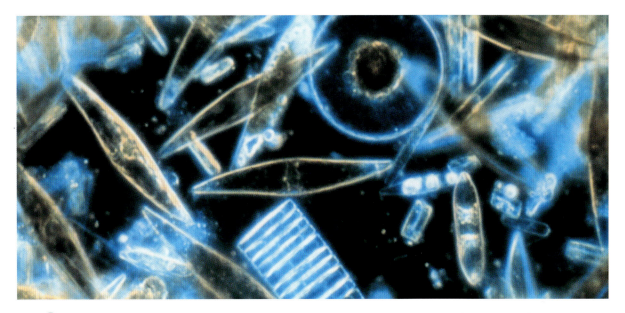

Figure 10-27. Assorted diatoms as seen through a microscope. These specimens were living between crystals of annual sea ice in McMurdo Sound, Antarctica. These tiny phytoplankton are encased within a silicate cell wall. The fact that their structures are so distinct (circles, diamond-shaped, etc.) is due to their having lived between ice crystals.

unusual, because, unlike in most eukaryotes, the chromosomes are compact and visible all the time. In the typical eukaryote, the chromosomes are only visible during the process of mitosis.

The population of some dinoflagellates increases dramatically on certain occasions. Water can turn red from large numbers of dinoflagellates. This phenomenon is known as a red tide. The species which causes red tides produces powerful nerve poisons, and high concentrations of these poisons may kill fish. Shellfish such as clams and oysters are unaffected by the poison; however, the poison builds up in their bodies. Humans who eat these shellfish may be paralyzed or even killed. Many dinoflagellates have the ability to produce light as the result of a chemical reaction, which is known as **bioluminescence**. If boats, fish, or waves disturb these types of dinoflagellates at night, the water seems to flash in spots with blue or green light. This flashing of light can be likened to numerous glow sticks that have been placed in the water, though the flashes have a very short life.

3. Phylum Chrysophyta

Most members of this phylum are photosynthetic. The predominance of other cell pigments gives these organisms their characteristic gold or brown color. Most chrysophytes do not have flagella. There are three Classes of chrysophytes: Diatoms (Bacillariophyta), Golden-brown algae (Chrysophyceae), and Yellow-green algae (Xanthophyceae).

Diatoms (dī-*a*-tŏm) comprise most of the 6,500 species of chrysophytes. These organisms live in both fresh and salt water. Diatoms have cell walls that contain two overlapping halves. The cell wall is composed of a gelatinous material called **pectin**. Most diatoms also contain silica in their cell wall. Silica is a glassy material that contains the elements silicon and oxygen, or silicon dioxide (SiO_2). Did you know that sand is composed of silica, and the glass that you look through every day is made from a refined version of this material?

Diatoms usually reproduce asexually. Following mitosis, the cytoplasm divides so that each daughter cell receives half of the old cell wall. The new daughter cell will then produce a new half cell wall. Occasionally diatoms reproduce sexually. In this process, the parent cell undergoes meiosis. One haploid nucleus survives the process, and the other three degenerate. A gamete with a flagellum breaks out of the old cell wall and swims until it encounters another gamete of the same species. These cells fuse to form a diploid zygote. The zygote will then develop an entirely new shell.

> **Section Review — 10.5 A**
>
> 1. What are Protists?
> 2. Where is Euglena found?
> 3. Why were biologists having difficulty classifying Euglena?
> 4. Which are the most common pyrrophytes?
> 5. What is bioluminescence?
> 6. Most members of Phylum Chrysophyta are _____.

B. Animal-like Protists

The animal-like protists are single-celled eukaryotes. They are classified as animal-like because of the way they move and feed.

1. Phylum Mastigophora: the Flagellates

This phylum contains the animal-like protists that move by means of flagella. These organisms may have one or more flagella. A few species of mastigophora live in fresh or salt water, but most live in the bodies of larger organisms.

One of the best-known flagellates is called *Trypanosoma*. This organism is a human parasite that causes **African sleeping sickness**. This parasite is found in the blood of many wild and domestic animals throughout much of Africa.

The trypanosome parasite is transmitted from one animal to another by the bite of the bloodsucking tsetse fly. Humans can easily acquire

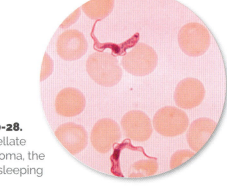

Figure 10-28. The Flagellate Trypanosoma, the cause of sleeping sickness.

African sleeping sickness, but the bites of the tsetse fly can easily be prevented by the use of repellants and clothing; for example, lightweight long sleeved shirts and trousers. The term **disease vector** is given to an insect such as a tsetse fly, which carries disease from one host to another.

Some flagellates living in animals are not parasitic but live symbiotically with their hosts. An example is the *Trichonympha*, which is extremely small and lives in the gut of termites. Trichonympha contain bacteria known as endosymbionts that help them digest cellulose. We typically think of the termite as eating wood, but it is the Trichonympha and the bacteria that they

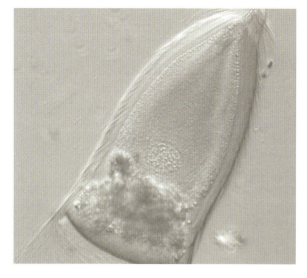

Figure 10-29. The Flagellate Trichonympha that lives symbiotically in the intestines of termites.

CHAPTER 10: SINGLE-CELLED ORGANISMS AND FUNGI

Figure 10-30. The Amoeba proteus.

harbor which make it possible. If antibiotics are placed in the diet of termites, the bacteria die, then the protists, and finally the termites.

2. Phylum Sarcodina

This phylum consists of the protists that move and obtain food by using **pseudopods** (sū-dō-pŏds), or "false feet." Pseudopods are temporary extensions of the cytoplasm. Some of the cytoplasm flows forward when a sarcodine moves, thus extending the cell membrane. All sarcodines feed by the process of **phagocytosis** (fă-gō-sī-tō-sĭs). In this process, an organism surrounds its prey with pseudopods, from which forms a food vacuole. Digestive enzymes that are released into the food vacuole help break down the prey. The cell then absorbs the resulting food molecules. The most well-known sarcodine is called *Amoeba proteus* and is more commonly known as the amoeba.

Amoebas live in ponds and crawl along the bottom of the pond or over plants. They have contractile vacuoles that remove excess water from their bodies. Amoebas also have several food vacuoles, each of which contains food at various stages of digestion. Amoebas reproduce asexually by mitosis and cell division.

3. Phylum Ciliophora: the Ciliates

The organisms in this group use cilia to move and capture food. **Cilia** (sĭl-ĭ-*a*) are short flagella that are composed of eleven microtubules. **Microtubules** are linear polymers and act as conveyer belts within cells. They help transport various items within cells and are associated with external movement of the cell.

Each single cilium is made up of a bundle of fibers called an **axoneme**. The bundles contain nine double microtubules around a center of two single microtubules. The cilia are very numerous and often cover the entire surface of the organism. Ciliates also have a pellicle like the euglenoids. Most ciliates live in fresh or salt water. A few species will attach to a surface by a stalk and will stay in place. Other ciliates live inside the bodies of host animals. The *Paramecium* is the most commonly studied ciliate.

Paramecia live in ponds. The cilia that surround the body of this organism move in a coordinated pattern, which allows movement and helps propel them through the water. The *Paramecium* also has stinging organelles called **trichocysts**, which are thread-like structures that are barbed at the end. These trichocysts are used for hunting and for self-defense. *Paramecia* are capable of responding to

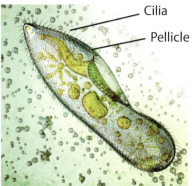

Figure 10-31. The Paramecium is the best known of all ciliates. Bubble-like structures throughout the cell are vacuoles. The entire surface is covered in cilia, which are blurred by their rapid movement.

Biology For Life

stimuli. For example, they swim toward food after detecting the presence of chemicals in the water. They may also swim away from harmful chemicals such as acids. There is a network of fibers beneath the cell membrane of the *paramecium* that helps to control the beating of the cilia necessary for motion.

Paramecia also use their cilia to capture food. Currents are created by the cilia, which allow food to be swept into the **oral groove** on the side of the organism. The oral groove leads to a gullet, and food vacuoles are formed at the base of this gullet. Any undigested food is removed from the organism when the food vacuole fuses with the cell membrane at the **anal pore**. A **contractile vacuole** removes excess water from the organism.

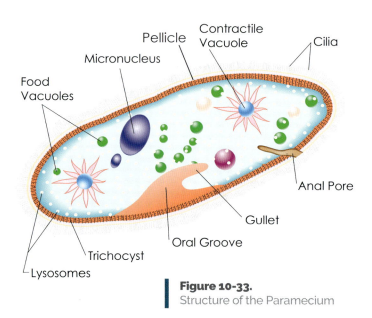

Figure 10-33. Structure of the Paramecium

Paramecia reproduce asexually by mitosis and cell division. These organisms also undergo a type of sexual reproduction known as conjugation (see the section on Bacteria). In this process the two individuals combine genetic information.

4. Phylum Sporozoa

All members of this phylum are parasites and have no means of locomotion. They depend on carriers or vectors to move from host to host, from which they acquire food and complete their life cycle. Sporozoa reproduce by the formation of spores. In this process, the nucleus of the parent cell undergoes mitosis numerous times. The spores are released when the parent cell bursts.

The best known sporozoans are those that cause **malaria** in humans. Malaria is a very harmful – often fatal – disease that is widespread in tropical and subtropical regions, including parts of the Americas, Asia, and Africa. *Plasmodium vivax* is one of the four species of Sporozoa that cause Malaria in humans. The life cycle of Plasmodium is quite complex.

Figure 10-34 summarizes this complex life cycle.

1. These parasites enter a person as sporozoites by the bite of an infected Anopheles mosquito.

2. The sporozoites travel to the liver where they infect liver cells. In the liver cells the sporozoites develop into merozoites.

3. At some point, the infection enters red blood cells, erythrocytes. In the red blood cells, the parasites continue to duplicate and spread the infection to other red blood cells.

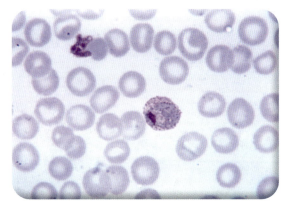

Figure 10-32. Plasmodium vivax is one of four species of malarial parasite that commonly infect humans and is seldom fatal. It is less virulent than *Plasmodium falciparum*, which is the most deadly of the four types.

CHAPTER 10: **SINGLE-CELLED ORGANISMS AND FUNGI**

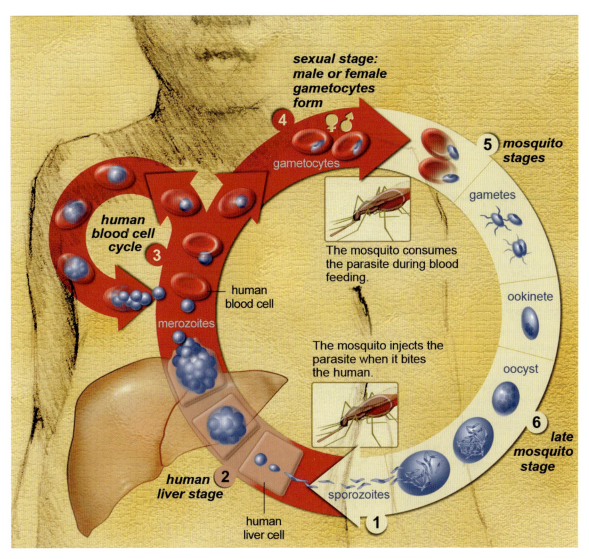

Figure 10-34. Life Cycle of Malaria in humans

4. In the red blood cells a small percentage of the parasites change into gametocytes, which are necessary for the sexual stage of further reproduction. If another mosquito bites this host, it ingests some of these gametocytes.

5. The mosquito's ingested human blood and the gametocytes go to the mosquito's midgut. There, the gametocytes develop into gametes, both male and female, and these join in fertilization to form a zygote.

6. The zygote undergoes further developmental stages which eventually form new sporozoites that travel to the mosquito's salivary glands, from which they can infect a new host through this mosquito's bite.

The parasite is relatively protected from attack by the body's immune system because for most of its human life cycle, it resides within the liver and blood cells and is relatively invisible to the immune system. However, circulating infected blood cells

Biology For Life

are destroyed in the spleen. To avoid this fate, the parasite displays adhesive proteins on the surface of the infected blood cells, causing the blood cells to stick to the walls of small blood vessels, thereby sequestering the parasite from passage through the general circulation and the spleen.

Widespread spraying of insecticides has been used to control mosquitoes in tropical countries where malaria is common. Unfortunately, the Anopheles mosquito has developed a partial resistance to the pesticides, and malaria continues to be a problematic disease. During the last twenty years, many villages throughout the world have begun to use bed-nets that have been impregnated with a repellent. As you can imagine, these repel numerous mosquitoes that are readily found in remote dwellings that lack windows and screens. Treatment of malaria involves supportive measures such as controlling the mosquito population, as well as specific antimalarial drugs. When properly treated, someone with malaria can expect a complete recovery.

Section Review — 10.5 B

1. What Flagellate is the cause of African sleeping sickness and how is it transmitted?
2. What is a pseudopod?
3. Name the phyla of the animal-like Protists.
4. Describe the life cycle of Malaria.

10.6 Kingdom Fungi

Section Objectives

- Describe the structural characteristics of fungi.
- Explain how fungi obtain their food.
- List the features of reproduction in fungi.
- List similarities and differences between cellular and acellular slime molds.
- List some of the characteristics of the Dimastigomycota.
- Summarize sexual and asexual reproduction in Zygomycetes.
- Give examples of some Basidiomycetes, and list the characteristics of this group.
- Describe how Deuteromycetes are different from the other true fungi.
- Describe some of the characteristics of lichens.

There are many varieties of mushrooms, all of which belong to the Kingdom Fungi. The part of the mushroom that we see is only the reproductive structure. The remainder of the mushroom is composed of a very large network of tiny threads (or mycelia) that grow through the decaying matter on the forest floor. Some types of fungi contain deadly poisons, but others are considered delicacies.

Members of the Kingdom Fungi

Figure 10-35. From left to right: Amanita muscaria, a basidiomycete; Sarcoscypha coccinea, an ascomycete; black bread mold, a zygomycete; a chytrid; a Penicillium conidiophore.

A. Characteristics of Fungi

For many years, **Fungi** were classified in the Plant Kingdom. Most fungi have a cell wall surrounding the cell membrane. Also, most fungi are sessile and do not actively move around. These characteristics of fungi make them similar to plants. However, there are important differences between plants and fungi. The majority of plants make chlorophyll, and fungi do not. Because fungi lack chlorophyll, they are unable to make their own food. Fungi also do not contain the specialized tissues found in most plants. Today, biologists classify fungi in their own kingdom.

B. Fungi Structure

Fungi are eukaryotes. As the fungi grow, their cytoplasm forms long tube-like extensions that are called **hyphae** (hī-fē). These hyphae are surrounded by cell walls. In several types of fungi, the cell wall is made of a complex carbohydrate known as **chitin** (kī-tĭn), the same hard material by which insects are covered. The hyphae branch to form a tangled network of threads called **mycelia** (mī-sē-lĭ-a, singular – **mycelium**).

C. Fungi Nutrition

All fungi are heterotrophs and feed by secreting enzymes into their surroundings. These enzymes break food into small molecules, and then the cells of the fungi absorb these molecules. This mode of obtaining nutrients is called **absorption**.

Most fungi are **saprophytes** (săp-ro-fĭts). A saprophyte is an organism that absorbs its food from dead or decaying organic matter. Fungi are often found growing on rotting leaves, wood, or animal wastes. The enzymes secreted by fungi help to break down these organic materials. Fungi are very important in the decomposition of plants and animals.

Consider for a moment, a very large tree that has been struck by lightning in a forest. It has

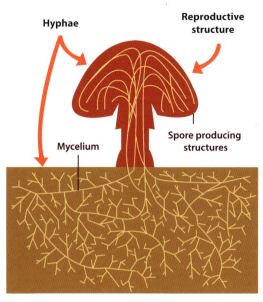

Figure 10-36. The Basic Structure of Fungi

Biology For Life

CHAPTER 10: SINGLE-CELLED ORGANISMS AND FUNGI

Figure 10-37. Slime Mold

had its bark ripped open, resulting in much tissue damages. Its defenses have been weakened. Within hours, fungi and insects that carry fungal spores find their way into the damaged tissues of the tree. The tree is beginning to die. Within a period of years, fungi and other organisms will consume the entire tree. That is, all of the stored energy in the form of cellulose will be broken down and consumed by fungi. Now, consider for a moment someone harvesting a same-sized tree to acquire firewood. The tree is reduced to piles of cordwood and, when burned, may supply heat for more than a couple of years. In both examples, the energy that is liberated from the dead tree will be about the same. The fungi capture this energy, then grow, reproduce, and produce spores. Nutrients are returned to the soil and shared among organisms. The homeowner liberates this energy and warms his family's house. Imagine how much energy is stored in a large tree. Clearly, this example shows the efficiency of nature and the designs of an omniscient creator God.

Some fungi are parasitic. The parasitic fungi usually grow on plants. They extend their hyphae between the cell walls of their host plant and then absorb food from its cells. Some parasitic fungi attack important crops such as corn, wheat, or potatoes. A few fungi are parasitic to animals. Athlete's foot and ringworm are fungal parasites of humans.

D. Reproduction in Fungi

Most fungi reproduce asexually and produce reproductive cells known as **spores**. The spores have tough cell walls, which help retain moisture. These spores may be carried through the air or water. A new mycelium may grow if the spore lands in a suitable environment with food and moisture.

Most fungi may also reproduce sexually. The part of the fungi that is visible, such as a mushroom cap, is actually the reproductive structure. These reproductive structures are known as **fruiting bodies**. The fruiting bodies release spores that have been produced by sexual reproduction.

E. Variety of Fungi

Fungi have been classified into three subkingdoms. The subkingdom **Gymnomycota** is composed of the fungi known as slime molds. The subkingdom **Dimastigomycota** contains fungi called water molds. Most of the fungi belong to the subkingdom **Eumycota**. Mushrooms and yeasts are members of the Eumycota subkingdom. These three subkingdoms will now be described in some detail.

1. Subkingdom Gymnomycota

There are two main types of slime molds. **Cellular slime molds** produce spores and rarely reproduce sexually. **Acellular slime molds** also produce spores; however, these spores are involved in sexual reproduction.

2. Subkingdom Dimastigomycota

Many members of this subkingdom are aquatic; therefore, this group is often called the water molds. The water molds produce flagellated spores during **asexual reproduction**.

Some species in this subkingdom do not live in water, and several species are parasites of terrestrial plants. One particular species caused a disease in potatoes known as late blight. From 1846 to 1847, Ireland's entire potato crop became infected with this disease. This important historical event was

Figure 10-39. Potatoes infected with late blight are shrunken on the outside, corky, and rotten inside. The spores of this water mold over-winter; that is, they lie dormant but survive through the winter on infected tubers, particularly those that are left in the ground after the previous year's harvest, in waste piles, soil, or infected volunteer plants. Infestations spread rapidly with warm and wet conditions. This can have devastating effects by destroying entire crops.

known as the "Potato Famine." During this time many of the Irish people starved to death. The famine caused many of the Irish to flee to other countries.

F. True Fungi

The members of the subkingdom Eumycota are also called "true fungi." Most true fungi have hyphae that form a mycelium. They all have cell walls made of chitin. Except for one group, true fungi do not have flagella. The subkingdom is divided into phyla based on the type of reproductive structure present.

1. Phylum *Zygomycetes*

Most zygomycetes are terrestrial **saprophytes**. A few types live in water, and some are parasites. The common bread mold, known as *Rhizopus*, is one of the best-known zygomycetes. *Rhizopus* may grow on many different foods. A branching

Figure 10-38. Water Mold from a Stream

Biology For Life

Figure 10-40. *Rhizopus stolonifer* (black bread mold). It is commonly found growing on bread and soft fruits such as bananas and grapes.

network of hyphae grows into the food and forms root-like fibers known as **rhizoids**.

You can easily "make" your own bread mold. Obtain a small piece of bread, about a 1-1/2 inch square. Put no more than 3-4 drops of water on it and place in a small, clean jar, and seal it tight with the lid. Next, place the jar in a warm area, perhaps near a water heater or clothes dryer. Within a few days, the bread will take on a blue-green color as bread mold begins to appear on the surface. Eventually, the mold will become black. It is truly amazing that these bread mold spores are everywhere!

2. Phylum *Ascomycetes*

The members of the **ascomycetes** have a saclike reproductive structure, which is called an **ascus**. The ascomycetes are also called sac fungi. These organisms have a mycelium made of hyphae with chitin walls. Many of these fungi are important to humans. The Neurospora fungus is used in many laboratories for genetic research. Truffles and morels are edible, delicious ascomycetes.

Yeast is an ascomycete used to help make bread. In nature, some yeasts grow around or within the roots of plants. They help the plants absorb minerals from the soil. In turn, products made by the plant are used by the yeasts. Many sac fungi are parasitic and cause considerable damage. American elm trees are parasitized by a sac fungus that causes Dutch elm disease. This disease is carried by beetles and enters the elms through wounds in the bark.

Before people began making bread with prepackaged yeasts, it used to be a common practice for those who made bread to leave behind a small piece of dough in the mixing bowl. It was called a "starter" and contained spores of natural yeasts that came from handling and that drifted through the air. Every time that dough was prepared and kneaded, a piece of starter was left behind. Today, this bread is made the same way and is called sourdough. I am familiar with one professional baker that has been using the same starter for over 20 years!

3. Phylum *Basidiomycetes*

Mushrooms are the most familiar of the Basidiomycetes. The mushrooms that appear suddenly after a rainstorm are actually the fruiting bodies of spreading networks of underground mycelia. The mushroom fruiting body is, in fact, a mass of hyphae. Beneath the cap of the mushroom are numerous thin **gills**. Special cells called **basidia** line the gills. These function to release spores when the mushroom is mature.

Bracket fungi, which grow on decaying trees, are also Basidiomycetes. Puffballs are another type of fungi in this phylum. If a mature puffball is disturbed, it sends clouds of spores in the form of yellowish dust through an opening in the top.

Parasitic rusts are also an important group of Basidiomycetes. Different rust species damage valuable crops such as wheat, corn, oats, and barley. Wheat rust spends part of its life cycle in the wheat plant and another part of it in the barberry bush. One of the best ways to control wheat rust is to destroy barberry bushes in the farming area.

4. Phylum *Deuteromycetes*: Imperfect Fungi

Deuteromycetes have never been known to reproduce by sexual means. If a sexual stage is ever

Figure 10-41. Truffles and Morels are highly edible ascomycetes.

CHAPTER 10: **SINGLE-CELLED ORGANISMS AND FUNGI**

Figure 10-42. The button mushroom is one of the most widely cultivated mushrooms in the world.

Figure 10-43. Bracket Fungus

observed with a fungus in this group, that particular fungus must be reclassified and put into a new group. *Penicillium* and *Aspergillis* are two well-known Deuteromycetes. These organisms are among the many common blue and green molds that you may have seen growing on food. The antibiotic penicillin is produced from the *Penicillium* mold.

G. Lichens

A **lichen** is a combination of two organisms. When two organisms live in association with and help each other, they are said to be symbiotic. Each lichen is a symbiotic association of a species of fungus and a photosynthetic organism.

The organism that is photosynthetic is usually a **cyanophyte**, or algae. The algae supplies the lichen with food, and the fungus keeps the lichen from drying out and attaches it to its surface. This symbiotic relationship between the cyanophyte and the fungus enables lichens to grow in places where other organisms may not. For example, lichens may grow on the surface of bare rocks or on the bark of trees. Lichens play a very small role in the process of weathering, that is, where natural processes break down rocks into components of soil. Rock tripe is a type of lichen that grows on rocks, which, in the past, has been used by explorers and soldiers in North America when food supplies were limited. It is also known as a famine food.

Section Review — 10.6

1. What is the main difference between plants and fungi?
2. What is the complex carbohydrate that, in several types of fungi, makes up its cell wall?
3. What is a saprophyte?
4. What are the types of slime molds? What is the main difference between them?
5. What are the different classifications of phylum in the subkingdom Eumycota or True Fungi?
6. What is Lichen?

Figure 10-44. Lichen

Biology For Life

CHAPTER 10: SINGLE-CELLED ORGANISMS AND FUNGI

Chapter 10 Supplemental Questions

Answer the following questions.

1. Know the content of Figures 10-21, 10-22, 10-25 and 10-33.
2. What is "taxonomy?"
3. Name the parts of the biological classification scheme, in order. Be able to provide a brief description of each level.
4. List differences between prokaryotic and eukaryotic cells.
5. Give an explanation of the process of nitrogen fixation.
6. Describe conditions needed to grow bacteria in the laboratory.
7. Discuss the reproduction of bacteria (prokaryotes).
8. Explain the processes of conjugation and transformation.
9. List examples of how humans are affected by bacteria.
10. List the characteristics used in the classification of bacteria into two divisions.
11. Describe the characteristics of schizophytes and cyanophytes.
12. To which Kingdom do viruses belong?
13. Identify: bacteriophage, decomposer, anaerobe, facultative anaerobe, and obligate anaerobe.
14. How are viroids similar to viruses, and how are they different? List at least one similarity and one difference.
15. Describe the structure of a typical virus, and discuss the two different reproductive cycles of a virus.
16. List the characteristics of the protists.
17. In no more than 3 sentences, define the terms autotrophic, heterotrophic, parasitic and symbiotic.
18. Describe a flagellum, its purpose and structure.
19. Give two differences between euglenophytes and other plant-like protists, like pyrrophytes and chrysophytes.

Chapter 10 Supplemental Questions

20. List the characteristics of dinoflagellates.

21. Describe the differences between aerobic and anaerobic bacteria.

22. What are saprobes?

23. Describe some of the characteristics of lichens.

24. Study all diagrams in this chapter. Be able to draw and label all diagrams.

CHAPTER 11
The Plant Kingdom

Chapter 11
The Plant Kingdom

Chapter Outline

11.1 Reproduction in Plants
    A. Asexual Reproduction
    B. Sexual Reproduction

11.2 Non-vascular Plants
    A. Algae
        1. Division Chlorophyta: Green Algae
        2. Division Rhodophyta: Red Algae
        3. Division Phaeophyta: Brown Algae
    B. Division Bryophyta: Mosses and Liverworts
        1. The Mosses
        2. The Liverworts

11.3 Vascular Plants
    1. Adaptations to Life on Land
    2. Tissues In Vascular Plants
    3. Organs of Vascular Plants
        A. Spore-Bearing Vascular Plants
            1. Whisk ferns
            2. Club Mosses
            3. Horsetails
            4. Ferns
        B. Gymnosperms
            1. The Seed
            2. Alternation of Generations in Gymnosperms
            3. Ginkgoes and Gnetophytes
            4. Cycads
            5. Conifers
            6. Life Cycle of a Pine
        C. Angiosperms
            1. Characteristics of Angiosperms
            2. Monocots and Dicots
            3. Use by Humans

11.1 Reproduction in Plants

Section Objectives

- Describe several forms of asexual reproduction in plants.
- Describe the life cycle of plants in terms of alternation of generations and discuss the haploid and diploid phases.
- Explain how gymnosperms undergo alternation of generations.

There are two main divisions of the Plant Kingdom, the vascular (fluid-carrying vessels) and non-vascular plants.

In this chapter, we will learn about the fascinating variety among the divisions of the Plant Kingdom. In addition, we will investigate the multicellular algae, which traditionally have been placed in the Plant Kingdom. The algae are included with plants because of their similarity to plants, especially their ability to use pigments by which they capture sunlight and to make their own food. Within this chapter, we

Vascular Plants	Non-Vascular Plants
Possess specialized vascular tissues that conduct nutrients, water, and synthesized compounds	Lack specialized vascular tissues; some possess specialized tissues for conducting water
Contain leaves, stems, roots, and flowers	Do not possess leaves; have structures for attachment called rhizoids
Numerous and diverse	Less numerous and not as diverse
Include flowering plants, conifers, ferns, horsetails, cycads, whisk ferns, and club mosses	Include algae, mosses, and liverworts

Biology For Life

CHAPTER 11: THE PLANT KINGDOM

will review the divisions of plants and algae that are common to most botanists.

Plants are eukaryotic organisms, that is, organisms with cells which have a nucleus. Plants are able to manufacture their own food through **photosynthesis**. The word "photo" refers to light, and "synthesis" means to make; therefore, plants use light, specifically sunlight, to grow and to make food. A vast array of organisms is found in the Plant Kingdom. These range in size from single-celled algae to the giant redwood trees in the forests of the Pacific coast.

All plants contain **chlorophyll**, a green pigment that is used to capture the energy of the sun. Chlorophyll is found in tiny plant organelles called **chloroplasts**. The cell walls of plants are composed of cellulose. Cellulose supports and stiffens the cell wall and prevents the cells from taking in too much water and bursting. The cell walls that cover the outside of some plants also provide some protection against abrasion or injury.

Plants cannot move, which means that they are non-motile, or stationary. In order to overcome this, plants have structures that bring together the materials and energy needed for photosynthesis.

The lack of motility, or mobility, in plants presents another problem in terms of uniting the gametes for sexual reproduction. (**Gametes** are mature reproductive cells, eggs or sperm, that combine DNA to form a new zygote.) However, this problem is solved by various methods. Some plants use water to transport the male gametes to the female gametes. Other plants use wind, insects, birds, bats, and other indirect means to unite the gametes. In this case, the male gamete is delivered in a small, sticky substance known as **pollen**. Bees

Figure 11-1. The incredible diversity of the Plant Kingdom

270 Biology For Life

CHAPTER 11: **THE PLANT KINGDOM**

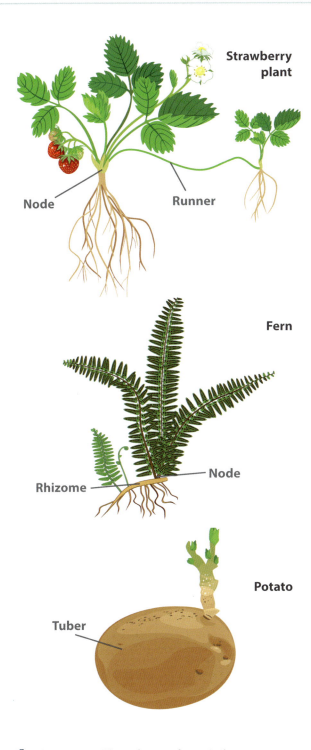

Figure 11-2. Three forms of vegetative reproduction. A strawberry plant (top) sends out runners. Grasses (center) produce underground stems called rhizomes. A potato (bottom) is a tuber and can sprout new plants.

are particularly important to plants for this mode of delivery, known as pollination. God designed bees to be completely covered with hairs, which enable the bees to easily collect and transport pollen. Once pollen has been delivered to the female reproductive structure of the same species, fertilization occurs.

A. Asexual Reproduction

Many plants are able to sexually and asexually reproduce within the plant. Any plant produced by **asexual reproduction** has the exact same genes as its parent. Vegetative reproduction is a type of asexual reproduction found in plants, where new independent individuals are formed without the production of seeds or spores. It occurs naturally in any of six kinds of plant structures: runners, **rhizomes** (which are underground stems), tubers, bulbs, food-storing roots, and leaves.

Blackberries and raspberries are known to reproduce by a process known as cane tip rooting. As these plants grow and colonize an area, the canes, or main stems, grow long, and form arches. The tips of these stems extend toward the ground and form roots when they touch the ground. This process is instrumental in forming a thicket and is essential in stabilizing habitats.

Other examples of vegetative reproduction include the formation of plantlets on specialized leaves (for example, in kalanchoe, the "maternity plant"), the growth of new plants from runners (for example, in strawberry) or rhizomes (for example, in grass), or the formation of new bulbs (for example, in tulips). Figure 11-2 shows examples of three types of asexual reproduction.

Commercially important plants are often propagated by asexual means in order to maintain desirable traits, such as specific color for a flower or stronger flavor in a vegetable or resistance to disease. One means for propagating is using cuttings, which may be taken from a parent plant and then rooted. Another means is grafting. **Grafting** involves attaching a section, sometimes called a scion, of a plant to another plant that is already in the soil. Grafting is widely used to

Biology For Life

CHAPTER 11: THE PLANT KINGDOM

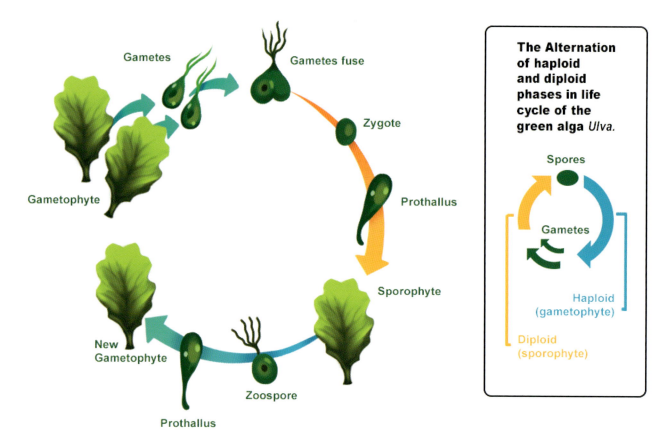

Figure 11-3. The life cycle of Ulva, a green alga, which involves an alternation of generations.

propagate a specific variety of shrub or tree. All apple varieties, for example, are propagated by grafting, putting a cutting onto a stock. The root system of the stock furnishes the attached cutting with water and minerals and provides support. In turn, the graft delivers nutrients to the stock, thus enabling its roots to grow.

Another type of asexual or vegetative reproduction is **fragmentation**. Fragmentation, or **gemmae**, is a type of vegetative reproduction found in multicellular algae. In this process, pieces of plants that are torn apart by animals, or pieces which the ocean currents may cause to drift elsewhere, grow into new plants. How good our God is to us, that we have an abundance of food provided by the earth. He provides these plants with the ability to fulfill their requirements for survival and reproduction in order to provide us with our earthly needs.

B. Sexual Reproduction

In sexual reproduction, the combining of material from two parents produces a new individual. In plants, as in animals, a sperm moves towards and unites with an egg. Any organism produced by sexual reproduction inherits genes that are similar to, but slightly different from, than its parents.

Most plants reproduce sexually. This method involves two processes: **meiosis** and **fertilization**. During meiosis, diploid cells divide to produce haploid gametes. The resulting eggs or sperm contain one-half of the genes of their respective parent cells. Fertilization unites the male and female gametes to form a diploid zygote. The genetic information of the zygote is complete, with one-half of its genes coming from the father and one-half from the mother. However, as would be expected from our omniscient

Biology For Life

creator God, He has provided many "twists" that show His infinite wisdom.

In many plants, the cells produced by meiosis do not immediately undergo fertilization. A haploid spore may grow into a multicellular plant body. The haploid organism will eventually produce gametes, which are capable of fusing with other gametes to form a zygote. In this situation, meiosis and fertilization are separated by a generation. The term for this type of life cycle is **alternation of generations** and is the means by which *all* terrestrial plants and many algae undergo sexual reproduction.

The diploid phase of a plant that reproduces by alternation of generations is called a **sporophyte**. The sporophyte undergoes meiosis to produce haploid spores, and the spores develop into haploid plants. Multicellular haploid plants are called **gametophytes**. The gametophyte produces haploid gametes that fuse with other haploid gametes to reproduce sexually.

An example of the above comes from *Ulva* (sea lettuce), an alga. The life cycle involves two distinct forms of the organism although they look identical. When gametes from two *Ulva* in the gametophyte stage fuse, they form a diploid zygote. All cells that

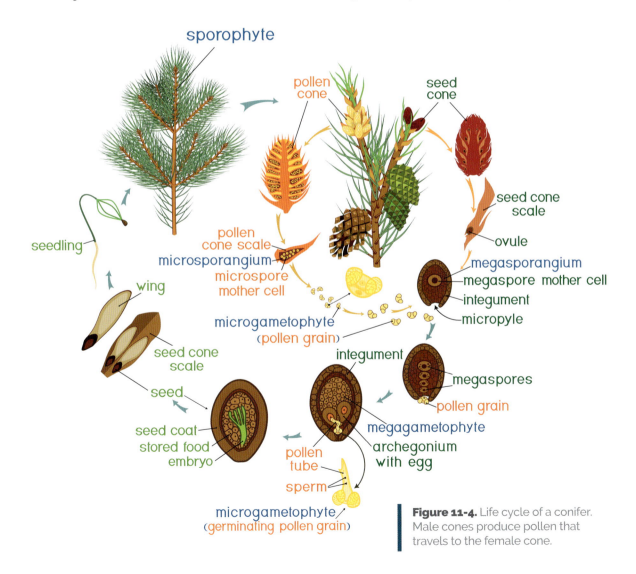

Figure 11-4. Life cycle of a conifer. Male cones produce pollen that travels to the female cone.

CHAPTER 11: **THE PLANT KINGDOM**

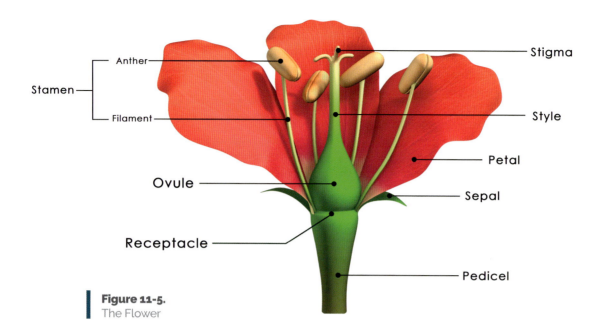

Figure 11-5.
The Flower

develop from the zygote are diploid, and the resulting plant body becomes the sporophyte. Its cells undergo meiosis and produce spores that develop into gametophytes.

In most conifers, a type of tree, sexual reproduction involves separate male and female cones that grow on the same tree. The male pollen cones produce microspores that develop into pollen grains, each of which is a male gametophyte. Within the female seed cones, megaspores develop into female gametophytes that contain egg cells. When the egg cells begin to form, the female cones secrete a sticky sap that traps pollen, which has drifted by on breezes. As the sap begins to dry, it draws the pollen towards the egg cells. The pollen grains then produce male gametes or sperm, which then fertilize the egg cells. The resulting diploid zygote develops into a conifer embryo contained within the seed. When the seed germinates or sprouts, it marks the beginning of the next generation.

In flowering plants, the flowers are essential for the reproductive cycle – see Figure 11-5.

In the flower, male gametes form in the **anthers**, and female gametes form in the **ovules**. These structures represent the gametophyte (haploid) generation for the plant. Following pollination, the fertilization of the egg by the sperm produces a diploid zygote, which is the plant's diploid, sporophyte stage.

Section Review — 11.1

1. Compare and contrast vascular and non-vascular plants.
2. What does the term "vascular" mean?
3. What are gametes?
4. _____, or gemmae, is a type of vegetative reproduction found in multicellular algae.
5. What is a sporophyte?
6. In the flower, male gametes form in the _____ and female gametes form in the _____.

Biology For Life

CHAPTER 11: **THE PLANT KINGDOM**

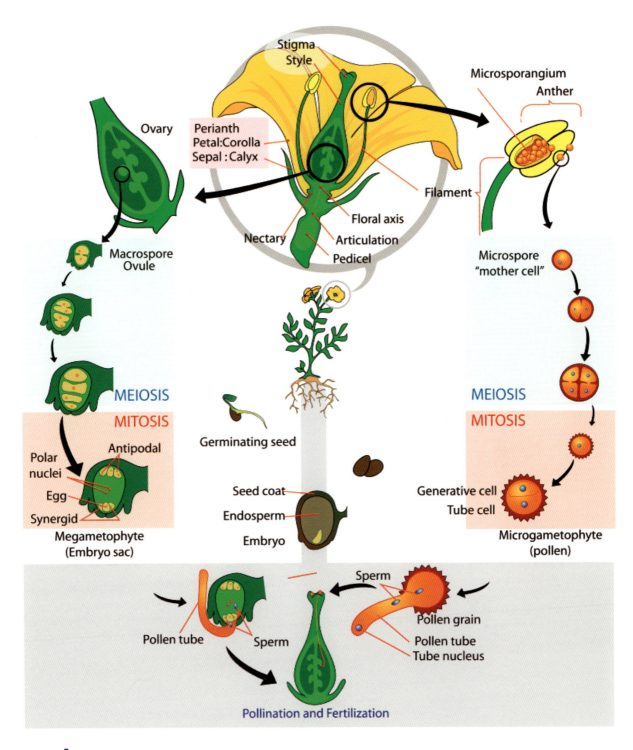

Figure 11-6. The life cycle of flowering plants. In the male part of the cycle, microspores form pollen grains. In the female part, a megaspore forms an embryo sac. Pollen grains are dispersed, land on the stigma, and form sperm that fertilize the egg.

Biology For Life

CHAPTER 11: THE PLANT KINGDOM

11.2 Non-vascular Plants

Section Objectives

- Describe the characteristics of the three divisions of multicellular algae and list a member from each division.
- Describe the characteristics of and compare sexual reproduction in the following organisms: *Chlamydomonas, Volvox, Spirogyra, Oedogonium.*
- List some characteristics of bryophytes, which enable them to live on land.
- Describe the structure and the life cycle of mosses and liverworts.

1. Division Chlorophyta: Green Algae

The Division **Chlorophyta** is the most diverse group of algae. Some of the green algae are single-celled, some live in colonies or groups, and some are multicellular. Their major pigment is chlorophyll, but they also contain yellowish-orange pigments called carotenes. Green algae are more commonly seen in lakes or ponds, though some are marine and live in brackish water or oceans.

Figure 11-8. Chlorella

Chlorella is a type of green alga that is unicellular (see Figure 11-8). This single-celled organism is bounded by a cell wall, contains a nucleus, and contains a chloroplast. *Chlorella* reproduces only by asexual means. Non-motile spores are formed within the parent cell. When these mature, the cell wall disintegrates, and the spores are released.

Chlamydomonas are flagellated unicellular algae that live in fresh water. They have cells with light-sensitive eyespots that direct the organisms toward light, and thus aid in photosynthesis. These organisms reproduce sexually. The gametes appear

The non-vascular plants lack vascular tissues, including **xylem** (zī-lĕm), **phloem** (flō-ĕm), **parenchyma** (pa-rĕn-kĭ-ma), and **cambium** cells; these non-vascular plants include the algae, mosses, and liverworts. The multicellular body of the non-vascular plant is called a **thallus**.

A. Algae

All algae contain chlorophyll; however, algae also contain other pigments. These pigments capture light at various wavelengths and pass this energy on to chlorophyll. Algae are classified according to the types of pigments they contain, as well as their biochemistry and methods of reproduction.

Figure 11-7. Green Algae. There are about 6,000 species of green algae. Many species live most of their lives as single cells, while other species form colonies or exist as long filaments.

Biology For Life

CHAPTER 11: **THE PLANT KINGDOM**

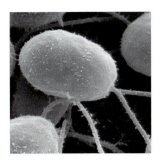

Figure 11-9. Chlamydomonas. Image is magnified 10,000 times. Notice that its shape resembles a grape.

to be identical and are known as isogametes. The process of **isogamy**, which means reproduction by isogametes, is the simplest form of sexual reproduction. Since both gametes look alike, they cannot be classified as "male" or "female." Instead, organisms undergoing isogamy are said to have different mating types, most commonly noted as "+" and "-" strains. Fertilization occurs when "+" and "-" gametes fuse to form a zygote having its own unique genetic structure.

Volvox is a colonial green alga. A **colonial** organism is multicellular, consisting of individual cells that are similar in structure. There is no specialization of cells or division of labor in colonial algae. Each Volvox colony is shaped like a hollow ball. Individual cells are enclosed in gelatin

Figure 11-10. Volvox colony. The mature colony possesses about 1000-3000 cells.

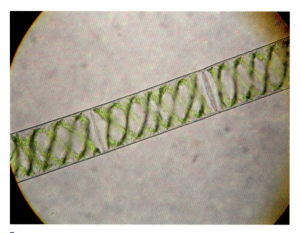

Figure 11-11. Single Spirogyra cell. Image is magnified 40x.

and are connected to each other by strands of cytoplasm. Each Volvox cell has two flagella.

The type of sexual reproduction in Volvox is called oogamous. In the process of **oogamy** (ō-ä-gă-mē), two types of gametes are formed. The egg and sperm fuse in the center of the colony and form a zygote. It is released when the parent colony disintegrates. The zygote begins to grow after a dormant period. Haploid spores called **meiospores** are formed during the process of meiosis. Each meiospore eventually produces a new Volvox colony.

Spirogyra is filamentous green algae that live in ponds and streams; see Figure 11-11. The cells of these organisms are arranged in long strands called **filaments**. Each cell of Spirogyra is a transparent cylinder that contains ribbon-like chloroplasts. Embedded in the chloroplasts are structures called **pyrenoids** (pī-rĕ-noid), which function to store starch. All green plants produce starch, for eventual use as an energy store, especially when the ability to conduct photosynthesis is limited.

Spirogyra reproduces sexually by a process known as **conjugation**. During conjugation, two filaments from separate Spirogyra align so that they touch each other. Small bumps grow outward from each cell and form bridges between the two cells. The cell wall, which divides the bridges, dissolves and leaves a passageway

Biology For Life

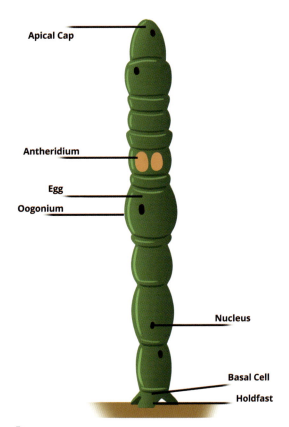

Figure 11-12. Oedogonium. It is about 250 μ or about 0.01 in. in length.

Oedogonium are capable of both sexual and asexual reproduction. In asexual reproduction, any cell, with the exception of the **holdfast** cell, may produce a **zoospore**. The zoospores swim away from the parent filament and attach to a suitable spot, so that they may grow into a new filament by cell division.

Two types of specialized cells are involved in sexual reproduction in the Oedogonium. The cell that produces the egg is called the **oogonium**, and the cell that produces the sperm is called the **antheridium** (an-tha-ri-dē-um). A zygote is formed after fertilization. When the zygote is released from the oogonium, it develops a thick wall and becomes a **zygospore**. Prior to undergoing meiosis, the zygospore enters a period of dormancy that can last several months. The process of meiosis results in four zoospores, each of which may start a new filament.

The evidence of God as an infinite Intellect becomes more obvious as we discover the incredible variety and complexity of the tiniest of creatures created by Him. All these creatures were created specifically for us. We may not know all the reasons now why these tiny creatures are so important to us, but as scientists study more and more, God will give them the graces to come to Him, and perhaps, discover how these creatures can make life better on earth.

between the pair of cells. The contents of the cell from one filament may enter the cell of the other filament. Gametes are transferred from one strand to the other, resulting in two groups of filaments. One group contains zygotes, and the other contains empty cells. The individual zygotes separate from each other, and each develops a thick cell wall. The cells then become **diploid zygospores**. When conditions are favorable, each zygospore undergoes meiosis. The haploid cells that result will grow into new filaments of *Spirogyra*.

Oedogonium is a filamentous freshwater alga that is more complex than *Spirogyra*. There is some specialization of cells in this organism. The filaments are attached to rocks or other solid objects by specialized cells called **holdfasts**. You can discern some of these features of the Oedogonium in Figure 11-12.

Figure 11-13.
Close-up of the red alga Laurencia, a marine seaweed from the regions around Hawaii.

2. Division Rhodophyta: Red Algae

The Division **Rhodophyta** contains the **red algae**, most of which live in oceans and brackish water. Most of these live in tropical regions, although they are common along rocky coastlines in cooler waters. Red algae may grow as long as one meter, and they attach to rocks, shells, or other surfaces. The larger species attach by holdfasts. The thalli (the multicellular bodies) of red algae often have a complex branched structure (see Figure 11-13).

The red algae contain **phycoerythrin**, a reddish pigment. The presence of this pigment allows some algae to live in very deep water. Phycoerythrin is able to capture the energy of blue light, which is the only wavelength of light that penetrates deep water. Phycoerythrin then transfers the energy to chlorophyll, which conducts photosynthesis.

Many species of red algae take up calcium from ocean water and deposit it in their cell walls. When these algae die, they leave behind calcium salts, especially calcium carbonate, the principal ingredient of limestone. In some cases, the numbers of dead algae are so immense that they form deposits of limestone that are as much as 300 meters thick. All red algae reproduce sexually by forming oogonia (eggs) and antheridia (sperm). The ocean currents carry the sperm to the eggs.

In parts of Asia, red algae are an important food source. Some red algae contain a substance known as carrageenan, which is used in puddings, preserves, and ice cream. Red algae are also used to make **agar**, which is a material that is used in laboratories for growing bacteria and fungi.

3. Division Phaeophyta: Brown Algae

The Division Phaeophyta contains about 1,000 species of **brown algae**. Most brown algae are marine and are found in coastal areas, especially in cold water. Brown algae are the largest type of algae, sometimes reaching fifty meters in length, which is the length of one-half of a football field! Sometimes swimmers become tangled up in brown algae, especially when there are strong waves that follow storms. The algae seem to wrap around the necks, arms, and legs of swimmers.

The pigment found in brown algae is called **fucoxanthin**. The branched thallus of brown algae contains **air bladders**, which function to keep the plant afloat. The broad leaves, called **blades**, are connected to a tough stalk. This stalk is known as the **stipe**. The reproductive organs are called **conceptacles** and are located at the ends of the blades. Sexual reproduction is oogamous, with separate egg and sperm cells.

In Asia, kelp and other brown algae are often used as a food source. In northern Europe, brown algae are used both as animal feed and fertilizer.

Figure 11-14. The Brown Algae Laminaria Sea Kale

Brown algae contain algin, a chemical that is often used in the manufacture of various consumer and industrial products, such as latex, ceramic glazes, cosmetics, and ice cream.

> **Section Review — 11.2 A**
>
> 1. How are algae classified?
> 2. Where are green algae found?
> 3. How do red algae conduct photosynthesis in deep water?
> 4. What type of algae do swimmers sometimes get caught in?

B. Division Bryophyta: Mosses and Liverworts

Bryophytes (brī-ō-fīts) are small terrestrial plants that lack vascular tissue. These plants do not grow in water but live in moist areas. Examples of these Bryophyta include the mosses and liverworts. The Bryophytes are well-specialized for living on land. The thallus is protected against desiccation (drying out) by an outer layer of epidermis and a waxy covering called the **cuticle** (kū-tĭ-k'l). The Bryophytes are anchored to the ground by root-like structures called **rhizoids**.

Bryophytes are restricted to living in moist areas of land, such as the base of waterfalls and rocky seepage areas on hillsides. They lack specialized conductive tissue and can transport food or water only a few centimeters. Water from the moist ground on which they grow diffuses into the thallus. Bryophytes lack supportive tissues or stems. Consequently, these plants grow close to the surface on which they attach. They are typically no more than several centimeters long, but they often grow in small, dense stands. Finally, the Bryophytes require water for sexual reproduction. As you can see, God has made water vital to all forms of life, and the Bryophytes are no exception.

1. The Mosses

There are about 14,000 species of mosses – it is obvious that God likes variety. They grow in shady places such as rocky ledges, on the sides of trees, along the banks of streams, and on seemingly sterile ground. Occasionally, mosses colonize large

Figure 11-15. Moss

areas and resemble a patchy carpet of many plants. You most likely have encountered these if you have ever gone camping or hiking in remote areas.

The moss life cycle begins when a haploid spore germinates (see top right of Figure 11-16). The spore will grow into a filament, which looks like a green alga. This stage is called a **protonema** (prō-tŏ-nē-ma) – see Figure 11-16. The protonema then develops into a small gametophyte that produces multicellular sex organs. These structures are the **archegonium** (är-kĕ-gō-nĭ-*u*m) and **antheridium**. The former produce eggs, and the latter make sperm (see bottom left of Figure 11-16). Mature sperm swim to the eggs through a thin film of water (see top left of Figure 11-16). The fertilized egg develops into an embryo, which is an immature diploid organism. The embryo grows a stalk that remains attached to the archegonium. A small **capsule** that produces spores is located at the upper end of this stalk.

Mosses require a thin layer of soil to grow and typically grow towards places where other plants cannot, such as in the cracks between rocks. Mosses are helpful towards the formation of soil. When mosses die, they release organic material into the soil, which builds up the soil so that larger plants may colonize that area. Mosses also prevent soil from drying out and blowing away. Sphagnum is a type of moss that grows in acid bogs. Sphagnum decays to form peat, which gardeners use to lower the pH of the soil and improve its texture.

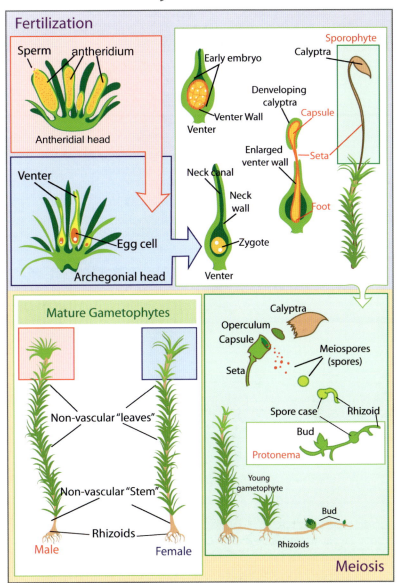

Figure 11-16. Life cycle of a typical moss. Note the mature archegonial and antheridial heads of the female and male gametophytes. The mature sperm swim to the egg through a thin film of water.

2. The Liverworts

Liverworts may be found along the banks of streams. Some liverworts are shown in Figure 11-17. They have a flattened thallus. They do not have

Figure 11-17. Liverwort

leaves; what are shown in Figure 11-17 are properly called lobes.

Liverworts anchor to ground with rhizoids and often form a carpet-like layer of small plants. Like mosses, they are capable of both sexual and asexual reproduction. The archegonium and the antheridium grow on stalks that project above the thallus (seen in the bottom of Figure 11-18). Mature sperm swim to the eggs through a thin film of water (see top of Figure 11-18). The fertilized egg develops into an embryo, which is an immature diploid organism. The embryo grows a stalk that remains attached to the archegonium. A capsule that produces spores forms at the upper end of the stalk.

Some liverworts have an unusual method of asexual reproduction. These liverworts have short stalks with little cups on the ends (see bottom of Figure 11-18). Tiny, flat structures called **gemmae** (je-may; singular **gemma**) are located in these cups. After the gemmae mature, raindrops and other disturbances may dislodge them from their cups. If a gemma lands on a suitable area, it develops rhizoids and grows into a gametophyte, which is genetically identical to the parent plant.

> ### Section Review — 11.2 B
> 1. What are the different divisions of non-vascular plants?
> 2. What are Bryophytes?
> 3. How are mosses and liverworts similar? How are they different?

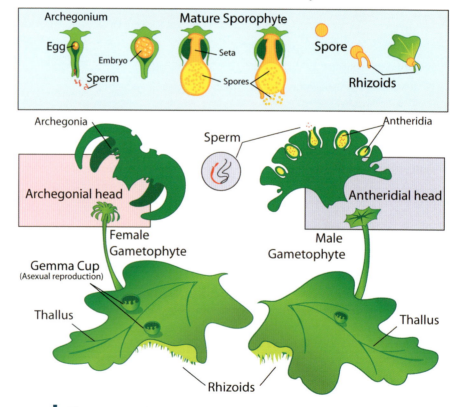

Figure 11-18. Liverwort Life Cycle. Note that the mature female gametophyte has an archegonial head and the mature male gametophyte has an antheridial head. The mature sperm swim to the egg through a thin film of water.

11.3 Vascular Plants

Section Objectives

- Describe the following adaptations of land plants: adaptations for conserving water, adaptations for providing support for tissues, adaptations for transporting gametes.
- Name and describe the following types of tissues in vascular plants: meristematic tissue, epidermis, parenchyma, supportive tissue, vascular tissue.
- Describe the characteristics of whisk ferns, club mosses, and horsetails.
- Describe the life cycle of the pine tree.
- Describe the structure and function of the seed.
- Describe angiosperms and their use by humans.

Vascular plants are known as the **Tracheophyta** (trā-kē-o-fīt-a). The word "vascular" means they have a system of vessels which carry nutrients. They have unique features that allow them to live on dry land. This capability differs from algae that live in water and mosses that must live in moist conditions.

The two major groupings of vascular plants are the spore-bearing plants and the seed-bearing plants. The latter group is known as the **Spermatophyta** and is divided into the gymnosperms and the angiosperms. First, we will discuss some of the characteristics of vascular plants, and then we will examine the major divisions.

1. Adaptations to Life on Land

The largest difference between life on land and life in the oceans is the availability of water. Algae are able to absorb water directly from their surroundings; however, it is more challenging for land plants to obtain water. To overcome this challenge, vascular plants possess structures that enable them to obtain water and dissolved nutrients from the ground.

Terrestrial (land) plants obtain most of their water through the roots. Water moves through the roots into the xylem. Minerals also enter the roots, but by a different mechanism. Water is absorbed by both active and passive means. The passive means is by osmosis, when the water saturation level in the soil is greater than in the root. ATP, which is produced by cellular respiration in the roots, is used to power the active transport mechanism when the water pressure in the soil is less than that in the root. In active transport, the ATP provides the energy to move water molecules across the root membranes in spite of the inverse pressure gradient. Once the water has moved into the xylem, it is pulled upward by transpiration and cohesion. Transpiration is the evaporation of water from the open stomata of the cell. The loss of this water from the leaf creates a negative water vapor pressure (vacuum) that causes water in the xylem to be pulled towards the leaf. Another force that pulls water upward in the xylem is cohesion. There are no pumps to push water up the plant tubes (xylem). However, the cohesive properties of water enable water to travel through the xylem tissues in

Figure 11-19. Vascular Plants

CHAPTER 11: THE PLANT KINGDOM

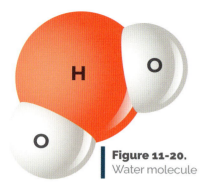

Figure 11-20. Water molecule

plants. Recall from chapter 2.6.B that it is the polar nature of water molecules which results in the cohesive properties of water. Despite what we know about the processes of water movement, it remains a mystery how water travels in an unbroken column from the roots to the crown of a tree.

In times of drought, known as water stress, a plant may appear to wilt or droop. The plant must be able to reduce water loss through its leaves. Wilting causes the stomata to close, so transpiration slows or stops. Wilting also causes temporary changes that occur in the plant's cells, as well as in the protective waterproof coverings. In leaves, this covering is known as a cuticle layer. It is very thick in desert plants and may also contain waxes and other compounds that help reduce water loss.

Terrestrial plants possess strong supportive tissues that enable them to grow to large sizes. They also have adaptations that allow fertilization to take place without water. Spores or seeds from terrestrial plants must rely on other agents than water, such as the air, for dispersal of seeds.

2. Tissues in Vascular Plants

Many of the vascular plants have specialized tissues that allow them to live on land. The major tissue types will be discussed.

Meristematic Tissue

The meristem is a type of tissue that contains unspecialized cells that are continually dividing and producing new cells with different and unique functions. These cells have thin walls and dense cytoplasm. The meristematic tissue found at the tip (called the apex) of a root or stem is called the **apical meristem**. A single layer of meristematic cells is also located between the bark and wood of a tree. This type of meristem is known as **vascular cambium**.

The cells produced in the meristematic tissue will eventually differentiate into other types of tissues. Some of the cell walls of meristematic tissue become thickened, while others elongate. Each cell becomes specialized for a specific function, such as absorption, transport, reproduction, or storage. Other meristematic cells differentiate or develop into other plant tissues. One might think of meristematic tissue as flour that is used to make many different types of foods, such as bread, cookies, or pastry for a pie.

Epidermis

The **epidermis** is a covering that develops from meristematic tissue. It functions to protect the internal cells and to regulate water loss. The epidermis is usually one cell thick and is covered by a waxy cuticle that is produced by specialized cells. The cuticle also functions to prevent water loss.

Parenchyma

Parenchyma develops from meristematic tissue, and is specialized for the storage of sugars and starches. The cells of the parenchyma

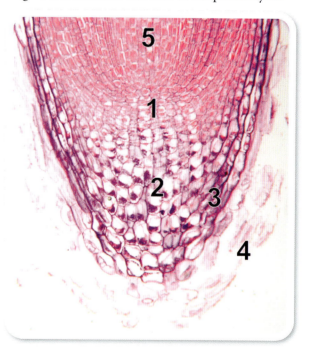

Figure 11-21. 10x microscope image of root tip with meristem. 1 - quiescent center; 2 - calyptrogen (live rootcap cells); 3 - rootcap; 4 - sloughed off dead rootcap cells; 5 - procambium.

Biology For Life

tissue are somewhat spherical. They have thin walls and contain large vacuoles. There are several specialized types of parenchyma cells. Chlorenchyma is a type of parenchyma that contains chloroplasts. Chlorenchyma is specialized for photosynthesis.

Supportive Tissue

Collenchyma is another specialized type of parenchyma cell. Collenchyma tissues provide strength and support. These cells are long and somewhat thickened near the corners. Collenchyma is often found in young stems and may be associated with vascular tissues. The long **sclerenchyma** fibers give plants their strength, mechanical support, and protection. Sclerenchyma cells have very thick secondary cell walls. The mature sclerenchyma cells often do not contain cytoplasm.

Cork

Cork cells are specialized for conserving water and protecting the plant. The cork cells are dead at the time of maturity and provide a waterproof layer around many stems. Perhaps you are familiar with cork cells that are commonly known as "cork." These are found on the tops of bottles and on drink coasters. This cork comes from a type of oak tree that is found in several Mediterranean countries, especially Portugal.

Vascular Tissue

Vascular tissue cells in plants are long and have thick cell walls. These tissues perform two important functions. First, they are used to support the weight of a large plant. Second, they conduct food, water, and dissolved minerals from one part of the plant to another. Although there is no direct comparison in animals, vascular tissue can be likened to an imaginary structure in humans, which would combine the transport functions of blood vessels and the structural support of bones.

There are two major types of vascular tissues, **phloem** and **xylem**. Phloem is composed of elongated cells that connect with each other. Its job is to transport starches and sugars from one part of the plant to another. Xylem is also

Figure 11-22. Cross section of celery stalk, showing vascular bundles, which include both phloem and xylem.

composed of elongated cells that connect with one another. Xylem tissue dies immediately after cell formation. Both phloem and xylem are hollow tubes. Phloem transports starches and sugars; xylem transports water and dissolved minerals. Xylem and phloem occur in strands referred to as vascular bundles.

3. Organs of Vascular Plants

An organ is composed of similar tissues that are grouped together to perform a specific function. Most vascular plants have roots, leaves, stems, and reproductive structures. These organs are connected by vascular tissues.

Consider for a moment a comparison of plant organs with the organs of the human body. This is tough to do, as plants and humans are two dramatically different organisms. However, just as the heart, brain, skin and other parts of your body perform specific tasks, so too do the various organs of plants.

Roots of plants are specialized organs that collect water and dissolved minerals from the soil. The roots also serve to anchor the plant in place and may extend far down into the ground. Roots are whitish in color, due to their starch content and lack of chlorophyll.

A **stem** is the organ that conducts water and minerals from the roots to other parts of the plant. The stem is also the conduit by which food from the leaves is transported to the rest of the plant. Many stems are able to store food, and some stems may conduct photosynthesis. The trunk of a tree is actually a stem.

The leaves are the main photosynthetic organs of vascular plants. The primary job of leaves is to produce glucose that will be distributed elsewhere in the plant. Leaves are typically flattened, which increases their surface area and allows more exposure of individual cells to the sun. The greater the surface area, the more there are cells that will be able to conduct photosynthesis. In some plants, especially small annual plants, the leaves are often curved upward to help channel rain downward, to the roots.

There are several kinds of reproductive organs among vascular plants. Spores are produced in multicellular organs, known as sporangia. The reproductive organs of seed-bearing plants consist of either cones or flowers.

There are three major types of vascular plants: spore-bearing, gymnosperms, and angiosperms.

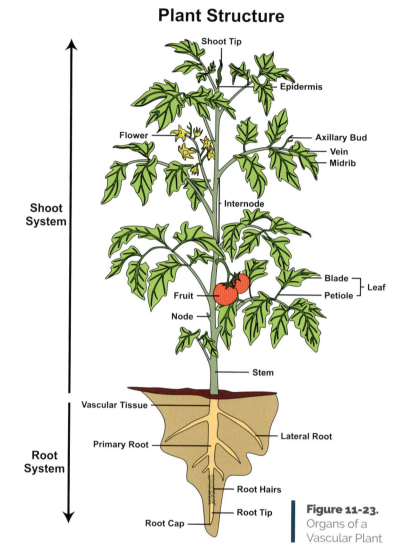

Figure 11-23. Organs of a Vascular Plant

Section Review — 11.3

1. The two major groupings of vascular plants are the _____ plants and the _____ plants.
2. What is the vascular cambium?
3. What are the phloem and xylem of a plant?
4. The three major types of vascular plants are: _____, _____, and _____.

A. Spore-bearing Vascular Plants

Spore-bearing vascular plants are placed in four subdivisions. These include the whisk ferns, club mosses, horsetails, and ferns. Within these plants, spores are released from the sporophyte and germinate to produce a gametophyte. The gametophyte begins its new life independent of the sporophyte. All mature gametophytes require water for fertilization. The sperm often use temporary water sources as a pathway for fertilization of the egg cells. These sources include dew and other water droplets that may collect from fog and clouds at high elevations.

1. Whisk ferns

The subdivision Psilopsida contains whisk ferns. These are the simplest of vascular plants, with only a few members of this subdivision still in existence. In North America, whisk ferns are found only in the state of Florida.

Whisk ferns are small plants that are rarely more than one meter tall. They lack both roots and leaves. Whisk ferns possess erect branches that look like bundles of green forked sticks. **Sporangia** are located at the intersections of some branches. Each sporangium releases many small spores that are carried and dispersed by the wind. If a spore falls on suitable land, it develops into a small gametophyte called a **prothallus**. The prothallus grows underground and sends out long brown rhizoids that penetrate the soil. The prothallus lacks chlorophyll and is unable to conduct photosynthesis. Accordingly, it relies on symbiotic fungi known as mycorrhizal fungi to provide it with food. This is made possible by fungal hyphae that penetrate its tissues.

The surface of the prothallus is covered by numerous antheridia and archegonia. The sperm are released by the antheridia and swim through moisture in the soil to reach the eggs of the prothalli. A simple embryo develops following fertilization. It eventually grows into a mature sporophyte and becomes the whisk fern.

Figure 11-24. Close-up of the whisk fern psilotum. Note the sporangia, which are the circular protrusions on the stems of the plant.

CHAPTER 11: THE PLANT KINGDOM

Figure 11-25. The club moss *Lycopodiella cernua* with close-up of a branch. Note that spores can grow at the bases of some branches.

Figure 11-26. The *Equisetum telmateia* (Great Horsetail or Northern Giant Horsetail). It is a horsetail with an unusual distribution, with one subspecies native to Europe, western Asia, and northwest Africa, and a second subspecies native to western North America. The biological name tells you that the organism belongs to the genus Equisetum and its species is telmateia.

2. Club Mosses

The subdivision **Lycopsida** contains simple vascular plants that are known as club mosses. Club mosses are common in tropical areas and on the forest floor in cooler climates. In North America, it is common to find large, dense stands of club mosses in undisturbed forests. If you happen to stumble on a large stand of these, it has taken about 100 years to develop!

The sporophyte of the club moss has tiny, scalelike leaves. Spore cases grow at the base of some of the branches. When the spores are released, they fall to the ground, then germinate and give rise to the gametophyte stage, or prothallus. Like the whisk fern, it grows underground and has a symbiotic relationship with mycorrhizal fungi that provide it with food.

Biology For Life

3. Horsetails

These plants usually grow in swampy or seepage areas, such as drainage ditches. Also, they are commonly found along railroad tracks. The only living genus is *Equisetum*, which means "horsetail" in Latin. Another name for horsetails is "scouring rushes." North American settlers gave this name to the plant many years ago. *Equisetum* is rich in silica (sand-like particles) and is very abrasive. Settlers used horsetails to clean pots and pans. Indians made use of its abrasive qualities to polish their arrows.

The stems and leaves of this plant are bushy like a horse's tail. Small leaves encircle the woody stem at each joint. Cones will eventually develop at the tips of some stems. These contain sporangia that produce and release spores. The spores germinate on the surface of wet soil and become a gametophyte, or prothallus. It is green in color, and is either branched or lobed.

4. Ferns

There are about 9,000 species of fern, which are distributed from the tropics to the Arctic. Ferns range in size from the tiniest of water ferns to the tropical tree ferns that often reach heights of twenty-five meters (about seventy-five feet). Most ferns in North America do not grow larger than one meter in height. Two common ferns that are found in the eastern United States are the Christmas fern and the cinnamon fern. The Christmas fern is common in many local parks and outdoor wooded areas. It stays green all winter, and it has been used in the past for Christmas decorations. Cinnamon ferns are large, with fronds growing up to six feet long and a foot wide. They grow in large clumps in moist woods, marshes, wet ditches, and stream banks. A cinnamon fern has two types of fronds: big green ones and smaller ones which begin as bright green and then turn a cinnamon color.

As you can see from Figure 11-27, ferns have true leaves, roots, and stems. The stems grow horizontally underground and are called rhizomes. The roots arise from these rhizomes and branch downward into the soil. The leaves of the fern are called **fronds** and grow upward from the rhizomes. The fronds of the sporophyte may die each fall; however, the rhizomes may live for many years. Young fronds are tightly coiled into a structure that is called a **fiddlehead**. The lower surfaces of some fronds possess brown spots known as **sori** (singular = **sorus**). Each sorus is a cluster of sporangia that contain many spores. Ferns show an alternation of generations in their life cycle; see Figure 11-28. The sporophytes are much larger than the gametophytes. Recall that the sporophyte is the diploid phase of the organism, and the gametophyte is the haploid phase. One amazing fact about ferns is that most of them produce a chemical which mimics an insect hormone. If parts of the plant are ingested by an insect, it would cause that insect to prematurely molt, or grow to the next stage, when it is not yet ready. This ultimately causes the death of any insect, and for this reason, most ferns do not have pest problems!

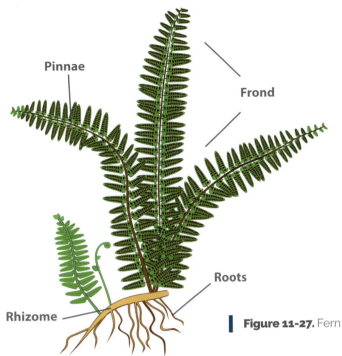

Figure 11-27. Fern

CHAPTER 11: THE PLANT KINGDOM

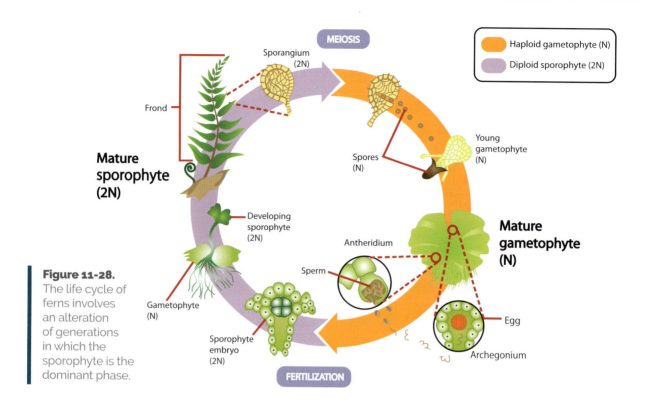

Figure 11-28. The life cycle of ferns involves an alteration of generations in which the sporophyte is the dominant phase.

Section Review — 11.3 A

1. What are the four subdivisions of spore-bearing vascular plants?
2. Where are whisk ferns found?
3. The subdivision _____ contains simple vascular plants that are known as club mosses.
4. The Latin word meaning "horsetail" is _____.
5. The leaves of a fern are called _____. Young fronds are tightly coiled into a structure that is called a _____. The lower surfaces of some fronds possess brown spots known as _____.

B. Gymnosperms

The largest subdivision of vascular plants is the spermatophytes, or seed plants. The seed is the means of reproduction in spermatophytes. They are divided into two major classes, the gymnosperms and the angiosperms (the flowering plants).

The gymnosperms (jĭm-nō-spŭrms) produce seeds in cones and generally keep their leaves or needles on throughout the year. There are about 700 species of gymnosperms, which make up about one-third of the world's forests. Modern gymnosperms include cycads, ginkgoes, gnetophytes, and conifers. Figure 11-29 shows the wide variety of seed placement in plants.

1. The Seed

The production of seeds is a common characteristic of all gymnosperms. Seeds are better adapted for dispersal than spores. Spores contain only one haploid cell, but seeds contain many cells. There are three parts to each seed: an embryo, a food supply, and a seed coat. The embryo is an immature diploid plant. The seed coat is a tough, waterproof covering which protects the seed. The seed coat allows the seed to lie dormant for many years without losing its ability to germinate, or begin to grow. When the seed germinates, it uses the food supply until it is capable of conducting photosynthesis.

Spores and seeds represent different stages in a plant's life cycle. A spore is a haploid and grows into a gametophyte plant. Conversely, a seed contains a diploid embryo and is the beginning of the sporophyte generation.

2. Alternation of Generations in Gymnosperms

The gametophyte generation of gymnosperms develops within the body of the sporophyte. Accordingly, the gametophyte does not live independently of the sporophyte. When mature, the gymnosperm sporophytes produce two types of spores. The small **microspores** give rise to the male gametophytes, which produce pollen. The larger **megaspores** give rise to the female gametophytes, which produce eggs.

Most gymnosperms do not require water for fertilization. The pollen is transported to the eggs mostly by the wind. The pollen grains are immature male gametophytes that complete their development when they reach a female gametophyte (egg).

3. Ginkgoes and Gnetophytes

The ginkgo, more commonly known as the maidenhair tree, is the only surviving species of a group that was once very large. The fan-shaped leaves of the ginkgo tree are shown in Figure 11-30.

The ginkgoes are native to China and are capable of growing very tall, often reaching heights greater than thirty meters. They are a popular ornamental tree and are grown in various regions throughout the United States. Ginkgoes are deciduous, which means that they lose their leaves in the winter. This term is used in contrast to the word coniferous (evergreen), which implies that trees keep their leaves on throughout the year (*section 5 below is about conifers*).

Traditionally, the Chinese have used extracts from the ginkgo tree to promote the "power of the

Different Kinds of Seeds

Fruits | Cones | Catkins
Seeds with Wings | Pods | Nuts

Figure 11-29. Different plants produce different kinds of seeds. Some produce seeds in a fleshy covering, such as a fruit or berry. Others produce seeds tucked in folds of cones or catkins. Others produce seeds with wings, and some produce seeds inside nuts or pods.

Biology For Life

Figure 11-30. Leaves of the tree Ginkgo biloba, also known as the Maidenhair Tree. It is a unique species of tree with no close living relatives. For centuries, it was thought to have been extinct, but was found growing in two small areas in Eastern China.

brain." Although there is some evidence that these extracts improve circulation to the brain, there is no evidence that they make one more intelligent.

Ginkgoes produce pollen and ovules (the female reproductive cell) on separate trees. As is the case with most pollen-producing organisms, the pollen is transferred by wind, insects, and birds.

The **gnetophytes** are a small group of plants that include a shrub known as *Ephedra*. This shrub is common in the deserts of the southwestern United States, where it is known as Mormon tea or squaw tea. The leaves of this plant are very tiny, and their green stems carry out photosynthesis.

Ephedra produce chemical compounds called alkaloids. These compounds include ephedrine and pseudoephedrine, both of which are found in popular, non-prescription cold remedies. These compounds are stimulants and are more commonly known as decongestants. They also tend to suppress appetites and are often sold as weight loss "gimmicks." While they may temporarily cause weight loss, once a person stops taking these drugs, the lost weight inevitably returns.

The overuse of *Ephedra* can lead to addiction and annually thousands of people require hospitalization for abusing this drug. *Ephedra* is like all of our plants, a gift from God. On one hand, it is beneficial to human health. On the other hand, it can be detrimental to it. This can be likened to our pursuit of virtue, where the grace of God helps us balance our excesses and deficiencies as we grow spiritually.

4. Cycads

Cycads are beautiful tropical plants that are often grown in greenhouses and arboreta. The cycads have un-branched stems that are crowned by long, leathery leaves. In the United States, cycads are found naturally in the southeastern states centered near Florida. Cycads grow in environments where their

Figure 11-31. *Cycas circinalis* with old and new male cones.

Biology For Life

rate of water loss through transpiration exceeds their gains through precipitation, except during one part of their growing season. Cycads produce pollen and ovules in specialized cones on separate plants. These cones are the "flowers" of these non-flowering plants. Pollination is made possible by specialized insect pollinators rather than by wind.

The method of fertilization in cycads shares some similarities with other plants. The pollen grains produce large sperm that are covered with cilia. A drop of water surrounds the cycad egg, and the sperm must swim through it before it unites with the egg. This is yet another vital use of water in the life of organisms.

5. Conifers

Conifers are very widespread and are able to grow in many different environments. The conifers form vast forests in northern Europe and in North America. There are about 600 species of conifers, which include pines, firs, spruces, cedars, hemlocks, and sequoias. A picture of a pine is seen in Figure 11-32.

The leaves of most conifers are called **needles**. These have a much smaller surface area than other leaves. The reduced surface area serves two purposes. First, it helps prevent desiccation during dry periods. Second, it reduces the chance of freezing during very cold winters. When we understand the reason for needles on conifers, we can see God's incredible wisdom at work.

Most conifers are known as evergreens, which means that they keep their leaves (needles) all year long. Some conifers, such as the larch, lose their leaves annually like deciduous trees.

Conifers have woody stems, which allow them to grow to large heights and widths. The redwood trees of California are capable of growing up to 380 feet, higher than a thirty-story building! Because conifers grow quickly, they are a major source of lumber and pulp for the production of paper. Pines are an important source of turpentine, an organic liquid that is used in paint thinners and varnishes. It is also used as a solvent for many industrial applications. Many birds and rodents use the seeds of conifers as a food source.

Figure 11-32. The conifer *Araucaria heterophylla*, also known as the Norfolk Island Pine, is endemic to Norfolk Island, a small island in the Pacific Ocean between Australia, New Zealand, and New Caledonia.

6. Life Cycle of a Pine

The **pine** is a typical **conifer**. Follow along in Figure 11-33 as we describe its life cycle.

Pine trees produce male pollen cones and female **seed cones** on the same tree. However, the male and female cones are usually produced on separate branches. The seed cones contain **megasporangia**. The cells in the megasporangia undergo meiosis and produce megaspores. These develop into female gametophytes within the cone. The female gametophytes are protected and nourished by the sporophyte. Female cones are open when dry but close in wet conditions.

The **microsporangia** in the male pollen cone produce microspores, which develop into male gametophytes. Before the gametophytes reach maturity, they are released as pollen grains. Each pollen grain contains a few haploid nuclei.

CHAPTER 11: THE PLANT KINGDOM

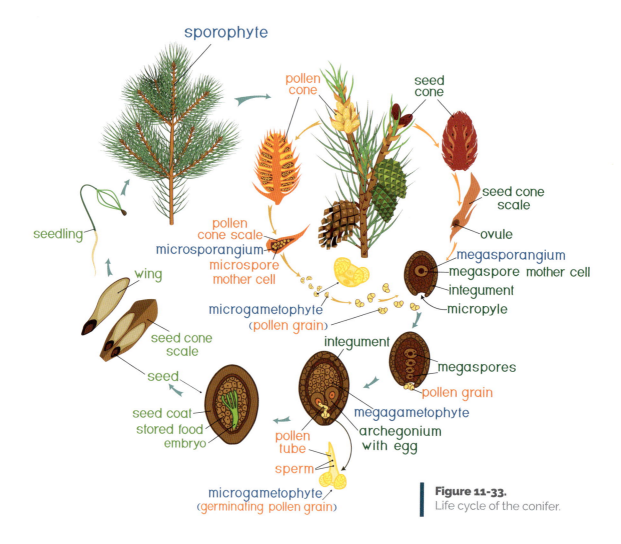

Figure 11-33.
Life cycle of the conifer.

The pollen grains are dispersed by the wind. The pollen drifts down between the open scales of the female cones in dry conditions. The process of **pollination** occurs when the pollen grains land near the female gametophytes. It is worth repeating that the female gametophyte is located inside the seed cones. After the pollen grain is situated inside the seed cone, it releases enzymes that digest tissues of the female gametophyte and forms a **pollen tube** while doing so. At this point in the life cycle, the male gametophyte is fully mature.

Each female gametophyte will produce eggs. Fertilization occurs when sperm pass through the pollen tube and unite with eggs. The fertilized egg is the first cell of the sporophyte, or diploid generation. It is important to recognize that pollination and fertilization are separate events. Sometimes a period of a year may pass between these two processes.

The female gametophyte absorbs food from the sporophyte and packages it into the seed, which is located inside the cone. During this time, a seed coat forms around the zygote and its food supply. When the seeds are ripe, the woody cone changes in color from green to brown and will eventually open, exposing the seeds to the environment. Each seed has papery wings, which allow it to be carried by the wind. Certain species of birds called crossbills have specially designed beaks that allow them to collect seeds from within the recesses of the cones.

CHAPTER 11: **THE PLANT KINGDOM**

C. Angiosperms

The angiosperms (ăn-jĭ-ō-spŭrms), or flowering plants, are the most abundant kinds of plants. There are over 250,000 species of angiosperms, which include trees, shrubs, cacti, herbs, grasses, vines, and floating plants.

1. Characteristics of Angiosperms

All angiosperms have flowers which produce seed-bearing fruit. The flower is a great advantage to seed plants because when pollen is spread by the wind, only a tiny percentage of pollen actually reaches the eggs of the same species. In plants containing flowers, insects transfer pollen from one plant to another. The insect flies directly from flower to flower, which increases the chances of

> **Section Review — 11.3 B**
>
> 1. The small _____ give rise to the male gametophytes, which produce _____.
> 2. The large _____ give rise to the female gametophytes, which produce _____.
> 3. What is a more common name for ginkgo trees?
> 4. Ephedra is an example of what kind of plant?
> 5. The leaves of conifers are called what?
> 6. Describe in brief the life cycle of a pine.

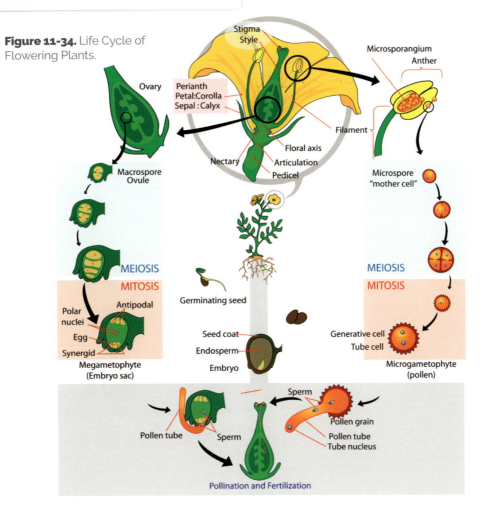

Figure 11-34. Life Cycle of Flowering Plants.

Biology For Life

pollination and makes it more precise. The life cycle of flowering plants is shown in figure 11-34.

As with gymnosperms, the gametophyte is small and is contained within the sporophyte. The reproduction of angiosperms is very similar to that of gymnosperms.

2. Monocots and Dicots

There are tiny embryonic leaves inside the seeds of angiosperms; however these are not true leaves, which will develop later. The leaves of the embryo are called **cotyledons** (kŏt-*i*-lē-d*u*ns). These are connected to the **plumule**, which is the embryonic stem of the plant. Figure 11-35 shows a dicot seed.

The cotyledons provide food in the form of starch for the young plant. In some angiosperms, the embryo has two cotyledons. These plants are called **dicotyledons**, or dicots. Angiosperms that have one cotyledon are known as **monocotyledons**, or monocots. Monocots have parallel veins in their leaves. Some of the common monocots are grasses, sedges, lilies, orchids, and palm trees. There are about 65,000 monocot species.

The dicot leaves have branched veins. There are about 170,000 species of dicots, which include the buttercups, peas, roses, sunflowers, and maple trees.

Figure 11-36. The flower of the magnolia, a typical dicot.

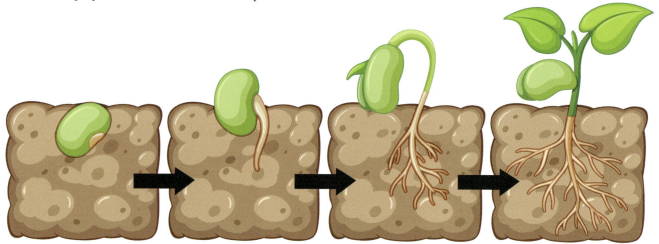

Figure 11-35. When a dicot seed, such as a bean, germinates, the expanding hypocotyl lifts the cotyledons out of the ground. Later, leaves form at the apical meristem, and the cotyledons wither. In a monocot, such as corn, the cotyledon remains underground as the plumule pierces the ground.

Biology For Life

3. Use by Humans

The majority of human food is produced by angiosperms. The fruits and vegetables in your refrigerator are angiosperms. The chicken or beef that you eat came from an animal that fed on grain from angiosperms. Angiosperms provide people with wood, cotton, oils, medicines, spices, and many other products.

The angiosperms also provide much of the beauty of the natural world around us. Perhaps you live close to an arboretum or botanical garden. These not only provide beauty, but often have plants that are exotic to the United States, and are not found growing outdoors. Plan a visit with your family and friends to a nearby arboretum or botanical garden. You can find books on famous botanical gardens at the library. It is a good opportunity to learn more about plants and will add to that which we have covered in this chapter.

Figure 11-37. The flower of Hemerocallis, a typical monocot known as the Daylily. The flowers of these plants are highly diverse in color and form, often resulting from hybridization by gardening enthusiasts.

Section Review — 11.3 C

1. What is the main difference between angiosperms and gymnosperms?
2. The leaves of the embryo are called _____.
3. In some angiosperms, the embryo has two cotyledons. What are these plants called?
4. What is the difference between monocot leaves and dicot leaves?
5. What are some things that angiosperms provide humans?

Chapter 11 Supplemental Questions

Answer the following questions.

1. Describe several forms of asexual reproduction in plants.

2. What is grafting?

3. Describe the life cycle of plants in terms of alternation of generations and discuss the haploid and diploid phases; explain how gymnosperms undergo alternation of generations.

4. Describe the characteristics of the three divisions of multicellular algae, and list a member from each division.

5. The alga Oedogonium has specialized filaments. What is their purpose and what are they called?

6. Explain the flower's role in reproduction.

7. Describe the characteristics of and compare sexual reproduction in the following organisms: Chlamydomonas, Volvox, Spirogyra, Oedogonium.

8. Which algae is used in the preparation of agar, the protein-rich medium used for growing organisms such as mold in a petri dish?

9. List some characteristics of bryophytes that have allowed them to adapt to living on land.

10. Describe the structure and the life cycle of mosses and liverworts.

11. What is unusual about the asexual reproduction of the liverworts?

12. Describe the following adaptations of land plants: adaptations for conserving water, adaptations for providing support for tissues.

13. Name and describe the following types of tissues in vascular plants: meristematic tissue, epidermis; parenchyma, supportive tissue, vascular tissue.

14. Discuss the functions of the following organs in vascular plants: roots, stems, rhizomes, leaves, sporangia, cones, flowers.

15. Compare and contrast xylem and phloem.

Chapter 11 Supplemental Questions

16. Compare and contrast monocot and dicot.

17. Compare and contrast angiosperm and gymnosperm.

18. What is the name of the gametophyte that develops from a whisk fern spore that falls on suitable land?

19. Describe the characteristics of whisk ferns, club mosses, and horsetails.

20. What does the word "flower" mean in describing the fertilization of cycads?

21. Conifers are called evergreens. Be able to name six varieties, and describe the main features of this class. Know the reproduction process.

22. Describe the life cycle of the pine tree.

23. Describe the structure and function of the pine seed.

24. Describe angiosperms and their use by humans.

25. Study all diagrams in this chapter. Be able to draw and label all diagrams.

CHAPTER 12
Invertebrates: Part 1

Chapter 12

Invertebrates: Part 1

Chapter Outline

12.1 Features of the Animal Kingdom
   A. Cell Specialization
   B. Animal Development
   C. Body Symmetry

12.2 Sponges and Coelenterates
   A. Phylum Porifera
      1. A Typical Sponge
      2. Reproduction of Sponges
      3. Sponge Diversity
   B. Phylum Coelenterata
      1. Polyps and Medusae
      2. The Hydra: a Typical Coelenterate
      3. Jellyfish, Sea Anemones, and Corals

12.3 Worms
   A. Flatworms
      1. Phylum Platyhelminthes: the Flatworms
         a. Class Turbellaria: Free-Living Flatworms
         b. Class Trematoda: the Flukes
         c. Class Cestoda: the Tapeworms
   B. Pseudocoelomates - Worms with False Body Cavities
      1. Phylum Nematoda: Roundworms
      2. Phylum Rotifera

   C. Phylum Annelida: the Segmented Worms
      1. Class Oligochaeta: the Earthworms
         a. Gas Exchange and Circulation
         b. Digestion and Excretion
         c. Movement
         d. Coordination and Senses
         e. Reproduction
      2. Class Hirudinea: the Leeches

12.4 Mollusks and Echinoderms
   A. Phylum Mollusca
      1. Development of Mollusks
      2. Mollusk Structure
      3. Class Gastropoda: Snails and Slugs
      4. Class Pelecypoda: the Bivalves
      5. Class Cephalopoda: Squid and Octopus
   B. Phylum Echinodermata
      1. Characteristics of Echinoderms
      2. Class Asteroidea: the Starfish

12.1 Features of the Animal Kingdom

Section Objectives

- Explain the function of cell specialization in multi-cellular organisms.
- List the germ layers of an embryo and describe their origin.
- Describe the types of body symmetry in living organisms.

Recall that all living things are currently classified into three Domains and six Kingdoms: Archaea, Bacteria, Protista, Plantae, Fungi, and Animalia. Also, remember that within the six kingdoms are further divisions, the first being phyla.

Animals are multi-cellular heterotrophic organisms. Heterotrophs must obtain food from the environment. Some animals ingest plants, some eat other animals, and some eat both. Most animals are mobile and move about in search of their food. Other animals, such as **sponges**, are called sessile, which means that most of their lives are spent attached to a single spot. Typically, the sessiles have a stage in their life cycle when they are motile (can move). Sessiles, like sponges, must capture food that comes near them.

Kingdom Animalia is often split into two generalized groups, the **vertebrates** and **invertebrates**. Vertebrates possess a backbone; invertebrates do not. Surprisingly, the invertebrates make up almost all of the divisions in Kingdom Animalia except for the phylum Chordata, which are those with spinal cords, or vertebrates. There are far more invertebrates than vertebrates in the animal kingdom. However, vertebrates are the organisms with which we are most familiar: dogs, cats, horses, and, of course, humans.

While scientists like to classify humans in the Animal Kingdom and the Phylum Chordata, the fact is that the difference between humans and all other creatures God has made is really immeasurable. There should be a special Kingdom for humans, because humans have an eternal soul and the ability to think, analyze, and choose good over evil. We can think of ourselves as belonging to the Heavenly Kingdom.

A. Cell Specialization

Animals possess many different types of cells, each of which performs a specialized function. Even though cells perform a special function, nevertheless, all of the different types of cells are interdependent. In other words, each cell within an organism depends on all other cells in the organism for its survival. Cells are grouped within specialized tissues and communicate with each other by hormones and chemicals or chemical messengers. Cell specialization is characteristic of both the plant and animal kingdoms.

Biology For Life

There are about thirty-five different phyla (categories) of animals, which are divided into two sub-kingdoms. This division is based somewhat upon the degree of interdependence among an animal's cells. In some phyla, there is little interdependence among the cells; in other phyla, there is a great degree of interdependence among the cells.

One sub-kingdom is called the **Parazoa**. Only sponges are in this sub-kingdom. Sponges possess specialized cell types, all of which are interdependent. However, each of their cells is not necessarily permanently specialized. Strange as it may seem, during the life of a cell within a sponge, specialized cells in the sponges can migrate to other tissues, transform, and take on new specialized functions. When pieces of sponge break away from the parent, amazingly, new groups of cells are capable of producing a new sponge, if the fragments of sponge contain the right group of cells.

The second sub-kingdom, known as the **Metazoa**, includes all animals except for the sponges. In this sub-kingdom, the cells are more highly specialized and do not have flexibility in function. The cell functions remain the same throughout the life of the animal.

The tissues and organs of animals are more highly differentiated than those of plants. In most animals, the tissues and organs are held together by a tough, stretchy, fibrous protein known as **collagen**. Did you know that gelatin, the key ingredient in various recipes in the kitchen, is created from collagen that has been extracted from pig or cow bones and connective tissues?

B. Animal Development

All members of the animal kingdom undergo sexual reproduction; however, many are able to reproduce asexually as well. An animal cell develops from one original cell, known as a zygote. The first cell divisions of the zygote are known as cleavages. After several cleavages, a small hollow ball of cells forms and is known as a blastula. The developing organism then becomes a gastrula. At this stage of growth, the cells begin to differentiate; each cell begins to take on its own special function.

Figure 12-1. Sponges are only in the sub-kingdom called the Parazoa.

The gastrula is marked by the formation of a tube that will eventually become the organism's digestive tube. The opening into the gastrula is called the blastopore. In some animals, the blastopore eventually becomes the mouth, and in other animals, the blastopore becomes the anus.

The cells begin to differentiate into specialized types, or functions, during the gastrula stage. It is during this time that the entire developing animal exists in three distinct layers of cells known as germ layers. The germ layers represent some of the first cells destined to contribute to specific types of tissue, such as muscle or blood in embryonic development. Cells that have this ability to develop into several specific kinds of tissue are termed **multipotent**. You may have heard of these first multipotent cells by their familiar name, stem cells.

The three germ layers of the developing organism are called the ectoderm, the endoderm, and the mesoderm. The outer layer of cells, or ectoderm, becomes the epidermis, or outer layer, of the animal. In mammals, this layer includes the hair, nails, and teeth. The ectoderm also forms the nervous system in most animals. The inner layer of cells is known as the endoderm. It gives rise to the epithelium that lines the respiratory passages, digestive tract, and multiple organs. Sandwiched between the ectoderm and endoderm is the third middle layer of cells known as the mesoderm. Mesoderm layer cells grow and diversify to give rise to the body's connective tissue, blood vessels, blood cells, and the epithelial cells of some organs and body cavities.

At the time when these three germ layers are forming, the developing organism is known as an embryo. In some animals, the embryo looks like a miniature version of the adult. In other animals, the embryo develops a form that looks very different from the adult. Sometimes these latter are called larva.

C. Body Symmetry

The bodies of most animals exhibit some type of **symmetry**, which is the balanced distribution of duplicate body parts or shapes. Organisms that lack any type of symmetry are called **asymmetrical**.

An organism that has **spherical symmetry** may be divided into equal halves. Such objects are shaped like spheres or globes. For example, a round ball exhibits spherical symmetry. The Volvox shown in Figure 12-2 provides a great example of spherical symmetry.

Several animal phyla have what is known as **radial symmetry**. These animals can be divided into equal halves by passing a plane through the central axis of the animal in any direction. These organisms resemble a pie where several cutting planes produce roughly identical pieces. An organism with radial symmetry exhibits no left or right sides. They have a top and a bottom surface only. Examples of radially symmetric organisms include starfish, anemones, and jellyfish. (Many flowers, such as buttercups and daffodils, are radially symmetric. Roughly identical petals, sepals, and stamen occur at regular intervals *around the center* of the flower.)

Animals that have **bilateral symmetry** may be divided into equal halves only along a single plane. A good example of bilateral symmetry comes from an airplane, where a plane passing through the center of the airplane from tip to tail would divide the airplane into two equal parts (on the external surface). Bilateral symmetry is found in insects, spiders, other invertebrates, and most vertebrates, including humans. (Some flowers, such as orchids and sweet peas, are bilaterally symmetrical. The leaves of most plants are bilaterally symmetrical.)

There are several terms used to describe animals with bilateral symmetry. The front of the animal is called the **anterior**, and the back of the animal is called the **posterior**. The top part of the

Figure 12-2. Types of Symmetry found in Nature

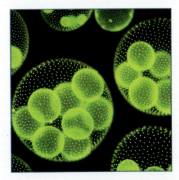
Spherical Symmetry

Radial Symmetry

Bilateral Symmetry

animal is called the **dorsal** surface, and the bottom (or belly side) is the **ventral** surface. The **lateral** surfaces refer to the sides of the animal.

In this chapter, you will learn about the invertebrates that belong to the following categories: sponges and coelenterates, worms, and mollusks and echinoderms. You will learn about their characteristics, methods of feeding, locomotion, and reproduction, as well as their habitats.

> **Section Review — 12.1**
>
> 1. Kingdom Animalia is often split into two generalized groups, the _____ and _____.
> 2. What is the difference between vertebrates and invertebrates?
> 3. What are the three different types of body symmetry? Give an example of each.

12.2 Sponges and Coelenterates

Section Objectives

- Identify characteristics of sponges.
- Explain sponge reproduction.
- Compare and contrast a polyp and a medusa.
- Identify and describe the function of specialized cells in hydra.
- List some of the types of organisms in the coelenterate phylum.

Sponges are the simplest group of animals. These organisms have several types of specialized cells that perform different functions, but as noted above, sponge cells are not permanently specialized. During certain conditions, cells that perform one function can change to another cell type. This is really incredible. It shows us that God is so concerned about the welfare of His creatures that He creates cells for some creatures which can perform different duties.

A. Phylum Porifera: Sponges

The Porifera is the only phylum, or division, found in the subkingdom Parazoa. It contains only the sponges. The body of a sponge has many interconnecting channels, which open to the outside through many tiny pores. Most sponges are asymmetrical. Sponges have very simple tissue development. Sponges do not develop the mesoderm, or middle germ layer, and, therefore, do not have organs.

We will study 1. A Typical Sponge, 2. Reproduction of Sponges, and 3. Sponge Diversity.

1. A Typical Sponge

Sponges have four basic types of specialized cells that are organized into two layers. The outer layer is called the epidermis and is composed of **flattened** cells. The epidermis is penetrated, however, by cylindrical cells known as porocytes. The porocytes permit water to enter the sponge's central cavity.

A jellylike material, known as the mesenchyme, sits below the epidermis. Hard, spike-like structures called spicules (spĭk-ŭlz) are embedded in the mesenchyme. The spicules serve two functions: to provide support for the cells and to help protect the sponge from predators.

The inner layer of tissue is called the **endoderm**. It contains a number of flagellated cells called **collar cells**, which line the interior cavity of the sponge. The collar cells create currents that draw water to the sponge's interior. The collar cells then withdraw food particles, such as algae, from the water. The collar cells supply all of the cells of the sponge with food in this manner.

Figure 12-3. Marine Sponge

A fourth type of cell in sponges is known as an *amoebocyte* (a-mē-bō-sīt) cell. These cells move around through the jelly-like mesenchyme by means of **pseudopods** (extensions which look like feet, so they are called "false feet," or pseudopods). The amoebocyte cells function to carry food particles from collar cells to epidermal cells and porocytes. Amoebocyte cells make the spicules in the mesenchyme of the sponge.

2. Reproduction of Sponges

Sponges are capable of both sexual and asexual reproduction. One form of asexual reproduction is called **budding**. In this method, small groups of cells, known as buds, grow out from the body wall of the adult sponge. The buds eventually break off, drift with water currents, and attach elsewhere.

Another form of asexual reproduction occurs in many freshwater sponges. During late fall, these sponges produce many *amoebocytes*, which carry food particles and are surrounded by a tough cell wall. These *amoebocytes* form new structures known as gemmules (jĕm-ūl). The gemmules are very hardy and can withstand winter freezing. The walls of these gemmule cells dissolve in the spring, and they differentiate, actually becoming a new sponge.

Most sponges, however, do not reproduce by budding or by producing gemmules. Most are hermaphroditic, which means that these sponges produce both eggs and sperm. The egg and sperm cells are produced by the *amoebocytes* or by collar cells that undergo meiosis (a type of cell division which produces both sperm and eggs). The sperm are released into the water and may enter porocyte cells of another sponge of the same species. After the sperm enters the sponge, it becomes surrounded by an amoebocyte cell. It is then carried to and fertilizes an egg. The larva that develops settles on a surface (such as a rock) and grows into a sessile adult sponge. You may recall that sessile means stationary or attached to a certain spot.

3. Sponge Diversity

Most of the 10,000 species of sponges live in salt water. One such species is shown in Figure 12-3.

Sponges have a variety of shapes, sizes, and colors and are classified according to the type of material found in their skeletons. The most common sponges have mats of flexible fibers in the jelly-like mesenchyme. These fibers are composed of a protein known as spongin. It is related to the collagen protein that holds tissues together in other animals.

CHAPTER 12: INVERTEBRATES: PART 1

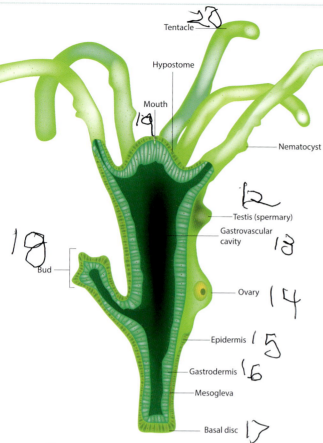

Figure 12-4. The Hydra: a Typical Coelenterate

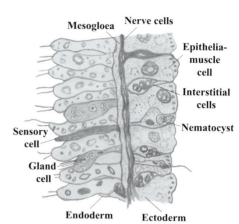

Figure 12-5. Structure of the Body of the Hydra. Note that the nematocyst has been discharged.

Sponges marketed for commercial use contain spongin. The sponges harvested from the ocean floor must be dried and pounded to remove the cell material and spicules. The spongin skeleton is collected and used for its water-absorbent properties. These "sponges" are sold in stores and tend to be expensive, in comparison with synthetic sponges that are made from cellulose or plastic polymers.

B. Phylum Coelenterata: Jellyfish, Hydra, Sea anemones, Corals

Animals such as the transparent jellyfish, flowerlike sea anemones, and the corals belong to the Phylum Coelenterata. (Notice the "c" is pronounced like an "s.") The **coelenterates** (suh-len-te-rāts) have radial symmetry and possess two distinct cell layers. In comparison to the sponges, the coelenterates have highly specialized cells. Most coelenterates are marine animals. They are carnivorous, and they feed on prey that ranges in size from tiny plankton to fish and animals that can be larger than themselves. However, they do not actively chase their prey. Instead, the coelenterates trap food organisms that inadvertently collide with the stinging cells of the coelenterata. Anyone who has been stung by a jellyfish knows just how powerful their sting is!

1. Polyps and Medusae

The coelenterates have two distinct body forms. One form is called a polyp. Polyps are shaped like a vase and are sessile, that is, attached to one location. The polyps attach to a surface with their mouth facing up. In contrast, the bell-shaped medusa (mĕ-dū-sa) is the other body form of the coelenterates. The medusa is a free-swimming form. It is not sessile.

Many coelenterates have a life cycle that alternates between the medusa stage and the polyp stages. The medusae reproduce sexually. After fertilization, the zygote develops into a blastula. It elongates to form a ciliated larva known as a planula, which attaches to the seabed and grows into a polyp. The polyp reproduces asexually by forming medusae, one on top of the other. The cycle is complete when the medusae form eggs and sperm. In other words, the medusa grows into a polyp, and the polyp becomes a medusa.

2. The Hydra: a Typical Coelenterate

The hydra is a freshwater coelenterate. The hydra is a polyp body form and has no medusa stage. Hydras are very small, measuring only about 0.5 centimeters in length. See Figure 12-5.

The body of a hydra is cylindrical in shape and contains two cell layers. The inner layer is called the gastroderm, and the outer layer is known as the epiderm. Between these two layers is found a jellylike material called **mesoglea** (měs-ō-glē-a). Nerve cells found in the mesoglea coordinate movement of the hydra.

The body of the hydra attaches to a surface (thus it is a sessile) by a structure called a **basal disk**. A number of tentacles project from the top of the hydra. The tentacles trap food organisms that float or swim near the hydra and direct these to the mouth.

The space in the interior of the hydra is called the **gastrovascular cavity**. Gastrovascular refers to anything to do with digestion and the distribution of nutrition to the body. Food organisms are pushed into this cavity through a single opening. Digestive enzymes released by specialized cells in the endoderm break down the food into food molecules. These move from within the gastrovascular cavity to all parts of the organism. Indigestible waste products are expelled back out of the same gastrovascular cavity opening.

The hydra have elongated extensions, called tentacles, which flow or stream out from their bodies. When organisms touch the tentacles of the hydra, the organisms are instantaneously stung with poisonous barbs. These tentacle-barbs are called nematocysts (něm-a-tō-sist). These are housed in specialized cells called cnidocytes (nī-dŏ-sīts). Cnidocytes are especially concentrated on the tentacles but are also scattered over the epidermis and gastrodermis of the hydra.

Cells that function like muscle are in the ectoderm of the hydra. When these cells contract, they make the hydra's body and tentacles longer. This way the tentacles can catch more organisms to digest. The contraction of other muscle-like cells will shorten the hydra's body.

During most of the year, the hydra reproduces asexually by budding. However, when winter approaches, hydras reproduce sexually. Both male and female reproductive glands, that is, the testes and ovaries, develop from the ectoderm. The sperm released from the testes of one hydra swim toward the ovaries of another hydra, where fertilization can occur.

3. Jellyfish, Sea Anemones, and Corals

All of the approximately 200 species of **jellyfish** live in oceans and spend most of their lives in the medusa form, that is, free-swimming form. Jellyfish swim with a gentle pulsing motion. Swimming is made possible by a ring of muscles that surrounds the edge of the medusa. These contract and push against the water. These contractions serve to move the jellyfish upward as the water pushes back with an opposite force. When these muscles relax, the jellyfish slowly sinks. Food organisms that brush against the tentacles of the jellyfish are stung, then paralyzed, and then pushed into the mouth of the jellyfish.

The **sea anemone** has a number of different compartments. The gastrovascular cavity is used for digestion and circulation. Stinging cells and tentacles line the outer walls of these compartments. Sea anemones feed on a variety of small animals, including fish.

Figure 12-6. Pacific Sea Nettle Jellyfish

Figure 12-7. Sea Anemone

Figure 12-8. Close-up of polyps of Montastrea cavernosa (Great Star Coral). Tentacles used to trap prey are clearly visible.

Corals are marine organisms that exist as small sea-anemone-like polyps, typically in colonies of many identical individuals. This group includes the important reef builders that are found in tropical oceans, which secrete calcium carbonate to form a hard skeleton.

Most corals are small animals that have skeletons that grow together into large masses. These large masses, or coral reefs, are merely new generations of coral skeletons that are built on older generations. Corals grow in a wide variety of colors and shapes. They play a very important role in maintaining fish and other animal communities that are found in the oceans. Figure 12-9 shows such a multigenerational construction.

Coral animals often occur in such large numbers that they may form ridges called **coral reefs**. Coral reefs can be found around islands and near coastlines. The Great Barrier Reef, which is located off the coast of northeastern Australia, is almost 2,000 kilometers (1,240 miles) long!

Figure 12-9. Pillar corals (dendrogyra cylindricus), a type of hard coral, which live in the western Atlantic Ocean. Pillar corals can grow to be up to 2.5 m (8 ft) tall. They grow on both flat and sloping sea floors at a depth of between 1 and 20 m (65 ft). They are one of the few types of hard coral whose polyps can commonly be seen feeding during the day.

Section Review — 12.2

1. What are collar cells?
2. How do sponges reproduce?
3. Coelenterates have two distinct body forms; they are _____ and _____.
4. What are three differences between polyp and medusa coelenterates?
5. _____ are marine organisms that exist as small sea-anemone-like-polyps, typically in colonies of many identical individuals.
6. Coral reefs can be found around _____ and near _____.

CHAPTER 12: **INVERTEBRATES: PART 1**

12.3 Worms

Section Objectives

- List the main characteristics of flatworms.
- Summarize the characteristics of each of the three classes of flatworms.
- Describe the life cycle of flukes and tapeworms.
- List some characteristics of parasitic worms.
- Contrast the roundworms from the other worm phyla.
- Summarize the hookworm's life cycle.
- Explain the usefulness of body segmentation as an adaptation.
- Name and describe the function of each organ system found in an earthworm.
- Describe some characteristics of leeches.

If you have ever worked in a garden or noticed the ground after a heavy rainfall, you have seen earthworms. Also known as night crawlers, earthworms help to make the soil suitable for the growth of plants. Earthworms are busy underground, digging tunnels through the soil. These tunnels create pathways for air and water to get to the roots of plants.

In the process of making tunnels, earthworms ingest dead animal and plant materials, along with soil particles. The food particles are digested, and the waste products are excreted. These waste products are high in nitrates, which are important nutrients for plants. If you have ever seen a compost pile, then you most likely have seen many earthworms at work. They are vital to the successful conversion of leaves, grass clippings, and other organic material into the dark, nutrient-rich compost material that gardeners cherish for their plants.

Earthworms are members of the Phylum Annelida in the Class Oligochaeta. Recall that the sequence **Domain – Kingdom – Phylum – Class – Order – Family – Genus – Species** is an organizational method used to show the progression of organisms from general (Domain) to specific (Species).

There are three major Phyla [plural for Phylum] of worms: flatworms, roundworms, and annelids. These worms are different from each other but share some common characteristics. All worms have bilateral symmetry, as well as organs and organ systems for carrying out their life's functions. Unlike the sponges and coelenterates, the tissues of worms develop from three germ (original) layers: ectoderm, mesoderm, and endoderm.

A. Flatworms

The flatworms are relatively simple bilateral, un-segmented, soft-bodied invertebrates. Unlike other bilaterals, they have no body cavity, and do not have specialized circulatory and respiratory organs. Therefore, they were designed with flattened shapes

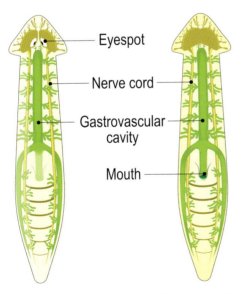

Figure 12-10. Diagram of a flatworm showing the eyespots, pharynx, and mouth. Note the "auricles" that project from the side of the head. These contain chemoreceptors that are used to find food.

Biology For Life

Figure 12-11. The planarian Dugesia that is often found in ponds and streams

that allow oxygen and nutrients to pass through their bodies by diffusion. The digestive cavity has only one opening for both the ingestion (intake of nutrients) and egestion (removal of undigested wastes); as a result, the food cannot be processed continuously.

1. Phylum Platyhelminthes: the Flatworms

The Platyhelminthes, or flatworms, have no spaces between the tissue layers. Since the worm is flat, all of its cells are close to the environment in which it lives. Oxygen diffuses directly from the environment and into its cells. Conversely, carbon dioxide diffuses directly from the worm into its surrounding environment. This is a very efficient method of respiration, or gas exchange.

The senses of sight and smell are located in the head of the flatworm. Nerve cells are concentrated near these sense organs. The nerve cells receive information from the sense organs and send the information to other parts of the body. The flatworm has a brain, which is a collection of nerve cells located in the anterior end. As is typical with animals that have a brain, the brain of the flatworm acts as a control center.

There are three different Classes of flatworms: **Turbelleria**, **Trematoda**, and **Cestoda**.

a. Class Turbellaria: Free-living Flatworms

The worms known as Planarians are representative of the Turbellarian worms. The planarian is usually less than one or two centimeters long. Most turbellarians live in salt water; however, the planarians live in fresh water. Planarians, like most other turbellarians, are free-living, which means that these organisms are not parasites. You will see later that some worms are parasites, living off other organisms.

Planarians have a nervous system, which consists of a brain composed of a large collection of nerve cells. These nerve cells work together to coordinate information that comes from the sense organs. Signals going to and from the brain are carried along nerve cells that form a ladder-like system along the entire body of the planarian. The head of the planarian has two light sensitive spots that resemble eyes. These eyespots sense the difference between light and dark. Planarians tend to move away from light. There are also projections on the sides of the planarian's head, which are sensitive to touch and water currents.

The digestive system of the planarian has a single opening, called the mouth, which is on the ventral surface near the middle of the animal. The worm obtains food by extending a muscular tube, known as a pharynx, from its mouth. The pharynx connects to the digestive cavity, or gut, of the worm. The gut has one anterior branch, two posterior branches, and many side branches. These serve to increase the surface area of the digestive system and bring the gut cavity close to many cells of the body.

Planarians eat protists and small organisms. The pharynx, the "throat," breaks up food into small pieces and pushes the pieces into the gut. The food particles are then enclosed in vacuoles, where they are broken down. The resulting food molecules then diffuse directly from the gut to the other cells of the body. Undigested food is expelled through the pharynx and mouth. These steps provide a very efficient means of nutrition and waste removal in planarians.

The planarian has a water-regulating system which functions to eliminate any excess water. Water is taken in by the process of osmosis, that is, the movement of water due to concentration differences between the inside and outside. Special ciliated cells are extensions that wave back and forth to move water along. These are called flame cells; they help to rid the planarian of excess water.

Planarians have two methods of locomotion. One method of movement involves the epidermal cells on its ventral, or underneath, side which have cilia extensions. The epidermal cells also produce mucus. As the cilia beat, the planarian is able to glide over the mucus. The second form of movement involves the three layers of muscle cells in the mesoderm layer. These muscles allow the planarian to stretch or shorten its length and even wrap itself around food.

Planarians are hermaphroditic, which means that they are capable of producing both eggs and sperm. During sexual reproduction, two planarians will exchange sperm. A process known as cross-fertilization occurs, which means that the sperm from one worm fertilizes the other worm's eggs. The fertilized eggs are released from the planarian in capsules. Each capsule has one or more eggs from which the young worms will emerge in a few weeks.

Planarians may also reproduce asexually. They are able to pull themselves apart to create two pieces. They can split into an anterior and posterior portion. Incredibly, each part will grow into a complete worm. Furthermore, the planarian has the ability to regenerate from pieces. This means, if the planarian worm is cut into pieces, each piece is able to develop into a new, complete, individual worm!

b. Class Trematoda: The Dangerous Parasite, the Fluke

The trematodes, more commonly known as flukes, are parasitic flatworms. Parasitic means they live off other organisms. Flukes have organs and organ systems similar to those of the planarians.

The fluke is characterized by its parasitic way of life. For example, the surface of a fluke is made up of a nonliving material called a cuticle. The cuticle serves to protect the fluke from digestive enzymes of the host organism. This means the organism which the fluke is living off cannot eat and digest the fluke. The host organism certainly has trouble getting rid of a fluke.

Most flukes have two sucker-like disks. The worm uses these suckers to attach itself to its host organism. In fact, the fluke usually attaches itself inside the digestive tract of the host organism! This is one case where the host organism would

Figure 12-12. *Fasciola hepatica*, also known as the common liver fluke or sheep liver fluke.

prefer not to be a host. The fluke does not need to digest its own food; it directly absorbs the already-digested food from the intestine of its host.

These parasitic worms, or flukes, often have more than one host during their life cycle. Each stage of the life cycle usually takes place in only one specific host organism. If a parasite is to survive, it must encounter the proper host at the right time. Parasitic worms usually produce a large number of eggs, so that a few eggs may find a host and continue the life cycle.

The life cycle of a fluke that lives in the liver of sheep is shown below.

1. Adult fluke worms in infected sheep reproduce sexually.
2. Fertilized eggs of the fluke are released in solid waste material from the sheep.
3. If the eggs are deposited near a marsh or pond, they hatch into larvae (miracidia).
4. The larvae must encounter a certain species of snail (lymnaeid snail) within eight hours, or the larvae will not survive. Each larva burrows into the snail.
5. In the snail, the larva reproduces asexually.

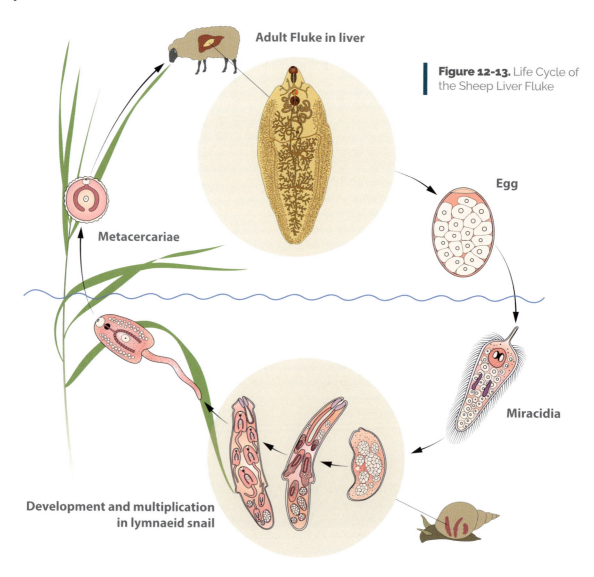

Figure 12-13. Life Cycle of the Sheep Liver Fluke

6. After several weeks, the new larvae leave the snail. The larvae swim until they reach blades of grass near the edge of the water.
7. Each larva forms a hard protective covering, called a cyst, around itself. The larva can last for months in this cyst form.
8. The larva will break out of its protective shell if a sheep eats the grass.
9. The fluke is carried to the sheep's intestine and eventually burrows into the sheep's liver.
10. The fluke grows to full size in the sheep's liver and produces a new generation of eggs.

The Blood Fluke

The blood fluke, known as **Schistosoma**, is a parasitic flatworm, and its biology is similar to the sheep liver fluke. However, it infects humans and causes a miserable disease. In some parts of Africa, it is not uncommon to find 75% of people with this disease, especially where people spend much of their time working or spending time in slow-moving water.

While you do not need to memorize the following details, this information gives you an idea of the life cycle of the blood fluke. Depending on the species of fluke, known as a schistosome, eggs are released from an infected human host in either the urine or feces. The eggs hatch into another stage of growth upon contact with water. In this stage, the parasites are capable of swimming and attempt to find a species of snail by which to attach themselves to.

Upon contacting a snail, the larvae quickly penetrate the snail's foot and eventually transform into another stage of growth. These parasites then migrate to an organ of the snail and transform themselves. They eventually produce thousands of parasites. Eventually, these parasites emerge from the snail and swim through the water in an attempt to find a human host. They have only 24 hours to find a human host, or they die. Upon finding a host, they penetrate the human skin and reach another stage of growth. They pass through the capillaries of the human bloodstream, then migrate to the lungs, and then move to the liver. The parasites become adults and live in either the urinary bladder or rectum. Male and female pairs

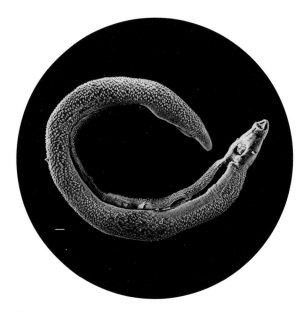

Figure 12-14. Electron micrograph of an adult male *Schistosoma*, also known as Blood Fluke. The bar represents a length of 500 μm.

mate, and the female may produce and release as many as 300 eggs per day.

People who suffer from this horrible disease often suffer organ damage and have a variety of health problems. Drugs are available to treat this disease, though many people often become infected again when spending time in the water. In some areas, there have been attempts to destroy populations of snails by increasing the populations of ducks and crayfish, both of which feed on the snails. In addition, scientists have developed extracts of plants that are known to kill the snails. These plants have been cultivated for this purpose and there has been much success. However, as with other health-related problems, these parasite diseases are difficult to manage, especially in remote areas where people are uninformed about the dangers of slow-moving and stagnant water.

c. Class Cestoda: the Tapeworms

The cestodes, commonly called tapeworms, are also parasitic flatworms. Tapeworms are more specialized than flukes. There are hooks on the head of the tapeworm that attach to the intestine of the unfortunate host. The tapeworm has a body that is divided into many sections, called proglottids (prō-glŏt-tids). The tapeworm will continue to make new proglottids, or sections, just behind the head. The tapeworm continues to grow in length by adding on more sections.

The tapeworm has no digestive system. As with the flukes, the tapeworm absorbs food directly from the host's intestine into the cells of the tapeworm. Nerves run through each proglottid section of the tapeworm. Each section has a flame cell system for eliminating excess water. Flame cells function like a kidney, removing waste materials and helping to maintain water balance.

Most of the proglottid section is filled with the reproductive organs. These sections contain both eggs and sperm. The proglottids that are distant from the head contain thousands of fertilized eggs, since they are older and therefore have had more time to produce eggs. The proglottids may break off and be excreted in the host's solid waste. In order to survive, these eggs must reach the proper host for the next stage of their life cycle.

The life cycle of the tapeworm may involve one, two, or three different host organisms. Infected pork, beef, or wild game that has not been properly cooked transmits several tapeworms that are human parasites. The meat contains a tapeworm larva that is encased in a cyst. The tapeworm larva can be rendered harmless, however, if the meat is cooked thoroughly and at a sufficiently high temperature. In the United States, most meat is inspected and does not reach the market if it is contaminated. Nevertheless, meat should be well cooked at high temperatures to prevent disease.

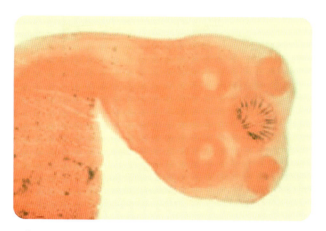

Figure 12-15. Head of the tapeworm Taenia solium; proglottids or sections are clearly evident.

B. Pseudocoelomates - Worms with False Body Cavities

Pseudocoelomates have false body cavities. They have a hollow, fluid-filled cavity called a pseudocoelom between the mesoderm and the endoderm. This is not a true body cavity formed from the endoderm, but it acts like one. This body cavity lack a complete epithelial lining from the mesoderm. However, this "false" cavity does allow some specialization of organ systems. Pseudocoelomates have a digestive tract with a mouth at the anterior end and an anus at the posterior end. The phylum Nematoda consists of roundworms. Most roundworms are free-living, but some are parasites of plants and animals. Rotifers are small animals, and most of them live as nonparasites in fresh water. The cilia surrounding their mouth sweep food into their digestive tract.

Pseudocoelomates have two major features that are characteristic of the phylum: a one-way digestive tract and a body cavity. These worms have a complete digestive tract, including a mouth where food enters and an anus, which expels undigested wastes. Food travels in one direction through this digestive tube.

Pseudocoelomates have a fluid-filled body cavity, which gives them rigidity. The fluid helps to circulate materials throughout the body. Included in the pseudocoelomates are the microscopic aquatic organisms known as rotifers, and the

Section Review — 12.3 A

1. What are three groups of flatworms? Which group is not parasitic?
2. Explain the life cycle of a fluke.
3. How does a tapeworm obtain nutrients, since it has no digestive system?

roundworms. Although rotifers are aquatic animals, they are known for their ability to withstand prolonged periods of desiccation, or extreme dryness, sometimes as long as nine years.

1. Phylum Nematoda: the Roundworms

The **nematode**, or **roundworm**, has a round, tube-like shape. The roundworm is extremely common; it has been estimated that there are over 10,000 species. Most roundworms are less than a few millimeters long. They live in soil and in most bodies of water. Most roundworms are harmless; however, some parasitic species do exist and may cause damage to plants, animals, and people. Nematodes are agricultural pests that can devastate crops if not controlled.

The outermost layer of the roundworm is called the **cuticle**. Under the cuticle lies the epidermis. Roundworms obtain oxygen and give off carbon dioxide by diffusion through the epidermis. There is a single layer of muscle that runs the entire length of the roundworm's body. The fluid-filled body cavity is below the muscle layer.

The digestive tract of the roundworm is located within the body cavity. A muscular **pharynx**, or throat, located just behind the mouth, sucks food into the mouth. The pharynx connects to the long intestine, where the food is digested and absorbed. Any undigested waste products are expelled through the anus.

The male or female sex organs are located inside the body cavity. The male and female roundworms are distinctly different organisms in most nematodes.

Parasitic Nematodes

The **hookworm** is a type of roundworm which is a parasite of humans. Infestations with hookworms can occur where sanitation is poor and where human wastes are used as plant fertilizer. For this reason, human waste is not typically used as fertilizer; however, in some poor parts of the world, it is used in agriculture and is often referred to as "night soil."

The human wastes from an infected person may contain hookworm eggs. When these eggs are deposited onto soil, they quickly develop into larval worms, and, upon contact with skin, quickly burrow into it. The worms then travel in the blood to various body parts, and cause severe damage to tissues.

The adult hookworms eventually settle in the intestine and will feed on blood and tissue fluids from the intestinal wall. When the adult worms reproduce, eggs pass out of the body with the waste material, and the entire cycle begins again. We all need to be aware of the importance of good personal hygiene and proper sanitation in the household so that parasites, like the hookworm, are not allowed to propagate.

The **trichina** is another roundworm parasite. At one particular stage in their life cycle, trichina worms form cysts in the muscles of pigs and other mammals. If humans eat undercooked pork that contains trichina cysts, the larvae will emerge and will infect human muscles. The disease is called **trichinosis**, and its symptoms include muscular aches and breathing difficulties. Until the natural human body chemicals kill the larvae, the person is really miserable, with flu-like symptoms. Due to the government inspection of meats, trichina infection is rare; however, all meat should be cooked thoroughly.

Another type of nematode is the filaria worm. The filaria worm resides in the lymphatic system and can produce a serious condition known as **elephantiasis** (ĕl-ĕ-făn-tī-a-sĭs). These worms, transmitted by mosquitos, are serious problems in Asia, Africa, parts of South America, and many of the Pacific Islands. A type of filaria worm that lives in dogs and cats can cause a problem known as heartworm and can be fatal. It too is transmitted by mosquitos.

A fourth type of nematode is known as the ascaris (ăs-ka-rĭs) worm. This parasitic roundworm feeds on partially

Figure 12-16. Hookworms

digested food and can reside in the intestines of its host. Ascaris infection can be prevented through cleanliness.

2. Phylum Rotifera

Rotifers are microscopic animals that can be found in many freshwater environments and in moist soil. The name "rotifer" is derived from the Latin word meaning "wheel-bearer"; this makes reference to the crown of cilia around the mouth of the rotifer. The rapid movement of the cilia in some species makes them appear to whirl like a wheel. Rotifers inhabit the thin films of water that are formed around soil particles. The habitat of rotifers may include still water environments, such as lake bottoms, as well as flowing water environments, such as rivers or streams. Rotifers are also commonly found on mosses and lichens growing on tree trunks and rocks, in rain gutters and puddles, in soil or leaf litter, on mushrooms growing near dead trees, in tanks of sewage treatment plants, and even on freshwater crustaceans and aquatic insect larvae.

Rotifers are multicellular animals with body cavities that are partially lined by mesoderm. These organisms have specialized organ systems and a complete digestive tract that includes both a mouth and anus. Since these characteristics are all uniquely animal characteristics, rotifers are recognized as animals, even though they are microscopic. Most species of rotifers are about 200 to 500 micrometers long. However a few species, such as Rotaria neptunia, may be longer than a millimeter. Rotifers are thus multicellular creatures who make their living at the scale of unicellular protists.

Since rotifers are microscopic animals, their diet must consist of matter small enough to fit through their tiny mouths during filter feeding. Rotifers are omnivorous. The diet of rotifers consists of dead or decomposing organic materials, as well as unicellular algae and other phytoplankton. Rotifers are in turn prey to carnivorous secondary consumers, including shrimp and crabs.

Reproduction in rotifers is unique; different types of reproduction have been observed. Some species consist only of females that produce their daughters from unfertilized eggs, a type of reproduction called parthenogenesis. In other words, these parthenogenic species can develop from an unfertilized egg, asexually. Other species produce two kinds of eggs that develop by parthenogenesis: one kind forms females and the other kind develops into degenerate males that cannot even feed themselves (sexual dimorphism). These individuals copulate, resulting in a fertilized egg developing within the rotifer. The males survive long enough to produce sperm that fertilize eggs, which then form resistant zygotes that can survive if the local water supply should dry up. The eggs are released and hatch in the water. If the egg develops in the summer, the egg may remain attached to the posterior end of the rotifer until hatching.

Phylum Rotifera is divided into three classes: Monogononta, Bdelloidea, and Seisonidea. The largest group is the Monogononta, with about 1500 species, followed by the Bdelloidea, with about 350 species. There are only two known species of Seisonidea, which is usually regarded as the most "primitive."

C. Phylum Annelida: the Segmented Worms

The Phylum Annelida is characterized by segmentation of the body of the worms, and the presence of a coelom. The **coelom** (sī-lŭm) is

Figure 12-17. Rotifers under a microscopic view

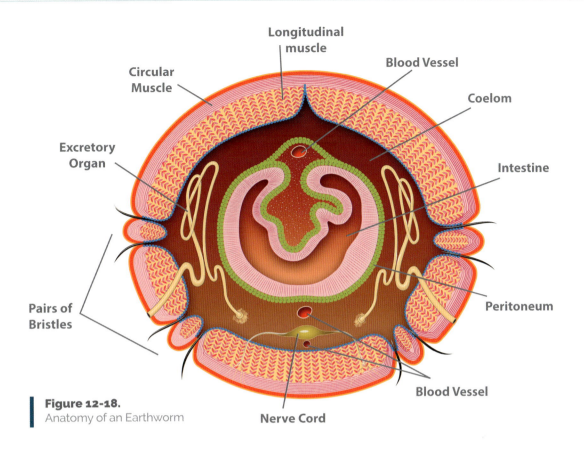

Figure 12-18.
Anatomy of an Earthworm

a fluid-filled body cavity that is surrounded by mesoderm. These worms have a membrane that originates from the inner mesoderm, which is called the **peritoneum**. The peritoneum suspends and supports the internal organs in the coelom. The coelom provides room for complex internal organs.

Animals with a coelom, or fluid-filled body cavity, have muscle tissue surrounding both the digestive tract and the body wall. These two muscle sets operate independently. The muscles of the body wall move the organism, and the muscles around the digestive tract move food through the digestive system.

Segmentation is the important characteristic of annelids for two reasons. First, an animal may increase in size by adding on identical segments. Second, different segments carry out specialized functions for these worms.

Figure 12-19. Earthworm

1. Class Oligochaeta: the Earthworm

There are about 2,500 species of oligochaete worms. The earthworm is one of the most familiar types of these worms and was discussed briefly at the start of this section. The following sections describe the life functions of the earthworm.

Biology For Life

CHAPTER 12: INVERTEBRATES: PART 1

a. Gas Exchange and Circulation

Earthworms have many more cell layers than either flatworms or roundworms. Blood carries gases to and from its body tissues. Gases are exchanged with the environment through the epidermis. Oxygen diffuses through the epidermis and into the nearby blood vessels. Carbon dioxide is carried from the body cells out to the epidermis, where it diffuses into the surroundings.

The blood carries molecules of digested food to the body cells. The movement of blood through the earthworm's body is called **circulation**. The earthworm has a circulatory system of blood vessels that pass to all parts of its body. The blood of the earthworm contains a red pigment, or protein, called **hemoglobin**, which carries and transports oxygen. Near the front of the worm, there are five pairs of hearts that contract and relax to move blood throughout the body. Blood moves from the hearts into a ventral vessel, which takes blood toward the body cells. A dorsal vessel takes the blood back to the heart.

b. Digestion and Excretion

The earthworm takes in food and soil particles through its mouth. Figure 12-20 shows the location of the mouth, its cavity, and other parts of the digestive system of the earthworm.

The mouth leads to a muscular pharynx, or throat. A short tube called the **esophagus** carries food from the mouth to a large storage organ called the **crop**. The food and soil are stored in the crop. Next, the food moves to the gizzard, where it is ground up. The organic material is broken into smaller pieces by sand particles. The smaller pieces move on to the intestine, where the food is digested and absorbed into the blood.

Sand particles and undigested food are expelled out through the anus (not shown in the illustration). This expelled material is called

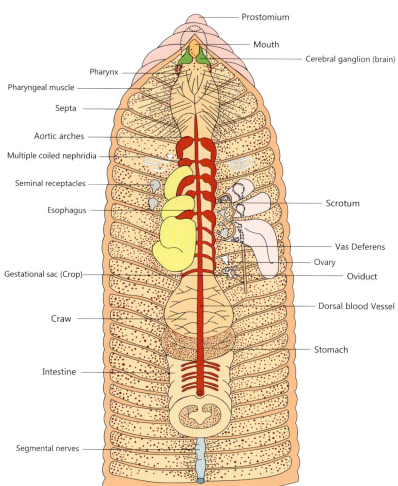

Figure 12-20. Head Anatomy of an Earthworm

castings, which actually fertilize the soil; this is the reason why earthworms are needed and welcome in the compost piles of gardeners. Earthworms have an excretory system that removes nitrogenous wastes. Every segment of the earthworm, with the exception of the first three and the last one, has a pair of structures called **nephridia** (nĕ-frĭd-ĭ-a). The nitrogen wastes from the blood diffuse into each nephridium and are excreted through an opening in the body wall of the earthworm.

c. Movement

Earthworms have muscle groups within each segment. These muscle groups provide the means for

Biology For Life

its movement. A layer of circular muscle is found just beneath the epidermis. There is a muscle layer that lies just beneath the circular muscle which extends the length of the segment. The earthworm moves by contraction and relaxation of these two muscle layers. Bristles, called setae, also aid the worm in movement. Except for the first and last segments, each segment contains four pairs of setae. The setae anchor each segment of the worm in the soil.

d. Coordination and Senses

The brain of the earthworm is located near the anterior end of the animal. Recall that the anterior end is the part of the animal that faces to the front; that is, it leads when the animal is moving forward. A **nerve cord** extends from the two halves of the brain and travels the entire length of the earthworm. The nerve cord contains swollen areas in each segment of the worm. Each swelling of the nerve cord is known as a ganglion, which refers to a grouping of nerve cells. These nerve cells send messages along the nerve cord and along nerves that branch into each segment. Nerves located in the epidermis detect touch and chemicals. Some nerves also detect light.

e. Reproduction

Earthworms reproduce sexually, but they produce both eggs and sperm. Eggs are produced in the ovaries, which are located in the thirteenth segment. The testes produce sperm and are found in the tenth and eleventh segments. When the earthworm mates, two earthworms line up side-by-side and the sperm move from each earthworm to an opening on the other worm's body where the sperm are stored temporarily.

After the earthworms part, an enlarged area of the body, called the clitellum, secretes a capsule. Eggs move out of the worm's body and into this capsule. The stored sperm are released and also travel to this capsule. The sperm and eggs unite in the capsule, and fertilization takes place. The capsule then slips off of the worm and is left in the soil. The fertilized eggs develop into small worms.

2. Class Hirudinea: the Leech

Leeches are annelids comprising the class Hirudinea. There are freshwater leeches, terrestrial leeches, and marine leeches. Like the earthworms, the bodies have both sperm and eggs, and they are fertilized in a capsule. Some, but not all, of the 300 species of leeches are hematophagous, or blood-feeding, and thus are parasites.

Figure 12-21. *Haemadipsa zeylanica*, a terrestrial leech found in Japan

The non-parasitic leeches feed on worms, snails, and soft-bodied insect larvae. The parasitic leeches attach to the outer surface of animals such as fish. The leech sucks the blood of its host and secretes anticoagulants. These compounds prevent the victim's blood from clotting and allow the leech to feed until it is satisfied. This may involve the leech acquiring a volume of blood that is many times its own weight in blood.

Leeches currently play an important role in medicine, as they have in the past. They are currently used in surgery, especially surgical techniques that involve the use of a microscope (microsurgery). Following surgery, capillaries quickly form clots that restrict blood flow and often prevent blood from flowing to areas that require blood. The compounds produced by the leech promote healing by preventing the formation of clots and by stimulating blood circulation. In this manner, blood is allowed to flow, allowing the delivery of nutrients and oxygen to repaired tissues.

> ### Section Review — 12.3 B & C
>
> 1. Name four types of roundworms (nematodes).
> 2. What are two reasons why segmentation is an important characteristic of annelids?
> 3. What do non-parasitic leeches feed on? What do parasitic leeches feed on?
> 4. Leeches are sometimes used to prevent blood clots. How is this possible?

12.4 Mollusks and Echinoderms

Section Objectives

- Describe and explain the significance of the trochophore larva.
- Describe the structure of a typical mollusk.
- List the major characteristics of gastropods.
- List the characteristics of bivalves that are different from other mollusks.
- Summarize the characteristics of echinoderms.
- Describe the major organ systems in the starfish.

The members of the phyla Mollusks and Echinoderms include the oysters, snails, squids, octopuses, starfishes, and sea urchins. In this section you will learn about the structure of these animals, as well as how they feed, move about, and reproduce.

A. Phylum Mollusca

The animals in the Phylum Mollusca include the oysters, snails, squids, and octopus (pl. octopi). Most of these animals are marine; however, there are many freshwater species. Some of the mollusks are also terrestrial, which means that they live on land.

1. Development of Mollusks

Following the fertilization of an egg, most mollusk species develop into a larva known as a trochophore (träk-a-fôr). A trochophore larva has its own digestive system, which has tufts of cilia located at each end of the larva. There is a ring of cilia around the middle of the trochophore. In many aquatic mollusk species, the trochophore larva is free-swimming and feeds on plankton.

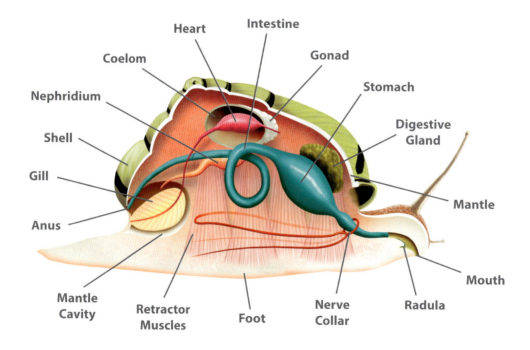

Figure 12-22. The Structure of a Mollusk. The mollusks have a large muscular foot that is used for locomotion. The foot secretes mucus, which helps it glide along a path. The head contains many specialized sensory organs that gather information from the environment.

2. Mollusk Structure

Body Structure of Mollusks

Mollusks have bilateral symmetry; the right half mirrors the left half. Mollusks lack internal skeletons. They have a **coelom**, a cavity within the body, and are soft bodied. A layer of cells called the mantle covers the body. The mantle secretes a material that forms the shell on the outside of the organism. Both the mantle and shell extend out farther than the body of the mollusk. At the posterior end, this extended section is so great that it forms a protected space called the **mantle cavity**. Water from the environment usually flows through this mantle cavity and serves several purposes. It helps deliver oxygen, obtain food, and remove wastes.

Digestive System of Mollusks

The digestive system of the mollusk begins at the mouth. The mollusk contains a feeding device known as a **radula** (răd-ŭ-la). It is a muscular structure with hard, tooth-like projections. The radula scrapes food from surfaces. The food travels to the stomach through the esophagus. The processes of digestion and absorption take place in the stomach with the aid of a nearby digestive gland. The intestine forms waste pellets that are excreted through the anus. The wastes enter the mantle cavity from the anus, and, with the aid of the water currents, pass out of the body.

Respiratory System of Mollusks

Gas exchange occurs through a pair of gills that are located within the mantle cavity. The gills are thin folded tissues which contain very small blood vessels and are in direct contact with water. These gills function to provide a large surface area for gas exchange: the gills absorb oxygen from the water and release carbon dioxide as a waste gas.

Circulatory System of Mollusks

Blood circulation in mollusks is different from those of annelids or earthworms. Recall that in the earthworm, the blood always stays in blood vessels as it travels around the body. This type of system is called a **closed circulatory system** and

Figure 12-23. The Caribbean Reef Squid (Sepioteuthis sepioidea), also known as just the Reef Squid, is a small (20 cm) torpedo-shaped squid with fins that extend nearly the entire length of the body and undulate rapidly as it swims. It is commonly found in small schools off the reefs throughout the Caribbean Sea and the coast of Florida.

is the type of circulatory system that is found in humans.

Most mollusks, however, have an open circulatory system. In an open type of circulation, the blood is not confined to vessels after leaving the heart. A two-chambered heart pumps the blood around the coelom, or cavity, and between the internal organs. The blood also travels through spaces between the tissues and organs. The blood contains an oxygen-carrying molecule, or pigment, known as hemocyanin. Unlike hemoglobin, which is red, hemocyanin is blue.

Excretory System of Mollusks

The coelom, or cavity, of the mollusk contains paired **nephridia**, or waste-discharging tubes, that remove nitrogen-containing wastes. Wastes are removed from the fluid in the body cavity and are excreted into the mantle cavity.

Reproductive System

Mollusks reproduce sexually. The sexes are separate in most mollusks; however, some species have both male and female cells within the same organism.

Figure 12-24.

Members of the Class Gastropoda

Helix pomatia,
a species of land snail

Marisa comuarietis,
a species of aquatic snail

Limax maximus,
an air-breathing land slug

Groups of mollusks have basic structures that are similar to those that have been described above. However, there are differences in these structures. For example, while the clams have paired shells to protect their soft body parts, snails have a single shell into which they can retract their bodies. The squid shell is different; it is essentially a supporting rod inside its body.

3. Class Gastropoda: Snails and Slugs

The gastropods are the largest group of mollusks, with over 35,000 species. They have a wide variety of forms and live in a diverse range of environments, more than any other mollusk group. They live in marine and freshwater habitats. Both the snails and slugs are the most common gastropods, and they live on land.

Snails lacking a shell or having only a very small shell are usually called slugs. There are really no appreciable differences between a slug and a snail except in habitat and behavior. A shell-less slug is much more maneuverable. Large land slugs can take advantage of habitats or retreats with very little space, that is, places that would be too small for similar-sized snails with shells. Slugs may fit under loose tree bark, under stone slabs, and under logs or wooden boards lying on the ground. Snails that have a broadly conical shell that is not coiled, or appears not to be coiled, are usually known as limpets.

The snail is representative of the Gastropod Class. The garden snail is an herbivore, which, as you will recall, means that it eats only plants. A problem for gardeners is that land snails and slugs may do extensive damage to agricultural crops. Some gastropods are carnivores, or meat eaters. Some gastropods are scavengers, which means that they feed on dead organisms. Finally, there exist a few gastropods that are parasites.

To search for food, the snail travels very slowly on its foot. It scrapes up food with its radula, or tooth-like projection. The snail has two eyes that are located on stalks, which look like antennae, in the head region.

Gastropods that live on land do not have gills. Instead, land snails have a simple lung that is involved in gas exchange, and so they are called lunged snails. The lungs consist of infoldings of the body wall that have a single, narrow opening to the outside. Because this opening is very small, it helps retain moisture in the lung tissues. In the lung, the gases diffuse between the air and the blood.

Gastropods that live in water have gills and are known as aquatic snails. One can easily tell the difference between both types of snails by observing the shape of the shell; if you look from the bottom foot area of the snail, the shell opens up leftwards.

All land snails produce both **spermatozoa** and **ova** in the same organism. However, most gastropods which live in water have separate sexes. Some species will release the eggs and sperm directly into the water; therefore, fertilization occurs in the water and not in the female's body. In other species, mating occurs between the two sexes, and fertilization occurs in the female's body. In most gastropods, a larva develops within the egg; however, this larva never leaves the egg. The free-swimming larva of some species develops into another larval form, and in other species, the free-swimming larva becomes a tiny adult snail within the egg.

4. Class Pelecypoda: the Bivalves

The mollusks that are pelecypods have two shells, or **valves**, hinged together. These animals are called **bivalves**. The two valves are hinged, and open and close due to the action of several large muscles. When the valves are closed, the animal is protected from predators. Animals that are bivalves include oysters, clams, scallops, and mussels.

The clam is a representative bivalve and has a specialized foot for burrowing in soft mud or sand. Most bivalves use their gills for feeding, as well as for respiration or breathing. The bivalve clam keeps its shells partly open so that water, which carries food and oxygen, is allowed to flow into the clam. The gills have cilia, hair-like structures, which move almost continually and create water currents in the mantle cavity. The mucus on the gills, which collect oxygen, also trap plankton in the water. The cilia sweep the mucus and food particles toward the mouth. Oxygen from the water diffuses into the blood, and carbon dioxide waste gas diffuses from the blood into the water.

Bivalves such as clams have an open circulatory system, which means the blood circulates through the body cavities and is not confined to blood vessels, as it is in a closed system. Bivalves also contain **nephridia**, which function in excretion to remove nitrogenous wastes and excess water. The nervous system of the bivalve is bilaterally symmetrical; that is, the right side is a mirror of the left side. The nervous system of the bivalve contains two pairs of long nerve cords and three **ganglia**, that is, masses of tissue which contain nerve cells. The edge of the mantle, the outer body wall, has sense organs that respond to touch, light, and chemicals.

In most bivalves, the sexes are separate. The sperm and eggs are shed into the water. Fertilization occurs outside of the body. The larva swims freely for some time, and then the larva settles to the bottom of the sea to develop into an adult.

Figure 12-25. A Giant Clam. It is about 4 ft (1.2 m) in length and weighs about 440 lbs (200 kg).

CHAPTER 12: INVERTEBRATES: PART 1

5. Class Cephalopoda: Squid and Octopus

The squid and the octopus are typical cephalopods and are the most highly specialized class of mollusks. The nautilus is the only living cephalopod that has an external shell. In the squid, the "shell" is an internal rod that runs the length of their bodies. The octopus has no shell.

The foot of this cephalopod has long arms that project from the head. The squid has ten arms, and the octopus has eight. The arms have suction disks on the bottom surface. Cephalopods use the arms to grasp prey and pull the food toward their mouths. All cephalopods are carnivorous and have hard, beaklike teeth in their mouths, which are used for biting and tearing prey.

Deep sea oil drills keep cameras in the deep sea to keep an eye on their drills. They are obtaining great film footage of deep sea creatures, which you can find sometimes on the Internet. The following is a description of such Shell Oil Company footage by National Geographic: "In a few seconds of jerky camerawork, the squid appears with its huge fins waving like elephant ears and its remarkable arms and tentacles trailing from elbow-like appendages."

The cephalopods have a closed circulatory system. This type of circulatory system enables the blood to carry and to efficiently deliver oxygen and food to all parts of the body through vessels. Closed circulation enhances an animal's ability to move rapidly. The cephalopods are the fastest moving mollusks.

The squid normally swims with graceful, wavelike motions. In times of danger, however, they are able to move quickly by using a nozzle-like siphon that is located on their ventral surface. Rapid muscular contractions force water out of the mantle cavity and through the siphon. The force of the water moving out of the siphon propels the squid and helps it escape from danger.

Cephalopods have an extensive nervous system with a large brain. In addition, they have complex sensory organs. The eyes of the squid and octopus are very similar in structure and function to the human eye.

B. Phylum Echinodermata

The echinoderms are a phylum of marine animals, which contain more than 5,000 species. Members of this phylum include the starfish, sea urchins, sand dollars, and sea cucumbers. The echinoderms have a coelom, that is, a cavity within the body, and a one-way digestive system like the mollusks.

1. Characteristics of Echinoderms

Most of the echinoderms have **pentamerous radial symmetry** (penta means five), which means they have five sections shaped like pieces of pie. Animals with this type of body symmetry are divided into five equal parts from a central point.

Figure 12-26. The Squid

326 Biology For Life

CHAPTER 12: **INVERTEBRATES: PART 1**

Figure 12-27. The Octopus

Echinoderms have two features that are unique to the invertebrates. The first is their system of tube feet. These are a series of suction disks that are used for both locomotion and obtaining food. The tube feet are powered by a system of water-pumping tubes.

The other unique feature of the echinoderms is the endoskeleton, or internal skeleton. It protects and supports the organism's soft tissues and attaches the muscles. The endoskeleton is composed of calcium compounds that form plates just beneath the outer epidermis. However, a number of spiny projections extend out from these plates through the epidermis. The name echinoderm actually means "spiny skin."

2. Class Asteroidea: the Starfish

The Class Asteroidea contains the starfish, which are also known as sea stars. The starfish has five arms, or rays, that project outward from a central disk. The outermost surface of the starfish consists of a layer of tiny little hairs, or ciliated epidermal cells. A nerve cell network lies just below the epidermis. The endoskeleton lies beneath the epidermis and nerve cell network.

Starfish

Sea Urchins

Figure 12-28. Echinoderms exhibit a wide range of colors

Biology For Life   327

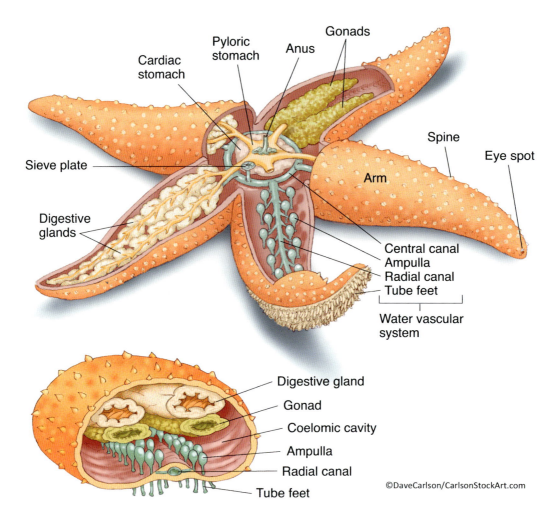

Figure 12-29. The dorsal surface of the Starfish. The mouth is located on the other side.

Starfish are usually bottom dwellers that move around by using their tube feet. The tube feet are located on the underside of each of the five rays. The tube feet are part of the water vascular system, a system of vessels which conduct water and nutrients for the starfish. In this system, seawater enters through a pore in the center of the starfish's upper surface. The water then moves through a series of canals and into the tube feet. Each tube foot moves due to the continuous filling and emptying of water.

Most starfish are carnivores; that is, they eat other animals. They take food into their mouth, which is located on the ventral surface of the starfish. Food travels through the esophagus and to the stomach, where digestion takes place. Digestive enzymes are produced by digestive glands in each of the five rays. Undigested waste products are eliminated through the anus, which is located on the dorsal surface of the starfish. Digested food molecules move to the animal's cells through the fluid in the coelom, or coelomic fluid. The coelom is lined with ciliated (with tiny hairs) cells that help to circulate the fluid. The coelomic or cavity fluid circulation actually functions as a circulatory system.

Starfish have no excretory system. Waste products diffuse from the cells into the coelomic fluid. The waste then diffuses out of the body through the tube feet. Specialized spines function as gills, and gases are exchanged between the coelomic fluid and the exterior by moving through the spines and the tube feet.

Figure 12-30.
A starfish is regenerating from an arm.

still attached to the arm. A few species of starfish can reproduce asexually by shedding arms, which then grow into another complete starfish. See Figure 12-30.

The starfish does not have a head, and its nervous system consists of a simple ring of nerve cells. The nervous system has branches that extend outward into the arms.

Reproduction in starfish is usually sexual. The males and females shed egg and sperm cells directly into the water, which is where fertilization occurs.

Starfish have a remarkable ability to regenerate. If a starfish is broken into pieces, many of the pieces will regenerate into another whole starfish! For example, a severed starfish arm can regenerate into an entire animal as long as a part of the central disk is

Section Review — 12.4

1. What animals are included in the Phylum Mollusca?
2. Hemoglobin is _____, and hemocyanin is _____.
3. What is a radula?
4. What animals are considered bivalves?
5. The squid and octopus are typical _____ and are the most highly specialized class of _____.
6. What animals are included in the Phylum Echinodermata?
7. What are two features do echinoderms possess that are unique to invertebrates?
8. What does echinoderm mean?
9. Are starfish herbivores or carnivores?

Chapter 12 Supplemental Questions

Answer the following questions.

1. Know the material in Figures 12-4, 12-10, 12-22, and 12-29.
2. Explain the function of cell specialization in multicellular organisms.
3. List the germ layers of an embryo and describe their origin.
4. Describe the types of body symmetry in living organisms; your answer should include appropriate drawings.
5. Explain sponge reproduction; your answer should discuss both asexual and sexual reproduction.
6. Compare and contrast a polyp and a medusa.
7. Identify and describe the function of specialized cells in hydra.
8. List some of the types of organisms in the coelenterate phylum.
9. List the main characteristics of flatworms. What is unique about their digestive system?
10. Summarize the characteristics of each of the three classes of flatworms.
11. Describe the life cycles of flukes and tapeworms.
12. List some of the characteristics of parasitic worms.
13. Contrast the roundworms from the other worm phyla.
14. Summarize the hookworm's life cycle.
15. Explain the usefulness of body segmentation as an adaptation.
16. Name and describe the function of each organ system found in an earthworm.
17. Describe some of the characteristics of leeches.

Chapter 12 Supplemental Questions

18. Describe and explain the significance of the trochophore larva.

19. Describe the structure of a typical mollusk.

20. List the major characteristics of gastropods.

21. List the characteristics of bivalves that are different from other mollusks.

22. Summarize the characteristics of echinoderms.

23. Describe the major organ systems in the starfish.

24. What does the word "sessile" mean? List at least two invertebrates that are sessile.

25. Be able to identify the parasites from this chapter that affect humans.

26. Study all diagrams in this chapter. Be able to draw and label all diagrams.

CHAPTER 13
Invertebrates: Part 2

Chapter 13

Invertebrates: Part 2

Chapter Outline

13.1 Phylum Arthropoda
  A. Introduction
  B. Arthropod Characteristics

13.2 Subphylum Chelicerata
  A. Characteristics of Chelicerates
  B. Class Arachnida: Spiders
  C. Reproduction in Spiders
  D. Other Arachnids

13.3 Subphylum Mandibulata
  A. Class Crustacea: Crayfish, Lobsters, and Shrimp
  B. Class Diplopoda: Millipedes
  C. Class Chilopoda: Centipedes

13.4 Class Insecta
  A. Characteristics of Insects
    1. The Insect Body
    2. Insect Development
  B. The Grasshopper
    1. Feeding, Digestion, and Excretion
    2. Senses, Nervous System, and Muscles
    3. Circulation and Respiration
    4. Reproduction and Development
  C. The Diversity of Insects
    1. Feeding Characteristics
    2. Appearance
    3. Reproduction
    4. Social Insects
    5. Insects and People

13.1 Phylum Arthropoda

Section Objectives

- List the characteristics of the Arthropoda.
- List the classes that belong to this phylum.
- State the differences between the two arthropod subphyla Chelicerata and Mandibulata.

A. Introduction

There are far more numbers of species in the phylum Arthropoda than in any other. These animals have a variety of means of locomotion, which include swimming, crawling, running, jumping, hitchhiking, and flying; some are sessile. Arthropods live in a variety of environments, including marine and freshwater, deserts, snow-covered regions, and even hot springs.

Insects, class Insecta, belong to one of the five classes of the Arthropoda. Almost 900,000 species of insects have been identified, and some entomologists estimate that there may be between 2-30 million that have neither been discovered nor described. The other four classes of arthropods are **Arachnida**, **Crustacea** (krus-tā-shē-a), **Diplopoda** (di-plō-pod), or millipedes, and **Chilopoda** (kī-lop-o-da), or centipedes.

In this chapter, you will learn about the characteristics, features, and diversity of the animals that belong to the **Arthropoda** of the Kingdom Animalia. Recall the classification scheme that is used in Biology to describe the hierarchy in a descending order towards the species level:

Domain → Kingdom → Phylum → Class → Order → Family → Genus → Species

The phylum Arthropoda contains about 1,170,000 described species. Some scientists have suggested that the arthropods account for 80% of all living creatures. Present-day arthropods include lobsters, spiders, scorpions, insects, mites, centipedes, and ticks, among others. Some organisms that are considered to be extinct, such as trilobites, are also included in this phylum. Figure 13-1 shows some members of this phylum.

There are many animals that belong to the Phylum Arthropoda; hence, they are divided into two subphyla, which are called the Chelicerata and the Mandibulata. The chelicerates (ki-lĭs-e-rāts) have pointed appendage-like mouthparts that are used to grasp food, and include arthropods such as the horseshoe crabs and spiders. The mandibulates have mouthparts that are used for biting, chewing, grasping, and cutting and holding food. These organisms include the crabs, centipedes, millipedes, and insects.

Biology For Life

Extinct and modern Arthropods
Figure 13-1

Trilobite

Crab

Butterfly

Millipede

Scorpion

B. Arthropod Characteristics

The name Arthropoda, in Latin, means "jointed foot." Legs or other movable extensions of the body are called **appendages**. Arthropods are characterized by having jointed appendages, a **segmented body** and an outer skeleton, or **exoskeleton**.

The main nerve cord of an arthropod is located on its ventral side, which is the same side that is closest to the digestive organs. This is in sharp contrast to that of humans, whose main nerve cord is on their dorsal surface and opposite of the digestive organs.

The arthropod has an open circulatory system with a tube-like heart on the dorsal surface. It pumps blood called **hemolymph** with rhythmic contractions. These contractions begin in the posterior half of the body and move the hemolymph towards the anterior half. In this system the hemolymph bathes the organs directly with nutrients and other important compounds. The hemolymph transports very little oxygen; that job is done by tracheae, or air tubes. The best way to understand the movement of hemolymph within insects is by the term "swishing around."

Muscular movements by the animal during locomotion can facilitate hemolymph movement, but diverting flow from one region of the body to another is limited. When the heart relaxes, blood is drawn back toward the heart through open-ended pores called **ostia**. This system is in contrast to the closed circulatory system. Closed circulatory systems, such as that which is found in humans, have the blood enclosed within vessels of different sizes and wall thicknesses. Furthermore, in closed circulatory systems, blood is pumped by a heart through vessels and does not normally fill body cavities. Both types of circulatory systems are shown in Figure 13-2.

An arthropod's skeleton is on the outside of its body. Therefore, this skeleton is called an exoskeleton. Recall that an organism with an endoskeleton has the skeleton within its body. The arthropod's exoskeleton provides support and protection for soft tissues. It also provides a site for muscle attachment. An arthropod's exoskeleton is made up of **chitin**, which is a strong, flexible

CHAPTER 13: **INVERTEBRATES: PART 2**

polysaccharide that is similar to cellulose. You may remember that a polysaccharide is formed of repeating units (either mono- or di-saccharides) joined together. If you took the dimensions of chitin for a given structure on an insect and then made that same structure with steel, you would find both of these have the same tensile strength. That is, our Creator gave insects an exoskeleton that is as strong as steel.

The arthropod's exoskeleton must be shed periodically in order for growth to occur. This process is called **molting**. After an arthropod molts, a short time is required before the new exoskeleton hardens, and during this time the animal grows significantly. However, having an exoskeleton does limit the maximum size to which the animal may grow, which is ultimately controlled by its genes.

All arthropods have flexible joints between body sections. These allow for movement and **flexion**, or **bending**. As mentioned above, arthropods have jointed appendages that extend outward from the main part of the body. A leg and an arm are two examples of appendages. Arthropods have pairs of muscles that enable joint movement. When one muscle contracts, the joint bends, and when the other muscle contracts, the joint straightens.

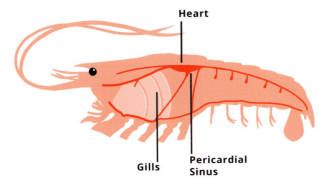

Open Circulatory System
Internal anatomy of a crayfish

Blood is pumped by a heart into the body cavities, where tissues and organs are surrounded by the blood

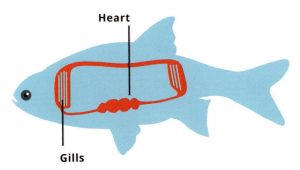

Closed Circulatory System
Internal anatomy of a fish

Blood is pumped by a heart through vessels, and does not normally fill body cavities

Figure 13-2. Open and Closed Circulatory Systems

Figure 13-3. A damselfly sheds its old skeleton in a process called molting. Note that the lower skeleton is now a cast skin, or exuvia. During this process the damselfly has re-absorbed 99% of the nutrients found in the outer tissue layers.

Biology For Life

Arthropods have segmented bodies; however, the segments are more highly specialized than the segments of the annelids (for example, earthworms or leeches). Many arthropods have segments containing appendages that function as jaws. Segments that are located just behind the head are often designed for locomotion, because they have paired limbs attached. The term tagma (plural tagmata) is used to denote a specialized grouping of segments with a common function, such as the head, thorax, and abdomen. You can see this segmentation in Figure 13-4.

The nervous system of arthropods consists of a brain and complex sense organs that are located in the head. The honey bee has the most brain cells of any arthropod, numbering around 80,000 cells. Arthropods have special organs that are used for the senses of touch, vibration, air pressure, temperature, humidity, and chemicals. Many arthropod eyes are very specialized. For example, a robber fly is an important predator, which is capable of seeing its prey, perhaps a grasshopper or wasp, at a distance of more than 45 feet.

Figure 13-4. A Tarantula Spider. Notice the jointed appendages and segmented body.

Section Review — 13.1

1. List the characteristics of an Arthropod.
2. The name Arthropoda in Latin means "_____."
3. What is molting?
4. Arthropods have special organs that are used for senses of what?

13.2 Subphylum Chelicerata

Section Objectives

- Describe the structure of an arachnid.
- Explain the uniqueness of reproduction in spiders.
- Name some arachnids other than spiders.

Like all arthropods, chelicerates have segmented bodies and jointed limbs, all of which are covered in an exoskeleton. The proper term for this covering is **cuticle**, a type of skin that is made of chitin and proteins. Chelicerate bodies are divided into two regions, the cephalothorax and abdomen. Mites differ from the other chelicerates in that they have no visible division between these sections. The chelicerae that give this group its name are the only appendages that appear before the mouth. In most chelicerate subgroups, the pincers function as gripping tools that are used in feeding. The chelicerae of spiders are known as fangs, which most species use to inject venom into their prey.

Typically, the Chelicerata within the Phylum Arthropoda contain the three classes: Arachnida (spiders, scorpions, mites, etc.), Xiphosura (horseshoe crabs), and Eurypterida (sea scorpions, extinct).

A. Characteristics of Chelicerates

There are two major parts to the chelicerate's body: the **cephalothorax** (se-fa-lo-thŏr-aks) and the **abdomen**. The cephalothorax is a fused section consisting of the head and thorax. As with the other arthropods, the brain and major sensory organs are located in the head, and the legs are found on the thorax. The abdomen consists of the posterior segments that contain most of the internal organs. Segmentation may or may not be present within each of these parts.

Most chelicerates have six pairs of appendages. The four pairs in the posterior of the cephalothorax are called the walking legs. The two anterior pairs are very specialized. The first pair is called **chelicerae** and aids the animal during feeding. They function like the jaws of other animals. The second pair of anterior appendages is the **pedipalps**. These are leg-like in structure and, depending on the kind of organism, perform a variety of functions that may include the manipulation of food. Often, the pedipalps serve as sensory receptors, picking up sensations of touch, taste, and smell. Pedipalps perform the same functions as the antennae of other arthropods.

B. Class Arachnida: Spiders

The spiders comprise the largest order of the class Arachnida, which includes the other orders such as the scorpions, ticks, and mites. The spider is composed of a cephalothorax and a segmented abdomen. Attached to the cephalothorax is one pair of chelicerae, one pair of pedipalps, and four pairs of walking legs. Like all arachnids, spiders have eight legs.

The Orb Weaver Spider

The chelicerae of spiders have piercing fangs with venom sacs that are designed to kill or immobilize their prey. All spiders have poison; however, only a few species pose a serious threat to humans. The brown recluse and the black widow spider are among the most dangerous spiders in North America. The brown recluse is capable of causing death. Most people are not aware of having been bitten by a brown recluse spider, but, rather, suspect the bite

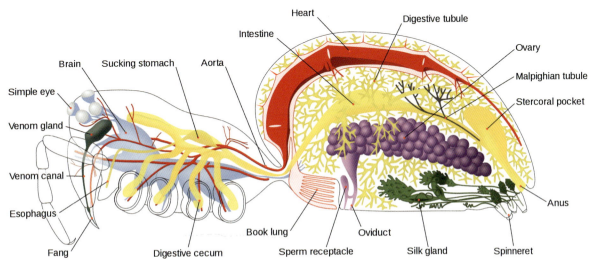

Figure 13-5. The major body parts of the spider

of a mosquito. In a matter of hours, the bite region grows in size and is accompanied by itching, pain, nausea, and vomiting. If a physician suspects that a patient has been bitten by a brown recluse spider, then emergency treatments are typically effective. The black widow rarely causes death but instead intense abdominal pain. Evidence for this comes from reports by women who have been bitten by the black widow and who previously have experienced labor pains during childbirth. They have said that the black widow bite causes more intense pain than that which they experienced during labor.

1. Digestion

Since spiders have no mandibles to chew food, they must digest food externally. The spider pumps digestive enzymes from its salivary glands directly into its prey. Enzymes then digest and liquefy the internal tissues of the prey. The resulting liquid-mixture is pulled into the esophagus by muscular contractions of the pharynx. Tubes branching from the stomach absorb food molecules into the hemolymph (the blood of the spider).

The mouth of the jumping spider can be seen in Figure 13-6.

Figure 13-6. Phidippus audax, jumping spider. The basal (the base of a stem) parts of the chelicerae are the two iridescent green mouth parts (shown in the zoomed detail). Recall that the chelicerae are the first anterior pair of appendages.

2. Silk Production

All spiders have silk glands in the abdomen. Silk is released from the body of the spider through nozzle-like openings called **spinnerets**. Many spiders make webs from their silk. Small insects become trapped in their sticky webs. Although all spiders make silk, not all spiders build webs. Some spiders use their silk to wrap their prey. Females often wrap their eggs in a protective silk case. Trap-door spiders use their silk to line their hiding places.

3. Respiration

Spiders are terrestrial organisms and have **book lungs** for gas exchange. This type of lung consists of folded membranes arranged in stacks, much like the pages of a book. With this arrangement, a large surface area of lung tissue is exposed to the air. The book lungs open to the environment through a single small passage to the outside, which is called a **spiracle** (spir-i-kal). In addition to book lungs, spiders also possess **tracheae**. Tracheae are tubes that bring oxygen close to the cells and circulating hemolymph of the spider. Air enters the tracheae through the spiracle in the abdomen.

4. Circulation and Excretion

In the circulatory system of the spider, hemolymph leaves the heart and moves through vessels to the body. Blood leaves the vessels and passes into open spaces around the tissues and organs. The excretory organs of the spider are called **Malpighian** (mal-pē-gē-an) **tubules**. These tubules have many branches and extend into the abdominal spaces. Nitrogen wastes are converted into uric acid and are eventually expelled from the body by the anus. Recall that nitrogenous waste in humans is urine. As we saw in Chapter 4, urine flows from the kidneys, through the ureter, to the bladder, and out through the urethra.

5. Nervous System and Senses

The nervous system consists of a brain and **ventral nerve cord**. Many spiders have well-developed sensory organs.

Most spiders have four pairs of **simple eyes**. A simple eye is a small organ that is light

sensitive and has a single lens; the human eye is an example of a simple eye. In the spider, three pairs are used for detecting light and brightness, both during the day and at night. The main pair of eyes that are located at the front of the head are capable of forming images. In many species, spider vision is very acute and perhaps better than that of dragonflies. This is somewhat surprising, as dragonflies and many other insects have compound eyes that may contain up to thousands of image-forming lenses known as **ommatidia**.

Spiders that build webs have a sensory organ in their legs that detects vibrations. When a trapped insect struggles to free itself from the web, it vibrates the web in the process. Hence these sensory leg organs help the spider know that its web has a meal waiting. A spider distinguishes vibrations from among its own young, another spider crawling on the web, or its trapped prey.

C. Reproduction in Spiders

1. Mating

The male spider produces sperm in the testes that are released through an opening in the abdomen. The male spider transfers sperm to the female by using its **pedipalps**. There is a small chamber in the female's

Figure 13-7. This jumping spider's main eyes (center pair) are very acute. The outer pair are "secondary eyes," and there are other pairs of secondary eyes on the sides and top of its head.

body called the **seminal receptacle**. After mating, the seminal receptacle holds the sperm cells within the female. Eggs are produced in the ovary of the female spider. When the ovaries release their eggs, the sperm too are released, which fertilizes the eggs.

Spiders have complex reproductive behaviors. In many spiders, the male is much smaller in size than the female. In some cases, female spiders are more interested in eating the males than in mating with them. To ensure successful mating and to avoid being devoured, males must approach females very carefully. Even if they are devoured during mating, the female will use the nutrients of the males to put into her developing eggs.

2. Courtship

In many species, the male spider offers the female a silk-wrapped insect. While the female is investigating this gift, the male mates with her and hastily retreats. In other species of spiders, the male strums the web in a special way, and the female, who recognizes these vibrations as special,

Biology For Life

allows the male to approach her. In species that do not spin webs, the male often performs an elaborate mating dance. After the eggs have been fertilized, the female usually places them in a silk wrapping until they develop into young spiders.

D. Other Arachnids

Scorpions are classified as a separate order within the Class Arachnida. The segments in the abdomen of the scorpion are distinct and not fused together like that of the spider. Accordingly, these moveable segments allow the scorpion greater flexibility.

Scorpions differ from spiders in a number of ways. The scorpion's chelicerae are small chewing structures and have no fangs or poison glands. The sting and poison gland of the scorpion is located in the last segment of its abdomen. Rather than serving a sensory function, the pedipalps of the scorpion are enlarged into pincers that grasp prey.

Perhaps scorpions are the most unique among the arthropods. They can withstand very high temperatures and drought, and live a very long time without eating. They eat a variety of organisms, including insects, other scorpions, and even mice. Following courtship and mating, the young hatch from eggs within a birth sack inside the mother; that is, they are born alive. The young scorpions instinctively crawl on to the back of the mother, who will protect them from between one to two weeks.

Figure 13-8. The tiny male of the Golden orb weaver (near the top of the leaf) is saved by producing the proper vibrations in the web. It may be too small to be worth eating by the female of its species, with whom it needs to mate.

The daddy longlegs, or harvestman, belongs to an order known as the Opiliones. Although daddy longlegs are arachnids, they are not true spiders. The cephalothorax and abdomen of these organisms are fused into a single rounded body. They do not have poison or silk glands. Daddy longlegs eat both plant and animal material, and they do not digest food externally.

Figure 13-9. A female scorpion carrying its young (white)

Mites are the most diverse of all arachnid orders. Most are less than one millimeter in length and are difficult to see. Mites are parasites of plants and animals, though many are scavengers. Some are pests of humans, including the **Demodex** mite, which lives in the hair follicles and sebaceous glands on the face of adult humans. A formidable but relatively harmless-looking mite is shown in Figure 13-10.

Ticks are larger than mites, but they have a similar body form. They are parasites of many terrestrial animals. Ticks attach to their host to feed on blood and eventually drop from their host. At some point, they will molt to another stage and wait for another passing host.

Most ticks find their hosts by climbing high on plants and grass and wait for passing hosts. This passive behavior is known as questing. They have a specialized grouping of cells or receptors that are located on the tip of their first leg. These are called the Haller's organ, and it is used to detect carbon dioxide that is emitted from potential hosts. As a potential host passes by a plant to which a tick is attached, minor air currents and vibrations stimulate the tick to a state of readiness. If the host brushes up against the tick or the tick detects carbon dioxide, it quickly attaches to the host.

Figure 13-11. A harvestman (a Phalangium opilio), showing the partially fused arrangement of abdomen and cephalothorax, which distinguishes these arachnids from spiders.

Ticks carry a number of human diseases, such as Rocky Mountain spotted fever, ehrlichiosis, and Lyme disease. People seem to have more fears concerning spiders and scorpions; however, mites and ticks may inflict far greater damage. The Middle East has one species of tick that has been reported to cause death by exsanguination, or severe blood loss. At least one account comes from the finding of two dead novice cavers, or spelunkers, in Israel. The two individuals ended their day by going to sleep in the cave. While they were sleeping, a large number of soft-bodied ticks came out of their hiding places. The ticks normally feed on bats but began to feed on the spelunkers. Because ticks normally feed without producing pain or itching, they fed and engorged themselves on the sleeping individuals.

Figure 13-10. The house dust mite, a common cause of asthma and allergic symptoms worldwide. It has been suggested that one gram of house dust contains 75 house dust mites. A typical house dust mite measures 420 micrometers in length and 250 to 320 micrometers in width. Both male and female adult house dust mites are creamy blue and have a rectangular shape. Dust mites can be transported in dust bunnies or airborne by minor air currents generated from normal household activities.

Section Review — 13.2

1. The two major parts of a chelicerate's body are the _____ and the _____.
2. Describe how a spider obtains food.
3. What are spinnerets?
4. How does a scorpion differ from a spider?
5. What is Haller's organ?

Biology For Life

13.3 Subphylum Mandibulata

Section Objectives

- Identify and state the function of the crayfish body parts.
- Describe how millipedes and centipedes differ from other arthropods and from each other.

The second subphylum of arthropods includes the mandibulates. This group of organisms has **mandibles**, or the mouthparts for chewing food. They also have **maxillae**, which are used for holding food and passing it to the mandibles and mouth. All mandibulates have **antennae**, which are segmented sense organs located on the head. The mandibulates also have three or more pairs of walking legs.

Insects are the largest class of mandibulates and will be discussed later in this chapter. There are three other major classes of mandibulates: the **crustaceans**, the **diplopods**, and the **chilopods**.

A. Class Crustacea: Crayfish, Lobsters, and Shrimp

The crustaceans include many animals used by humans for food, including crayfish, lobster, crab, and shrimp. *Daphnia*, the water flea, and barnacles are also considered crustaceans. Most crustaceans are marine animals; however, some live in fresh water. Sow bugs (woodlice, roly-polies, pillbugs) are the only terrestrial crustaceans that exist. They need moisture because they breathe through gills called pseudotrachea. They are usually found in damp, dark places, such as under rocks and logs. Roly-polies are easy to keep in terrariums. They require moist (not wet) soil to walk on and potato peelings for food. Typically, they congregate underneath these peelings.

The crayfish is a typical crustacean, with a cephalothorax and an abdomen. A single portion of exoskeleton known as a **carapace** covers the cephalothorax. In addition to chitin, the exoskeleton of crustaceans contains calcium and other minerals. The minerals make the shell very hard and inflexible. A crayfish is shown in Figure 13-12 – can you see the carapace?

1. Appendages

Crustaceans have paired appendages that are attached to each segment. The two pairs of antennae are found on the most anterior part of the animal. The mandibles are located right behind the antennae. Two pairs of maxillae are found slightly behind the mandibles. Slightly below the mandibles, there are three pairs of **maxillipeds**, which serve as sense organs. These also help pass food to the mouth.

The large pincer-bearing **chelipeds** (kē-lĭ-ped) are the next appendages. The crayfish uses its chelipeds to obtain food and to protect itself from predators.

Four pairs of walking legs lie directly behind the chelipeds. The appendages called swimmerets, which aid in swimming, are located on the abdominal

Figure 13-12. The Crayfish. Note the two plates just behind the head, which form the carapace.

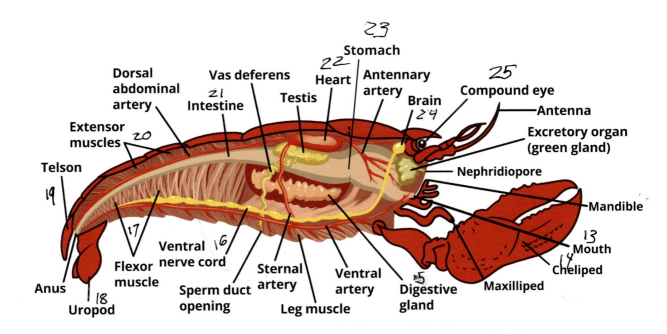

Figure 13-13. Diagram of the internal and external structure of a crayfish

segments of the crayfish. The swimmerets hold the fertilized eggs on the female crayfish. There are several flattened appendages called **uropods** (yùr-ō-pod) at the posterior end of the crayfish. The uropods lie on both sides of the flattened tail, called a **telson**, and are used for swimming.

2. Feeding and Digestion

Crayfish are scavengers, and they also feed on small animals such as snails, worms, and leeches. The mandibles function to shred food into small pieces. The esophagus, which leads to the stomach, has toothlike structures that are made of chitin. These toothlike structures grind food into fine particles. Digestion and absorption of food take place in the large digestive glands, which are located on both sides of the stomach.

3. Circulation, Respiration, and Excretion

The gills of the crayfish are located in the thorax. Water flows over the gills as the crayfish moves, and gases diffuse between the water and the blood. Crayfish have an open circulatory system, which consists of a heart and several arteries. Blood is pumped by the heart to the arteries and flows throughout the crayfish. Blood travels to the organs and other tissues and returns to the gills. Similar to fish, the gills function to provide gas exchange by absorbing oxygen and releasing carbon dioxide. The oxygenated blood then moves to a space around the heart, known as the pericardial sinus. Blood then enters the heart through the ostia. Crayfish excrete nitrogen wastes from two glands, known as **green glands**, which are located in the head. These glands filter wastes from the blood in the body cavities. The wastes pass through pores to the outside of the body.

4. Senses

The crayfish has a brain and a complex nervous system that receive stimuli from a variety of sense organs. The antennae are sensitive to both touch and chemicals in the water. The entire crayfish body is covered with fine, hairlike projections, which are sensitive to touch and to vibrations in the water. The crayfish has two large **compound eyes**. A compound eye is made up of many individual light-

Biology For Life

CHAPTER 13: INVERTEBRATES: PART 2

Figure 13-14. The Rusty Millipede showing the abundance of legs typical for many millipedes. Despite their name, these creatures do not have a thousand legs, although members of the rare species *Illacme plenipes* can have up to 750. Common species of millipedes have between 36 and 400 legs.

sensitive units. The crayfish's compound eyes sense motion and, perhaps, see crude images.

The crayfish has a pair of organs called antennules that it uses for balance. These are located near the base of the antennae. Antennules are small spherical cavities lined with many nerve endings. The crayfish inserts a grain of sand into each balance organ, and the sand is pulled down by gravity. The crayfish receives information about the position of its body from the nerve endings that are touched by the sand grain. As we learned in Chapter 5, it is the semicircular canals in our ears that help us maintain our balance.

5. Reproduction

Crayfish have separate sexes, and they reproduce sexually. During mating, sperm is deposited by the male into a receptacle within the female's body known as the spermatheca. The sperm stay there and are released when the eggs leave the ovary. Each egg is then fertilized as it leaves the female's body. The eggs stick to the swimmerets and stay there until they hatch.

B. Class Diplopoda: Millipedes

The term millipede means "thousand feet"; however the millipede does not literally have 1,000 feet. Some millipedes do have several hundred legs. A millipede has two groups of simple eyes, a pair of antennae, mandibles, and maxillae on its head.

The first four segments behind the head each bear a single pair of legs. The abdominal segments are actually two segments fused together. The abdominal segments have two pairs of legs per segment. Millipedes range in length from two millimeters to thirty centimeters (300 millimeters). There may be anywhere from ten to over 100 segments in a millipede. The millipede is an herbivorous organism, so you can go ahead and gently pick one up – it will not hurt you.

Along some woodland trails of the eastern United States, it is very likely that you may encounter a yellow and black millipede that is 1-1/2 to 2 inches long. Its name is *Pleuroloma flavipes*, and it defends itself by emitting cyanide when it is disturbed. If you happen to handle this species and smell your hand afterward, your hand will have a faint smell of cyanide, which has the aroma of almonds. It may have a pleasant fragrance, but this smell is repellent for many would-be predators.

C. Class Chilopoda: Centipedes

The term centipede means "hundred feet." Centipedes vary in size, ranging from a few centimeters to thirty centimeters in length. They may have between fifteen and 150 pairs of legs. The main difference between millipedes and centipedes is that

Figure 13-15. A centipede. The segmentation is clearly discernible, as are the appendages. Can you count them?

centipedes have simple, unfused body segments. Each segment of a centipede has only one pair of legs.

Centipedes are carnivores. They have a pair of anterior appendages called poison claws or fangs that are used to capture and subdue prey. Although centipedes have a reputation for being dangerous, only the very largest of the tropical centipedes may seriously harm people. The centipedes found in North America are harmless.

One species of centipede is commonly found in houses. It scares many people but is harmless. In fact, it is quite beneficial, as its diet consists of insect and spiders. If you happen to find some of these centipedes in your house, you may see them displaying a pet-like schedule. Day after day, you can see them showing up in the same locations, even at the same times. Large house centipedes can bite, but only if roughly handled.

> **Section Review — 13.3**
>
> 1. What are the three major classes of mandibulates besides insects?
> 2. A single portion of exoskeleton known as a _____ covers the cephalothorax.
> 3. What does the term "millipede" mean?
> 4. What is the main difference between centipedes and millipedes?

13.4 Class Insecta

Section Objectives

- Identify the major characteristics of the insects.
- Compare and contrast complete and incomplete metamorphosis.
- Name the external structures on the body of the grasshopper.
- Identify and describe the function of the feeding, digestive, and excretory structures of the grasshopper.
- Describe the circulatory and respiratory systems in the grasshopper.
- Describe the reproduction and development of a grasshopper.
- Give examples of how insect mouthparts are adapted to different food sources.
- List an example of camouflage and mimicry in insects.
- Describe some reproductive adaptations of insects.
- Explain the difference between social insects and other insects.
- Describe the different types of bees that exist in a bee colony (that is, worker, queen, drone).
- List ways in which insects are helpful and harmful to humans.

The Class Insecta forms the most diverse group of animals on the planet. The number of living species is in the millions, with almost one million species having been described. Insects represent more than half of all known living organisms and account for over 90% of the differing life forms on Earth. It has been suggested that if you took all of the people on the Earth and all of the insects and placed these on opposite sides of a balance, they would outweigh us by more than 12 times! Insects may be found in nearly all environments except the oceans. The arthropods that live within the ocean are crustaceans.

A. Characteristics of Insects

The insect is able to live in almost every habitat imaginable. Insects live in and on the soil, as well as within and on top of fresh water. Insects are found in, on, and around plants and animals.

CHAPTER 13: **INVERTEBRATES: PART 2**

Examples of Insects

Dragonfly

Yellowjacket Wasp

Western Honey Bee

Figure 13-16. Some examples of insects. Notice the clear three body sections of the yellowjacket wasp.

In His infinite wisdom, God has taken basic designs and exploited these to their maximum possibilities. Take the butterflies and moths, for example. These consist of a head, thorax, and abdomen. Nearly all adult forms have wings. However, some species lack wings or have wings but cannot fly (female gypsy moth). Butterflies and moths exhibit all the colors of the light spectrum. Most butterflies and moths display colors through the use of pigments, but many show color through the principle of structural interference. In these butterflies, light passes through the wing membrane, where it bounces off or is reflected from small crystals. Some colors are absorbed, and others are reflected. What we see are the reflected colors.

Perhaps the best examples of omniscience and infinity come from the wings. Wings come in a range of sizes and shapes: short and narrow, short and pointed, short and feathery, long and narrow, long and wide, long and pointed, etc. Some have tails, and some do not. All of these different wing sizes and shapes indicate unique flight characteristics for each species. Indeed, an experienced entomologist can tell the Family a butterfly belongs to by simply watching its flight behavior. The point is worth repeating. There is a single unique theme with an almost infinite number of variations. You could ask the world's best graphic artist to sketch all of the possible butterfly and moth wings in the world, both known and unknown, and the artist would not be able to accomplish this task in a single lifetime.

1. The Insect Body

The insect exoskeleton is made of chitin. Insects have jointed legs and a segmented body. However, insects have three major body sections, unlike most arthropods, which have two major sections.

These three divisions are the head, thorax, and abdomen. All insect heads have mouthparts, one pair of antennae, and eyes. Insects usually have a number of simple eyes and a pair of compound eyes. In general, the thorax contains three pairs of legs as well as two pairs of wings. The thorax contains muscles that operate the legs and wings, making locomotion possible. Insects are the only arthropods that have wings. The insect abdomen contains several internal organs, including those of reproduction. All insects have female and male sexes. The insect's thorax and abdomen also contain spiracles. Air enters the insect's body through the spiracles, and the oxygen is delivered to the cells of the insect by means of the tracheae.

2. Insect Development

In order to grow, insects must molt. The insect passes through several distinct stages during its development. **Metamorphosis** is the term used to describe the process of developing into an adult by progressing through different structural stages. Some insect species, such as dragonflies, grasshoppers, and aphids, have a three-stage or **incomplete metamorphosis**, as shown in Figure 13-17.

There are three stages of incomplete metamorphosis: the egg, nymph, and adult. The immature form is called a **nymph** and hatches from the egg. Nymphs resemble small adults; however, they do not have wings or reproductive organs. After passing through a series of molts, the nymph reaches its adult form and size. Recall that we saw a stage of molting in Figure 13-3. Which stage of incomplete metamorphosis was shown in the figure?

Insects such as butterflies, beetles, and ants undergo a four-stage or **complete metamorphosis** (see Figure 13-17). The four stages of complete metamorphosis are the following: egg, larva, pupa, and adult. The larva hatches from the egg. Insect larvae have segments and look much like worms. They are specialized for eating. The larval form does not have wings or reproductive organs. After completing several molts, the larva becomes a **pupa**. During the pupal stage, the body of the

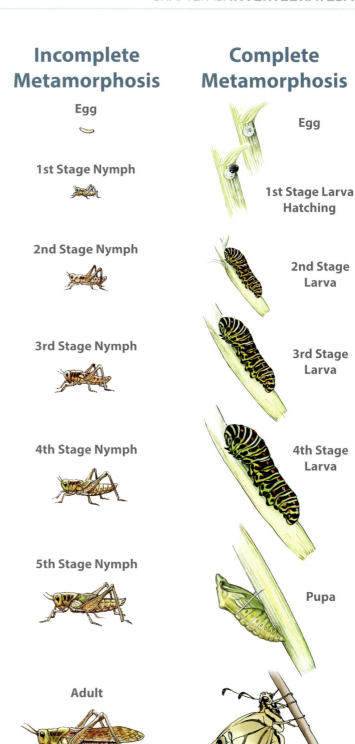

Figure 13-17.

larva is slowly transformed into an adult. When the adult emerges from the cocoon or other casing, it has wings and all of its reproductive organs.

In both "incomplete metamorphosis" and "complete metamorphosis" there is a stage of completion when the organism reaches the adult stage. Therefore the term "incomplete" may seem misleading. It is incomplete only in the sense that it has three stages and not four.

Figure 13-18. The Grasshopper

B. The Grasshopper

The grasshopper is a representative insect species. In the next section, the various features and characteristics of the grasshopper will be explained in detail.

The grasshopper, cricket, and locust all belong to the order Orthoptera, which includes 10,000 different species. These insects have enlarged rear legs that are specialized for jumping. It has been suggested that if we had the muscles and leg structure of a grasshopper, we would be able to jump the length of a football field in one leap! Figures 13-19 and 13-20 show the body parts of the grasshopper, some of which will now be discussed.

The head of the grasshopper contains the brain and sensory organs, which include a single pair of antennae, three simple eyes, and a pair of compound eyes. The chewing mouthparts are also on the head.

The thorax is divided into three segments, with a pair of legs attached to each segment. Two pairs of wings are also attached to the thorax. The first pair of wings is thickened, in order to protect the second set of wings that lies underneath.

The abdomen of the grasshopper is divided into ten segments. A round membrane, called the **tympanum**, is an organ used for hearing and is located within the first segment. Each abdominal segment has a pair of spiracles, along with the two posterior segments of the thorax. The spiracles open into the tracheae.

1. Feeding, Digestion, and Excretion

The mouthparts of the grasshopper are designed for its plant diet. The **labrum** is a flap of exoskeleton that covers the other mouthparts. It helps to hold food between the biting and chewing mandibles. One pair of maxillae, which holds food for chewing, is located behind the mandibles. The **labium**, which is located behind the maxillae, also holds and manipulates food. The **sensory palps** are organs that taste food, and they are found on the maxillae and labium.

The grasshopper has a complete one-way digestive tract that begins at the mouth. Chewed food passes

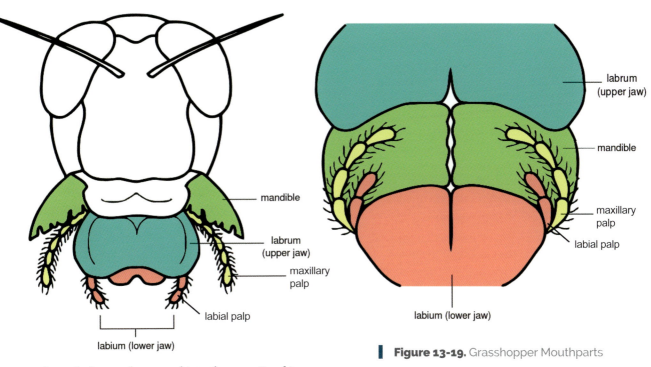

Figure 13-19. Grasshopper Mouthparts

through the esophagus and into the crop. Food is stored temporarily in the crop. Behind the crop lies the gizzard. It has hard, chitinous teeth for grinding food. Next, the food passes to the stomach, or **midgut**. Food is broken down in the midgut by enzymes. Many pouches, called **gastric caeca** (sing. caecum), branch from the stomach and extend into the body space. These increase the surface area for digestive enzymes to accomplish their task, as well as the ability to extract useful nutrients. Molecules of digested food move through the lining of the caeca and into the body cavity. Solid indigestible wastes pass through to the hindgut.

Like spiders, grasshoppers have **Malpighian tubules**, which remove nitrogenous wastes from the body. These tubes extend into the abdominal cavity. Nitrogen wastes leave the hemolymph and enter these tubules. As water is removed from these wastes, they begin to solidify. More water is removed in the rectum, and solid droppings consisting mostly of uric acid are eventually expelled from the body through the anus.

2. Senses, Nervous System, and Muscles

The large compound eyes of the grasshopper are able to detect shapes and movement and discern crude images. The smaller simple eyes detect changes in brightness and daylength.

The exoskeleton of the grasshopper is covered with tiny sensory hairs called setae, which detect slight touches. Setae are particularly numerous on the antennae. The antennae contain receptors and nerve cells that function to detect important airborne chemicals. This is the grasshopper's sense of smell.

The nervous system of the grasshopper consists of a brain and a double nerve cord, which extends ventrally into the body. The nerves branch to all body parts from the brain, and ganglia are found within each body segment. Nerves carry messages to various body parts, as well as to the muscles.

Pairs of muscles bend the joints of the exoskeleton in either direction. When one muscle contracts, the other muscle relaxes. The flight muscles move the wings by changing the shape of the thorax.

CHAPTER 13: INVERTEBRATES: PART 2

3. Circulation and Respiration

The grasshopper has an open circulatory system. The heart is a simple tube that is located on the dorsal (back) side of the abdomen (see Figure 13-20). The **aorta** is a vessel that extends forward from the heart to the head. The heart pumps hemolymph, or blood, through the aorta, and then it flows from the aorta into the body cavity. Hemolymph travels toward the posterior (rear) end of the grasshopper and finally returns to the heart. The hemolymph of the grasshopper is a clear, watery fluid that transports nutrients, hormones, waste molecules, and some gases.

Insects obtain oxygen by means of a tracheal system that opens to the outside of the body through spiracles. The tracheae of grasshoppers open into large, balloon-like **air sacs**. The air sacs allow the grasshoppers to move large amounts of air through the tracheal system. The grasshopper breathes by contracting and expanding its abdomen and by opening and closing its spiracles.

4. Reproduction and Development

The male grasshopper produces sperm in paired testes. During mating, the sperm cells travel through tubes to the penis. The male uses the penis to deposit sperm in the reproductive tract of the female. The female stores the sperm in a **seminal receptacle** known as the spermatheca.

Eggs develop in the ovaries of the female. Before fertilization can occur, the eggs leave the ovary and travel through tubes called **oviducts**. As these pass through the oviduct, the spermatheca releases sperm, each of which fertilizes an egg inside the uterus. These fertilized eggs pass to the outside of the grasshopper's body through the **ovipositor**, a hollow egg-laying organ. When the grasshopper is ready to deposit her eggs, the well-musculated ovipositor is used to dig a burrow into the ground. When this burrow has been sufficiently deepened, she deposits her eggs into it. Typically, grasshoppers mate in late summer and early autumn. The nymphs emerge in the spring.

C. Diversity of Insects

Insects exhibit a variety of characteristics. There are differences in body structure, the number and structure of wings, and the types of mouthparts. Another difference is the type of metamorphosis that an insect undergoes. Insects differ in behavior; some live as individuals, while others live in social groups. Various features of insects help to account

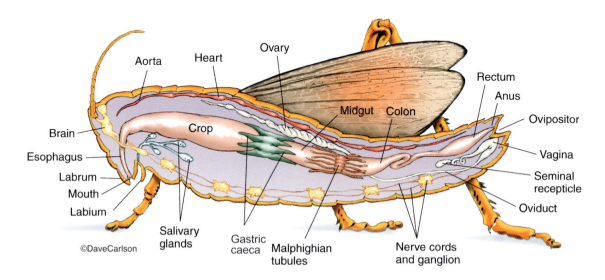

Figure 13-20. The Internal Structure of a Grasshopper

Biology For Life

CHAPTER 13: **INVERTEBRATES: PART 2**

for the large numbers that exist. We will discuss five of these features below.

1. Feeding Characteristics

The mouthparts of insects are designed for many different functions. For example, the mouthparts of the mosquito are a pair of hollow tubes used for puncturing the skin of an animal and sucking its blood. Insects that are predatory also use a variety of structures to help them capture prey. The praying mantis has forelegs that rapidly extend outward to ambush and capture other insects. Wasps use venom glands and stingers in their abdomen to subdue and paralyze their prey.

2. Appearance (as a defense mechanism)

Some insects use **camouflage** for protection against predators. Camouflage allows the insect to blend into its environment. Walking stick insects look like twigs and cannot easily be seen if they are on a small branch. You may have seen this uniquely designed insect near your home or in an insect zoo. This type of camouflage is called cryptic camouflage.

The praying mantis is the same shade of green as the leaves on which it lives, and so it blends in with its environment, making it hard to see. Katydids typically have detailed wings that look like leaves, even having leaf-like veins or venation. Our Creator spares no details! Three such insects using this form of camouflage are shown in Figure 13-21.

In some cases, insects benefit by attracting attention. For example, it is a general rule of thumb that brightly colored red and orange insects are distasteful, poisonous, or harmful to predators. Predators avoid such insects, either by instinct or by learning from one bad experience. This type of bright body coloring is called warning coloration. For example, ladybird beetles (ladybugs) produce repellent compounds that render them inedible. These beetles are usually brightly colored and spotted. Three insects that protect themselves by warning coloration are shown in Figure 13-22.

Mimicry is an insect characteristic in which one organism closely resembles another organism or an

Brimstone Butterfly

Geometer Moth

Grasshopper

Figure 13-21. Some insects protect themselves by blending in with their environment. Notice how hard it is to see these insects against their background.

Biology For Life

Figure 13-22. Some insects protect themselves by advertising. The monarch butterfly is foul-tasting and poisonous; the palm beetle discourages predators with a chemical discharge; and the yellowjacket is a ferocious adversary of its predators.

object in its environment. A predator will avoid the mimic organism because the mimic closely resembles a harmful species. The viceroy butterfly mimics the monarch, because birds avoid the monarch, which is very distasteful to the bird. Figure 13-23 shows the viceroy butterfly, which, as you can see from Figure 13-22, mimics the inedible monarch butterfly.

3. Reproduction

Insects have a short lifespan, but, despite this fact, they are very successful at producing large numbers of offspring. Insects have a number of features that allow males and females of the same species to recognize and attract one another. In order to attract males, many females release airborne chemicals called **pheromones**. These chemicals are very powerful. For example, the female Polyphemus moth is capable of attracting a male moth of the same species from a distance of up to five miles away. I once recall hatching a female Promethea moth from a cocoon in late spring. After several hours spent watching her unfold and harden her wings, we released her into our backyard. She flew into our maple tree and within less than 2 minutes she had "called" a male, which flew right next to her, with her pheromones. Some insects lay eggs that develop asexually into adults. The development of eggs without the process of fertilization is known as **parthenogenesis**.

Figure 13-23. Some insects protect themselves by mimicry. The viceroy butterfly mimics the inedible monarch butterfly; the sesiid moth and flower fly mimic hornets and wasps, which threaten potential predators.

4. Social Insects

Termites, ants, and most bees and wasps live in societies. An insect **society** is a group of insects that live together and exhibit a division of labor. Different individuals within the colony perform different tasks. The tasks of social insects include nest building, caring for the young, obtaining food, defending the colony, and reproducing.

The best-known social insects are the honey bees. There are 40,000 to 80,000 bees in a typical beehive. Figure 13-24 shows one lone honeybee doing work for its colony.

Each colony has three types of bees: the queen bee, worker bees, and drones. There is one **queen bee** per hive, and her job is to lay eggs. She also has two other important roles that stem from the pheromones that she produces. The pheromone prevents the worker bees from developing their own eggs, and it promotes colony cohesion. If something happened to the queen, chaos would erupt in the beehive until a new queen was produced.

The worker bees care for the young, gather food, and maintain the hive. The purpose of the male bees, which are called **drones**, is solely to mate with the queen bee. At the end of the growing season, when bees no longer collect pollen and nectar, the drones are "kicked out" of the hive. This is done primarily to ensure that the beehive has sufficient food resources to be used by the worker bees and their developing brood throughout the winter.

A new honey bee colony is formed when the queen bee leaves the colony with a large group of worker bees in a process called **swarming** (see Figure 13-26). Shortly thereafter, the old colony gains a new queen when an immature queen bee matures, leaves the hive, and mates with up to a dozen drones during a mating flight. The sperm received during this single mating flight will be used to fertilize all of the eggs that she lays in her 5-year lifetime. The queen then returns to the hive to begin laying eggs.

Figure 13-24. Honey Bee Collecting Pollen

The honeycomb is made of six-sided chambers called cells. The queen lays an egg in each cell. Larvae hatch from the eggs and feed on pollen that is provided by worker bees. When mature, the larvae molt and become pupae. After the pupal stage, the new bees emerge from their cells and join the colony. Many of the empty cells of the honeycomb are also used to store honey and pollen.

The sex of the bee is determined by fertilization. Drones develop from unfertilized eggs, and females develop from fertilized eggs. A queen bee develops from a worker larva that is fed a substance called royal jelly. It is rich in glucose and is secreted by glands within the mouths of workers. Female larvae that are not fed this royal jelly will become worker bees.

5. Insects and People

Many insects, especially caterpillars and beetles, cause damage to plants that are important to agriculture. Each year, the potato beetle, the European corn borer (caterpillar), and the boll weevil (beetle) do extensive damage to agricultural crops.

Some destructive insect species have inadvertently been brought to new locations where they have no natural enemies. For example, the gypsy moth was accidentally released into the United States in the 1800s. These insects strip leaves from thousands of square kilometers of forests each year.

Some insects, on the other hand, are beneficial to humans. For example, honey bees are raised for the honey that they produce. However, their most important activity is that of pollination. In any given fruit orchard, honey bees are indirectly responsible for up to 75% of the fruit that will be produced. Say, for example, that the typical harvest for an apple tree is ten bushels. In the absence of honey bees, the yield might only be 2 ½ bushels. For this reason, managers of orchards often have large numbers of beehives delivered by trucks when the trees are flowering. Believe it or not, this is an ancient practice. Even the ancient Egyptians transported beehives up and down the River Nile to take advantage of the bees' pollination abilities.

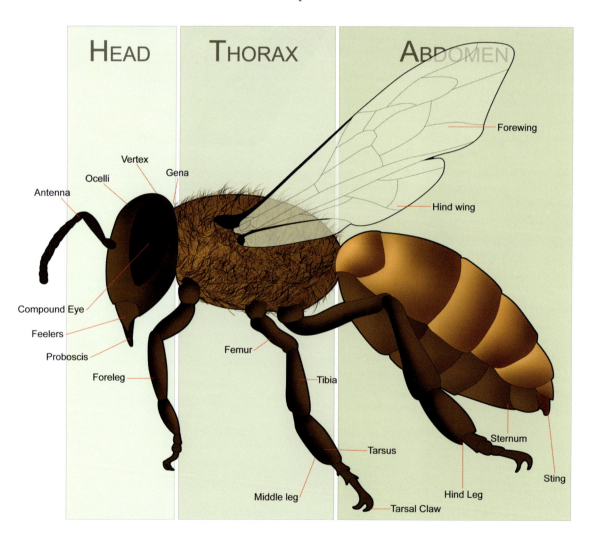

Figure 13-25. The Structure of the Worker Bee. Note the three distinct body sections characteristic of insects.

CHAPTER 13: **INVERTEBRATES: PART 2**

Some insects also eat harmful insects. For example, hornets and yellowjackets from one nest may capture hundreds of insects per day, including flies, crickets, caterpillars, and other pest species. The insects are chewed into a pulpy mass and fed to the hornet or yellowjacket larvae. This past spring, we had a developing hornet nest on our garage. We decided to leave it undisturbed until it became a problem. The back of the nest was attached to the movable garage door. Every time the door opened, it ripped open the back of the nest. While it was open, we could approach the nest slowly (no closer than 2 feet), and I would show my daughters how the worker hornets tended their larvae. The hornets eventually adapted to this situation and sealed off the back of the nest. In any case, the nest continued to grow until the queen died. Perhaps you will think twice about destroying hornet and yellowjacket nests, unless they are interfering with your ability to freely move around. Remember that all of God's Creation has been described as good.

Figure 13-26. A Honey Bee Swarm. Swarming is the way in which a colony reproduces itself. A majority of workers take the queen from the hive to seek out a new location in order to establish a new colony. A number of workers remain in the old hive with new queens that have emerged from their cells.

Section Review — 13.4

1. What is metamorphosis?
2. What are the stages of incomplete metamorphosis? What are the stages of complete metamorphosis?
3. What are some defensive mechanisms that some insects possess?
4. The development of eggs without the process of fertilization is known as _____.
5. What are some examples of social insects?
6. What do the tasks of social insects include?
7. How is a queen bee determined?

Biology For Life

Chapter 13 Supplemental Questions

Answer the following questions.

1. Know and be able to label Figures 13-5, 13-13, and 13-20.
2. Name the five classes of arthropods and give examples from each.
3. List the characteristics of arthropods.
4. Name the two arthropod subphyla and state the differences between them.
5. Describe the structure of an arachnid.
6. Discuss at least three uses of silk.
7. State the basic purpose of the malpighian tubules in a spider.
8. Explain the uniqueness of reproduction and digestion in spiders.
9. Name some arachnids other than spiders.
10. List four or five facts about ticks and mites, one of which should be a comparison to spiders and scorpions.
11. Identify six types of crustaceans.
12. Identify and state the function of seven of the body parts of a crayfish.
13. Describe how millipedes and centipedes differ from other arthropods and from each other.
14. Identify the major characteristics of insects.
15. Compare and contrast complete and incomplete metamorphosis.
16. Name the external structures of the grasshopper's body.
17. Identify and describe the function of the grasshopper's feeding, digestive, and excretory structures.
18. Describe the circulatory and respiratory systems in the grasshopper.
19. Describe the reproduction and development of a grasshopper.

Chapter 13 Supplemental Questions

20. Give examples of how insect mouthparts are adapted to different food sources.

21. List an example of camouflage and mimicry in insects.

22. Describe some reproductive adaptations of insects.

23. Explain the difference between social insects and other insects.

24. Describe the different types of bees that exist in a bee colony (that is, worker, queen, drone).

25. List ways in which insects are helpful and harmful to humans.

26. Study all diagrams in this chapter. Be able to draw and label all diagrams.

CHAPTER 14
Vertebrates: Part 1

Chapter 14
Vertebrates: Part 1

Chapter Outline

14.1 Introduction

14.2 Fish
    A. Class Agnatha: the Jawless Fish
    B. Class Chondrichthyes: the Cartilage Fish
        1. Skeleton and Movement of the Shark
        2. Shark Skin and Teeth
        3. Gas Exchange in the Shark
        4. Temperature Regulation in Fish
        5. Shark Reproduction
        6. Rays and Skates
    C. Class Osteichthyes: The Bony Fish
        1. Digestion in the Perch
        2. Excretion in the Perch
        3. Circulation and Gas Exchange in the Perch
        4. The Nervous System of the Perch
        5. Senses of the Perch
        6. Reproduction in the Perch
        7. Adaptations of Bony Fish

14.3 Amphibians
    A. Characteristics of Amphibians
        1. From Water to Land
    B. A Typical Amphibian: The Frog
        1. External Anatomy of the Frog
        2. The Frog Skeleton
        3. Digestion in Frogs
        4. Excretion in Frogs
        5. The Frog's Circulatory System
        6. Frog Respiration
        7. The Nervous System of the Frog
        8. The Frog's Senses
        9. Reproduction and Development of Frogs
    C. Other Amphibians
        1. Order Urodela: Salamanders
        2. Order Apoda: Caecilians

14.1 Introduction

In this chapter, you will learn about the fish and amphibians. These animals are vertebrates and belong to the Phylum Chordata. The spinal cord of the chordates is encased and protected in a backbone, which is positioned near the dorsal surface of the animal. The presence of a backbone in the vertebrates easily distinguishes them from the invertebrates. Recall the classification scheme that is used in biology to describe the hierarchy in a descending order towards the species level:

Domain → Kingdom → Phylum → Class → Order → Family → Genus → Species.

14.2 Fish

Section Objectives

- List an example of the Cyclostomata and describe its characteristics.
- Describe the characteristics of the Chondrichthyes.
- Explain the replacement of teeth in the shark.
- Contrast the shark's gills with those of bony fish.
- State how the internal temperature of fish is determined.
- Differentiate between shark reproduction and reproduction in other types of fish.
- List other members of the Chondrichthyes in addition to sharks.
- List the types of fins of bony fish and state their functions.
- Describe the functions of the perch's digestive and excretory organs.
- Discuss the pathway of blood flow in a bony fish.
- Explain the function of the gills and swim bladder in a fish.
- Name the major portions of the perch brain and state the function of each part.
- Describe the sense organs of fish.
- Detail the process of reproduction in a bony fish.
- Describe several adaptations of bony fish.

Biology For Life

Pacific Hagfish

Sea Lamprey

Figure 14-1. The Jawless Fish: Hagfish and Lamprey. The lamprey's mouth is modified to serve as a sucker-like device.

The design of a fish's body makes it well-adapted for its watery environment. The smooth contours of their bodies and the presence of fins instead of legs allow the fish to move easily and rapidly through water. Many types of fish live in large groups, known as "schools." This schooling behavior provides a source of protection for the group. For example, when one fish in the school becomes injured, the group may sense danger and move away. Also, an entire group of fish presents a more intimidating presence for potential predators.

A. Class Agnatha: The Jawless Fish

The class of jawless fish is also called the Cyclostomata, and sometimes referred to as the **cyclostomes**. The jawless fish, the hagfish and lamprey, are parasites and scavengers. You can see both in Figure 14-1.

The lamprey uses its mouth to attach to the bodies of other fish – see Figure 14-3. The file-like tongue of the lamprey cuts through the scales and skin and feeds on the blood of its host.

The lamprey has simple structures. For example, this fish does not have paired fins, as do most fish. The lamprey swims by snakelike movements of the body. It has multiple gill slits. The skeleton of the lamprey is a flexible, rod-like structure called the **notochord**. It is found along the dorsal surface of all vertebrates during their embryonic stage. The notochord is eventually replaced by a backbone, or **vertebral column**, when vertebrates reach the adult

Figure 14-2. The mouth of the lamprey

Figure 14-3. Lampreys attached to a Lake Trout

CHAPTER 14: **VERTEBRATES: PART 1**

stage. The lampreys are an exception and do not develop a backbone.

B. Class Chondrichthyes: the Cartilage Fish

The **Chondrichthyes** are jawed fish, which have an endoskeleton that is composed of cartilage. The best-known members of this group are sharks, skates, and rays. These fish are predators and have a movable jaw for grasping, chewing, and crushing their prey.

The shark is a representative member of the Chondrichthyes. The gray reef shark is shown in Figure 14-4. Notice the sleek, graceful look of this creature of God. It almost appears harmless, or at least non-threatening.

1. Skeleton and Movement of the Shark

The endoskeleton of vertebrates is initially formed from cartilage. The cartilage skeleton is eventually replaced by bone during maturation into adulthood. However, adult Chondrichthyes retain their cartilage skeletons for their lives.

An **endoskeleton** is an internal support structure of an animal. It may function purely for support (as in the case of sponges), but it typically serves as an attachment site for muscles. Appendages and other structures are attached by muscles, of which the opposing ends are attached to the skeleton. This arrangement makes possible a mechanism for transmitting muscular force, either by contracting or extending the muscles.

The cartilaginous endoskeleton provides flexibility for the shark. Sharks have two sets of paired fins (see Figure 14-5). Two **pectoral fins** are located in the front of the fish, directly behind the gills. Farther back and almost in the middle of the fish are the **pelvic fins**. These two sets of paired fins allow the shark to make turns or move up and down in the water column. The large tail fin pushes from side to side and propels the fish forward. If a shark stops swimming, it will sink, because the body of the shark is denser than water.

2. Shark Skin and Teeth

Shark skin is gritty to the touch and has a texture like sandpaper. The skin is covered by small, spiny projections called **placoid scales**. These scales are often referred to as dermal denticles and are similar in structure to teeth.

Sharks have very sharp teeth that are triangular and slant backward in the mouth. This unique design helps to prevent food from slipping out once

Figure 14-4. The Gray Reef Shark

Biology For Life

CHAPTER 14: VERTEBRATES: PART 1

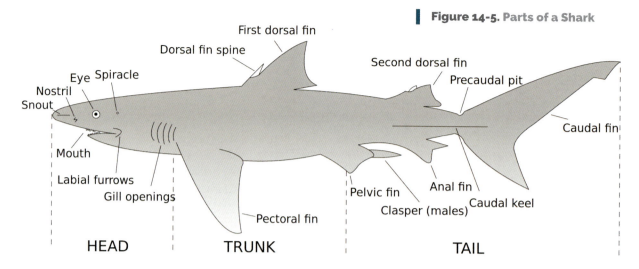

Figure 14-5. Parts of a Shark

Figure 14-6. Teeth of a Tiger Shark

the shark has bitten into it. Both the slanting and the triangular shape of the teeth of the tiger shark are shown in Figure 14-6.

The teeth of sharks grow in rows. As new teeth grow, the teeth in the very first row fall out and are replaced by other teeth that grow in from behind. Typically a shark has two to three working rows of teeth with 20 to 30 teeth in each row.

3. Gas Exchange in the Shark

Sharks have gills that remove oxygen from the water, and, in turn, release carbon dioxide. The gills are composed of many folded tissues that contain a host of blood vessels. These folds of tissue provide a large surface area for gas exchange. Each gill opens to the outside through a structure called a gill slit.

4. Temperature Regulation in Fish

Along with most fish, sharks are **cold-blooded**. An animal that is cold-blooded does not maintain a constant internal body temperature. Indeed, its body temperature changes with the changing temperature of its environment. Fish must remain in water that provides a suitable temperature in which to function. Even slight, rapid changes in temperature can shock fish. This is known as thermal shock and is quite capable of injuring or killing fish. Fortunately, water warms and cools slowly, much more slowly than air. This principle is known as thermal inertia and is due to the density of water.

Figure 14-7. Egg Case of Port Jackson Shark. These are often found washed up on beaches and are also known as mermaid's purses.

CHAPTER 14: **VERTEBRATES: PART 1**

Skate

Ray

Figure 14-8. The Other Chondrichthyes: a Skate and a Ray

5. Shark Reproduction

Sharks reproduce by internal fertilization; that is, males deposit sperm in the body of the females. The fertilized eggs develop and become enclosed in a leathery egg case (see Figure 14-7). The female deposits these egg cases in the water, though some species of shark release their eggs directly into the water. The young sharks develop on their own, without any parental care.

In many species of shark, the female keeps the eggs inside her body until the embryos have fully developed. Upon hatching, the young are born alive. Newborn sharks are independent and receive no care from their mothers.

6. Rays and Skates

Rays and skates also belong to the Chondrichthyes (see Figure 14-8). They have the same general characteristics as sharks; however, their bodies are greatly flattened. The pectoral fins are expanded and merge along the longitudinal axis of the body. Most rays and skates are adapted for living on the bottom of the ocean.

The mouths of the rays and skates open ventrally and are directed downward when lying on the ocean floor. Since their mouths are often buried in the sand, it is impossible to take water into the mouth without taking in a large amount of sand. The Creator has solved this problem by providing gills through two openings on the top of their head. As they rest on a sandy bottom, water passes through these openings to allow gas exchange in the gills. Water continues to pass through the gill slits that are positioned on the bottom of the fish.

Section Review — 14.2 A & B

1. Give an example of a cyclostome.
2. The skeleton of the lamprey is a flexible rod-like structure called the _____.
3. What are placoid scales?
4. Describe the characteristics of shark's teeth.
5. Along with most fish, sharks are _____.

Biology For Life

CHAPTER 14: VERTEBRATES: PART 1

C. Class Osteichthyes: The Bony Fish

You are probably most familiar with the bony fish, most of which belong to the Class Osteichthyes. These exist in a variety of shapes and sizes and include the goldfish, tropical fish, and those that are most often found in rivers, ponds, and lakes. The bony fish range in size from the tiny guppy to the 400-kilogram (about 880 pounds) tuna. In comparison with cartilaginous fish, the bony fish are adapted to a wider range of environments.

Bony fish begin life with a cartilaginous skeleton, which is eventually replaced by bone as the fish becomes an adult. The perch is a representative member of the bony fish (see Figure 14-9), and we will discuss its internal and external structure.

The perch has overlapping scales that are arranged much like the shingles on a roof. The scales vary in size and shape, both of which change with age. The main function of the scales is to protect the fish and to prevent water from entering or leaving the body.

The fish possesses the same number of scales throughout its life, because adult fish do not grow new scales. Annual lines form in the scales with growth, much as tree rings form on a tree. You can determine a fish's age by counting the number of dark, concentric "rings" (otoliths) on one of its scales.

Fish feel slimy to the touch because they have glands that secrete slippery mucus that covers the scales. The purpose of the mucus is to protect the fish against microorganisms. The mucus also gives the fish a smooth surface in order to reduce friction, which eases movement through the water.

The perch has several types of fins. These are composed of thin membranes and supporting cartilage rays. The perch has a pair of **dorsal fins** along the back, a **caudal fin** on the tail, and an **anal fin** on the ventral side. These fins are located on the midline of the fish and help the fish to swim in a straight line. There are also two pairs of lateral paired fins. A pair of **pectoral fins** lies near the bottom of the gill openings. Right behind the pectoral fins are two **pelvic fins**. The tail fin serves to move the fish forward through the water. The other fins help the fish to maintain balance and to change direction. Can you locate these types of fins on the perch?

1. Digestion in the Perch

The perch is carnivorous. Its small, sharp teeth help it to grasp and hold prey. The mouth leads to a short tube called the esophagus, which is connected to the stomach. Food is broken down into a soupy consistency in the stomach. It then enters the intestine, where enzymes reduce the food to small molecules that are absorbed into the bloodstream. There are several sac-like pouches, called **pyloric caeca** (singlular: caecum), that are located near the junction of the stomach and intestine. Digestion and absorption occur in these pouches. Food that is indigestible is expelled through the anus. The liver of the perch functions in digestion by producing bile, a substance that helps digestive enzymes break down fats. The liver also stores sugar for future energy requirements.

2. Excretion in the Perch

The **kidneys** remove nitrogen wastes from the blood. These wastes leave the body as urine through an opening just behind the anus. The kidneys also regulate the balance of water and salt in the tissues of the fish. This role of the kidney is particularly important for fish, since many live in salt water.

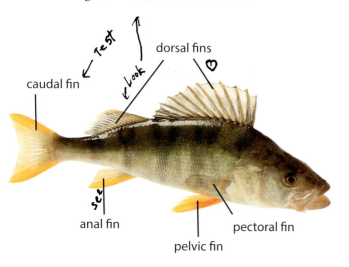

Figure 14-9. Yellow Perch

CHAPTER 14: **VERTEBRATES: PART 1**

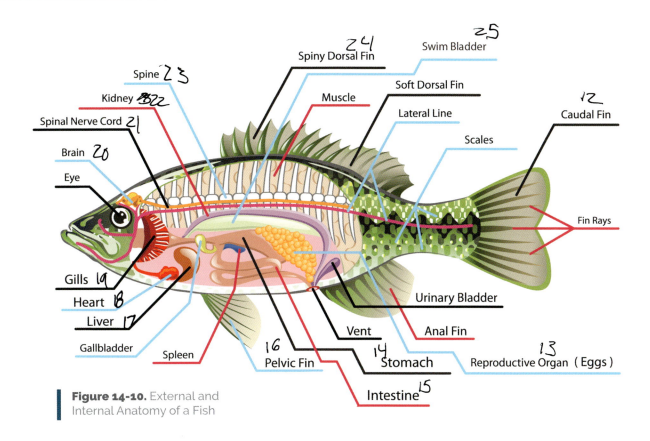

Figure 14-10. External and Internal Anatomy of a Fish

3. Circulation and Gas Exchange in the Perch

There are three types of blood vessels that carry blood through the fish: arteries, veins, and capillaries. As in mammals, arteries carry blood away from the heart, and veins return blood to the heart. Nutrients and oxygen diffuse from the blood of the capillaries into the cells of all tissues. These cells carry out cellular respiration and acquire nutrients. Simultaneously, carbon dioxide and nitrogen wastes diffuse out of these cells and into the capillaries.

Veins then return blood from the body and empty it into a sac, which is located behind the heart. This sac is called the **sinus venosus**. Blood then enters the heart itself.

The heart of the fish is a muscular pump that contains two separate chambers. The first and second chambers are called the **atrium** and **ventricle**, respectively. When the atrium contracts, it forces blood into the ventricle. The ventricle then pushes blood through a large blood vessel, which is called the **ventral aorta**. The blood travels from the ventral aorta to the gills.

The gills of a fish contain numerous capillaries that are in direct contact with the blood. The capillaries acquire oxygen from the water and also eliminate carbon dioxide. The oxygenated blood enters a vessel called the **dorsal aorta** and travels to all parts of the body. This type of circulation is familiar to you and is referred to as a **closed circulatory system**.

Respiration in the perch occurs in the gills. The perch takes water into its mouth, and then, by raising the floor of its mouth, the perch forces water over its gills. The water leaves the fish through a single slit on each side of the head. The fish has a protective covering for the gills on the side of its head, which is called an **operculum** (ō-pŭr-kū-lum). The operculum protects the delicate gill tissue from damage.

Biology For Life

The structure of the gills will be discussed next. The fish has four **gill arches** on each side of its head. Each arch is supported by a piece of cartilage. Folded tissues, called **gill rakers**, are located on the front of each arch. The purpose of the gill rakers is to keep food particles from passing through the gills and damaging the delicate tissue.

In addition to the gills, gases also enter and leave the blood in another organ called the **swim bladder** (gas bladder). It has no respiratory function, but, rather, it acts as a gas bag that functions to control the depth of the fish. The fish may change its depth in the water by changing the volume of gas in the swim bladder, by releasing or absorbing gas from the blood stream. When the fish descends to a lower depth, it decreases the volume of gas in its swim bladder. Conversely, when it begins its ascent, it increases the volume of gas.

4. The Nervous System of the Perch

The perch has a well-developed brain and spinal cord. There are five major parts to the brain:

1. The **olfactory bulb** brings information about smell from the nostrils to the brain via the olfactory nerve.
2. The **cerebrum** is composed of two lobes that are primarily involved in the interpretation of smell.
3. The **optic lobes** are located behind the cerebrum and process visual information. These lobes also send impulses to the muscles, so that the fish can respond to what it sees.
4. The **cerebellum** is located behind the optic lobes. This structure coordinates complex muscle movements.
5. The **medulla** is located beneath the cerebellum. It controls the internal organs of the fish.

The spinal cord runs from the medulla and travels down the vertebral column. The spinal cord is the major pathway for transmitting information between the brain and body. Nerves branch out from the brain and spinal cord. **Cranial nerves** branch out from

Figure 14-11. A school of fish

the brain, and spinal nerves branch from the spinal cord.

5. Senses of the Perch

The sense of smell is very important to the perch, and to fish in general. The perch has two olfactory sacs that are located inside the fish and near its mouth. These sacs are organs that are sensitive to chemicals dissolved in the water. Also, the fish has cells that are sensitive to taste. These are located in and around the mouth.

The perch does not see very well, in spite of the fact that its optic lobes and eyes are large. Fish see objects that are very close to them and most detect movement at limited distances.

A pair of semicircular canals is located near the back of the fish's brain. This organ is involved in the fish's sense of balance and hearing. The fish hears sounds that are transmitted through the body and skull bones to the semicircular canals.

All fish have a sense organ called the **lateral line**. This line runs along each side of the fish and is sensitive to pressure changes in the water. Fish use this organ to detect movement nearby. The fish that travel in schools use the lateral line to locate other fish in the school and to synchronize changes in direction. If you have ever seen the amazingly coordinated movement of fish in a school, you now know that these lateral lines are the reason why.

6. Reproduction in the Perch

Eggs are produced in the ovaries of the female perch, and sperm are produced in the testes of the male. The reproductive cells exit the perch's body through an opening just behind the anus. Fertilization in the perch is external. The female lays several hundred eggs in the water. The male releases **milt** near the eggs. Milt is the term given to

Figure 14-12. Salmon eggs found in a river

the fluid that contains the sperm cells. Fertilization occurs in the water, and the parents swim away, leaving the young unprotected. On the other hand, protection by parents occurs in many fish, including the smallmouth bass. The male guards the fertilized eggs and, upon hatching, the fry stay near the male for a couple of weeks. If he leaves the fry for even a few seconds, it is quite possible and even likely that predators will devour most of them.

7. Adaptations of Bony Fish

Fish have many unique adaptations to their specialized environment. Let us look at some of these amazing adaptations.

Saltwater Fish

Bony fish live in freshwater, saltwater, and brackish water. Some species, such as the steelhead trout of the western United States, spend several years living in saltwater, and, when mature, migrate to freshwater rivers to spawn. In the eastern United States, striped bass do the same. One can only imagine the difficulties faced by fish in their aquatic environment, especially when many live in saltwater for years, and then must adapt to living in freshwater.

There is a higher concentration of salt in saltwater than there is in the blood of a fish. Consequently, saltwater fish tend to lose water from their cells by osmosis, that is, the movement of water from a high concentration to a lower one.* Another way to consider this is that water tends to follow the movement of salts. Because there is more salt dissolved in salt water, there is a tendency for water to travel out of the fish's cells. This loss of water has the potential to be injurious to the cells and perhaps lethal to the fish. However, fish are able to maintain the salt balance in their bodies by a process called **osmoregulation**. This process involves the active transport of salts that are pumped out from the gills. For example, if the fish were located in saltwater, it would transport excess salt from its body and through its gills. Conversely, if the fish were dwelling in freshwater, it would need to hold onto salts by transporting them away from the gills and circulating these back into the body.

Scales, which are impermeable to water, help prevent water loss. The kidneys of saltwater fish excrete only small amounts of highly concentrated

Figure 14-13. Atlantic Salmon

urine, which also helps them to conserve body water.

(A good way to examine the effects of osmosis is to slice a carrot and put a piece in a container of saltwater for a few days. The salt moves into the cells of the carrot, and water moves out, thereby shriveling and wrinkling the carrot.)

Freshwater Fish

Fish that live in freshwater can adapt to changes in salt concentration of their surroundings by osmoregulation. The scales of these fish do not allow water to pass into the body. Also, the gills actively transport small amounts of salt from the water and into the fish. The kidneys help regulate salt balance by excreting large volumes of excess body water. Only a few bony fish, such as the salmon, steelhead trout, and striped bass, are able to travel from freshwater to saltwater and back. Amazingly, the gills and kidneys of salmon reverse their water and salt transport functions.

Life Cycles

Salmon have an amazing life cycle, illustrated by their ability to adapt to freshwater and saltwater. They hatch in freshwater streams and then swim to the sea, which is saltwater. An Atlantic salmon is seen in Figure 14-13.

Figure 14-14. The Great Barracuda with prey

Salmon live most of their adult lives in the ocean. When mature, they return to the freshwater place of their birth to reproduce and then die. The new generation returns to the same freshwater stream to reproduce and die.

Feeding

The barracuda has sharp teeth for catching prey – see Figure 14-14.

The streamlined shape of this fish is an adaptation for the swift pursuit of prey. There are many other examples of adaptation that help fish feed. For example, the hard, beaked tooth of the parrotfish helps this fish to scrape algae from rocks; the wormlike appendage on the head of the anglerfish attracts prey, much like a worm on a fishhook.

Coloration

Some fish have brilliant colors, an adaptation that allows males and females of the same species to recognize one another. Distinctive coloration may also be used for defense purposes. For example, enemies of the lionfish – seen in Figure 14-15 – have learned to recognize its elaborate fins and bright colors, and, therefore, its poisonous spines.

Figure 14-15. The side view of a Red Lionfish

Biology For Life

Figure 14-16. Lungfish

Also, the flounder has the ability to change its color for purposes of camouflage.

Deep Ocean Fish

Very little light reaches ocean depths below 150 meters. Hence, some fish that live at these depths adapt by generating their own light, through a chemical process called **bioluminescence**. Some of these fish have lights for attracting mates, and some have lights that lure prey.

Reproductive Behavior

There are many variations in the reproductive structures and behavior of the bony fish. Most bony fish carry out external fertilization; however, some fertilize internally and bear live young. Some fish breed by keeping their eggs in their mouths. After the eggs have been fertilized in the water, the female scoops them into her mouth and holds the eggs there until they are hatched. A few fish build nests for their eggs and guard the nest until the eggs are hatched.

Lungfish

Some fish have special adaptations for breathing air. The African lungfish lives in ponds that completely dry up for one season during the year. While in the pond, the lungfish uses gills for respiration. The fish buries itself in the mud before the pond dries up. The fish leaves a tiny passageway for air to enter the hole. The fish draws this air into a structure called a lung. The lungfish survive in the mud hole until the next rainy season fills up the pond with water. These fish are found in Africa, South America, and Australia.

> **Section Review — 14.2 C**
>
> 1. A fish has a protective covering for the gills on the side of its head; this is called an _____.
> 2. What is a pyloric caecum?
> 3. List the five major parts of a perch brain.
> 4. What is osmoregulation?
> 5. What is bioluminescence?

Biology For Life

CHAPTER 14: **VERTEBRATES: PART 1**

14.3 Amphibians

Section Objectives

- List the three amphibian orders.
- Describe the major identifying characteristics of the amphibians.
- Identify the external structures of a frog.
- Describe the functions of each part of the frog's digestive system and excretory system.
- Contrast the frog's circulatory system with the fish's circulatory system.
- Name the three structures used by the frog for gas exchange.
- List the major parts of the frog's nervous system.
- Identify the major senses of the frog.
- Distinguish the differences between tadpoles and frogs.
- Describe the features of salamanders and caecilians.

In this section, you will learn about a group of vertebrates called amphibians, which includes the frogs, toads, and salamanders. These animals are unique in that they share the characteristics of fish that live in water and reptiles that live on land. This is because amphibians, in most cases, spend part of their lives in water and part of their lives on land. All amphibians require a moist environment in which to survive and reproduce.

A. Characteristics of Amphibians

Amphibians are divided into three orders: **Anura** (frogs and toads), **Urodela** (salamanders), and **Apoda**. The Apoda (caecilians) is composed of a group of wormlike, legless amphibians that live in tropical regions. A member of the order Anura is seen in Figure 14-17.

1. From Water to Land

The term **amphibian** means "dual life." Amphibians begin their lives as eggs. These hatch into larvae that are capable of swimming in water. The larvae have fins for swimming, gills for gas exchange, and a circulatory system which is similar to that of a fish. Most amphibians undergo a metamorphosis to the adult form that will live on land and near fresh water. The adult amphibian has four legs for movement on land and lungs for gas exchange with the air. The circulatory system of

Figure 14-17. The western spadefoot toad (*Spea hammondii*) is a typical member of the amphibians that is found in North America. Adult toads are between 3.8 and 7.5 cm long. Populations of *Spea hammondii* are localized, or found in small isolated areas, but are found throughout a broad geographic range. The western spadefoot toad prefers living in grassland, scrub, and chaparral habitats, but is also found in oak woodlands. It is nocturnal, and its activity is limited to the wet seasons and during summer storms. In addition, it is active when the soils in which it dwells are very moist.

Biology For Life

CHAPTER 14: VERTEBRATES: PART 1

Figure 14-18. White's Tree Frog, or Dumpy Tree Frog (*Litoria caerulea*), is a species of tree frog that is native to Australia and New Guinea, with introduced populations in New Zealand and the United States. The adults reach 10 centimeters (4 inches) in length. White's Tree Frogs are docile and well-suited to living near human dwellings. They are often found on windows or inside houses, and they eat insects that are drawn by light.

the adult is also more complex than that which is found in the larva.

All amphibians have smooth skin with no scales. The skin is covered with mucus and must be kept moist. The skin of an amphibian is an important respiratory surface by which gases are exchanged.

All amphibians are cold blooded; therefore their internal body temperature is the same as the surrounding environment. Most species of amphibians live in areas where there is little temperature variation, since they cannot regulate their internal temperatures. Species that live in extremely hot or cold temperatures have special adaptations that allow them to survive.

B. A Typical Amphibian: The Frog

Frogs and toads have no tails. This characteristic distinguishes the order Anura from the other amphibians. They also have broad, flat bodies, with no distinct neck, and have very large hind legs that are specialized for jumping.

All adult frogs and toads are carnivorous. They eat worms, insects, insect larvae, and other small animals. The larval stage of a frog is called a **tadpole**, or polliwog. Tadpoles lack legs but have tails and gills. Tadpoles eat plants and small vegetable matter that is found in water.

1. External Anatomy of the Frog

The skin of a frog is both smooth and **permeable** to water. Frogs do not need to drink water, because they are able to absorb the water that they need through their skin. Frogs must periodically moisten

Biology For Life

their skin by entering water, because they are vulnerable to drying out in the air and sunlight. This is the reason why frogs must live near water.

The frog has glands that secrete mucus onto the skin. The mucus slows the evaporation of water to air, and also makes the frog slippery to predators. Because the frog's skin is loosely attached, it is difficult for some predators to hold onto the frog. Frogs and toads are similar animals; however, there are some differences between them. Toads live farther from water than frogs do. Toads have drier skin, but they must stay in humid places.

At the front of the head, the frog has two nostrils that open into nasal passages. The nasal passages are connected to the mouth cavity. The nostrils of the frog have two functions: the sense of smell and the passage of air. A frog can be submerged in water and breathe with only the nostrils exposed to the air. Located behind the nostrils are two large eyes that stick up above the head. When the frog is beneath the water, its eyes can still be above the surface of the water. Frogs have a protective covering over their eyes known as a **nictitating** (nik-tĭ-tāt-ing) membrane. This membrane covers the eye, protects it from water, and keeps the eye moist when the frog is out of water.

The male frog emits sounds to attract females during the period of courtship. In many frog species, the male pushes air from its mouth into vocal sacs that are located under the chin. These vocal sacs help to amplify the sound. Sound is received by oval, flat **tympanic membranes** that are located behind each eye. These are the organs of hearing in the frog.

A frog's front legs are smaller than its rear legs. The rear legs are large and muscular and are well adapted for jumping. The male frog has an additional pad on the area of the front legs that is like our "thumb." This padded area helps to push eggs out of the female's body during mating.

2. The Frog Skeleton

The skull of the frog consists of a lower jawbone along with a number of bones that surround the brain. As in fish, vertebrae surround the spinal cord. There are only a small number of vertebrae in frogs, and there are no ribs. A long bone extends from the end of the spine to the most posterior point on the body.

The **pectoral girdle** of the frog is a group of bones that almost surround the thorax and attach front limbs to the body. The **pelvic girdle** is located where the rear limbs are attached to the body.

Figure 14-19.
A frog's rear legs are long and muscular and are adapted for jumping.

Biology For Life

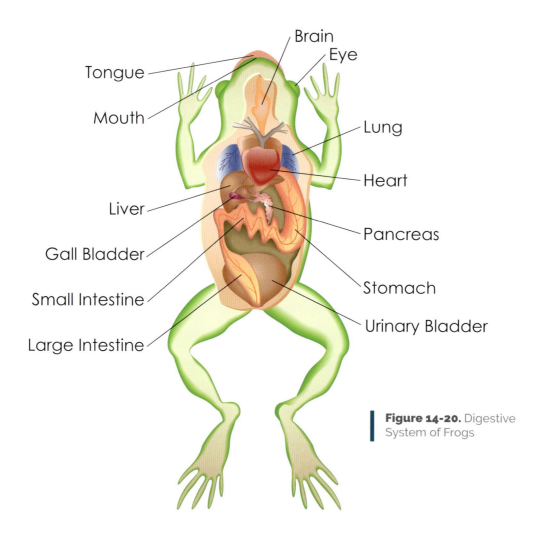

Figure 14-20. Digestive System of Frogs

3. Digestion in Frogs

The frog has many very small **maxillary teeth** that are located near the edges of the upper jaw. Two additional **vomerine teeth** are found on the inside of the upper part of the mouth, near the internal openings to the nasal passages. Frogs do not chew with their teeth, but they use their teeth to hold their prey. The frog has an unusual tongue, because the point of attachment of the tongue is in the front of the mouth. This allows the frog to flick its tongue far out in the front of its mouth, so that the frog may grasp flying insects or other prey.

Food is immediately pushed to the back of the mouth and into the gullet, which is the beginning of the esophagus. The esophagus leads to the stomach, where acid secretions convert the food into a soupy mixture. Food then enters the duodenum, which is the first part of the small intestine.

The liver is a large organ that is located near the stomach. It produces bile, which aids the frog in the digestion of fats. A small gallbladder stores bile that is eventually transported to the duodenum through a small tube known as the bile duct. Many of the digestive enzymes are produced in a gland near the stomach, which is called the pancreas. A tube called the **pancreatic duct** transports the enzymes to the duodenum. The molecules of food are absorbed into the blood through the walls of the small intestine.

The blood carries food molecules from the small intestine directly to the liver. In the liver, glucose is converted to glycogen and is stored. Between meals, the liver hydrolyzes, or breaks down, the glycogen into glucose. Glucose is transported to the body cells via the blood. Excess food that is not immediately needed for energy is converted to and stored as fat. It is stored in structures known as **fat bodies** that are located near the kidneys.

Food that is not absorbed in the small intestine is passed on to the large intestine. The large intestine removes excess water from the waste material. The final part of the digestive system is the **cloaca**. This structure is a common opening through which solid wastes, liquid wastes from the kidneys, and reproductive cells are expelled from the body.

4. Excretion in Frogs

The major organs of excretion in frogs are the **kidneys**, which are located on either side of the vertebral column near the dorsal body wall. The **renal arteries** carry blood to the kidneys. The kidneys remove nitrogenous wastes from the blood, and the blood then returns to the circulation through the **renal veins**. The nitrogenous wastes form urine when dissolved in water. The urine leaves the kidneys through a tube called the **ureter**. Urine is held in the saclike **urinary bladder** before it is expelled through the cloaca.

5. The Frog's Circulatory System

The blood of a frog contains **red blood cells**, each of which has a nucleus. The cytoplasm of the red blood cells contains the respiratory pigment known as **hemoglobin**. The red blood cells and hemoglobin work to transport oxygen from the respiratory surfaces of the frog to all of the cells of the body.

The white blood cells of the frog also have nuclei, with colorless cytoplasm. The white cells defend the frog against diseases, just as they do in humans and other organisms. The liquid portion of the frog's blood carries food molecules, the waste products of cellular metabolism, and proteins.

Similar to the other vertebrates, frogs have a closed circulatory system. In the frog, blood leaves the heart and travels to the lungs, where it acquires oxygen. The blood then returns to the heart and is pumped to the rest of the body. After giving up its oxygen to cells and tissues, the blood then returns to the heart. The circulation between the heart and lungs is called the **pulmonary loop**. The circulation between the heart and the rest of the body is called the **systemic loop**.

The heart of the frog has three chambers: a ventricle, a **left atrium**, and a **right atrium**. Deoxygenated blood from the body returns to the heart through two vessels, an anterior and a posterior **vena cava**. These veins empty into a large collecting area behind the heart, which is called the **sinus venosus**. The blood then enters the right atrium of the heart. The right atrium pumps blood into the ventricle, and the ventricle pushes blood upward. Some of the blood will enter the **pulmonary artery**, which takes blood to the lungs. Blood is oxygenated in the lungs and returns to the **left atrium** by way of the **pulmonary vein**. The oxygenated blood from the left atrium is pumped into the ventricle. Blood then passes through the **conus arteriosus** and enters the **aorta**, before traveling to the cells and tissues of the body.

6. Frog Respiration

The lungs of a frog are two air sacs that branch off from the gullet. These are slightly folded on the interior surface, which increases lung surface area. This results in more contact with blood vessels, thus allowing better exchange of oxygen between the air and the blood. Respiration is also enhanced by capillaries, which cover the outer surface of the

CHAPTER 14: VERTEBRATES: PART 1

lungs. These also function to exchange oxygen for carbon dioxide between the air and the blood in the capillaries.

Frogs also depend on two other surfaces for gas exchange: their moist skin and the moist membranes that line the inside of their mouths. The skin surface has a rich supply of capillaries. Blood is oxygenated when it passes near these. Gases are exchanged through the lining of a frog's mouth when the frog gulps air.

7. The Nervous System of the Frog

The **central nervous system** of the frog consists of the brain and spinal cord. The **olfactory lobes** are the most anterior parts of the brain. These lobes receive input from the nose for the sense of smell. An area called the **cerebrum** is located behind each olfactory lobe. The cerebrum is associated with memory and the initiation of body movements. The **optic lobes**, which are associated with vision, are located behind the cerebrum.

The **cerebellum** and **medulla** are located behind the optic lobes. The cerebellum is involved with the control of muscle movement, and the medulla controls basic body functions such as heart rate and breathing. A small piece of tissue, called the **pituitary gland**, is attached to a stalk of nerves at the base of the brain. This gland secretes many chemical hormones that control and regulate a variety of body processes.

The **peripheral nervous system** of the frog is composed of the nerves that branch off of the **central nervous system** and into the body. The frog has ten pairs of **cranial nerves** that branch off from the brain. A number of **spinal nerves** branch out from the spinal cord.

8. The Frog's Senses

The frog has a sense of taste and smell. The frog will accept or reject food based on the information received from these senses. The frog depends on a good sense of vision in order to see flying insects, which it rapidly captures with its tongue.

In the frog's ear, the **tympanic membrane** transmits sound to the inner ear. The inner ear sends information about sound to the brain. A tube called the **Eustachian** (yü-stā-shē-an) **tube** connects the space behind the tympanic membrane to the mouth. The Eustachian tube equalizes the air pressure on both sides of the tympanic membrane, just as it does in humans. An organ that senses balance is also located in the area of the inner ear.

9. Reproduction and Development of Frogs

Sperm cells are produced by the male frog in a pair of testes that are located near the kidneys. These cells then travel through many fine tubes known as **vasa efferentia**. These tubes merge with the urinary ducts in the kidneys. Sperm cells travel through the ureter to the cloaca. During mating, the sperm cells are expelled from the body through the cloaca.

The female frog produces eggs in the ovaries. These develop in the ovary and are eventually released into the abdominal cavity. The eggs are then swept into funnel-like tubes called oviducts that are lined with cilia. The cilia move the eggs toward an enlarged sac called the uterus. Each egg is surrounded and protected by a capsule of gelatinous material. The eggs accumulate in the uterus until mating, at which time they are expelled from the female frog's body through the cloaca. The key stages in a frog's life cycle are shown in Figure 14-21.

Frogs generally mate in spring. Males attract the females by croaking out specific mating calls. The male grasps the female around the abdomen and helps squeeze eggs out of her body. As the eggs are expelled from the body of the female, the male releases his sperm on top of the eggs. This type of fertilization, known as external fertilization, takes place in the water. The fertilized eggs appear as jelly-filled sacs that can often be seen near the bank of a pond or river.

Once the eggs begin developing within this gelatinous mass, it is possible to see the embryo in various stages of development. The tadpoles emerge from the eggs within a few days, the speed of which depends on water temperature. Upon hatching, the

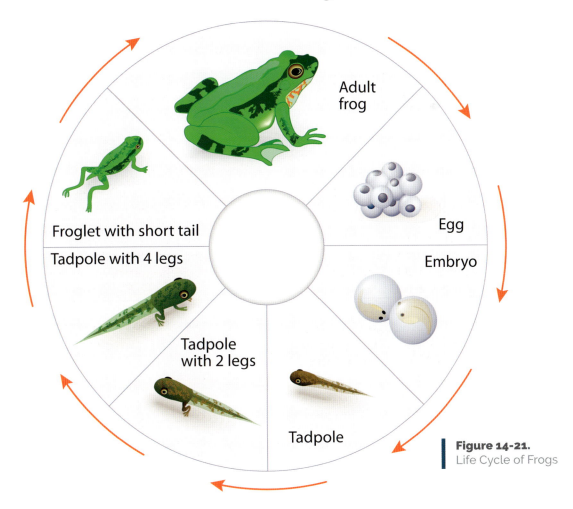

Figure 14-21. Life Cycle of Frogs

tadpole feeds, grows, and then undergoes a gradual metamorphosis to become an adult frog. The entire process takes about one to two months.

A number of frog species live in areas that may become extremely cold in winter and extremely hot in summer. Since frogs are cold blooded, they must be able to adapt to these extreme changes in weather in order to survive. For example, frogs that survive cold winter climates do so by hibernating. They dig into the mud at the bottom of a pond. The frog's metabolic activity slows down greatly, and they require very little oxygen. During the hibernation period, the frog lives on the stored food in the fat bodies of the abdomen. Frogs that live in hot areas, where ponds dry up, burrow in the mud and **aestivate**. **Aestivation** is a resting period during which the frog awaits the return of more favorable weather conditions.

C. Other Amphibians

Some of the other orders of amphibians include the **Urodela**, or salamanders, and the **Apoda**. A number of salamanders are found in the colder regions of the world. The apods, or caecilians, are legless amphibians and belong to the order Apoda. They tend to be a rare group and are not found in North America.

1. Order Urodela: Salamanders

A salamander has a different body plan than a frog. The salamander is long and slender and has a neck between its head and body. It has a long tail and resembles a lizard. However, keep in mind that lizards are reptiles, while salamanders are amphibians. The salamander has smooth, moist skin like the other amphibians. Three species of salamanders can be seen in Figure 14-22.

Some salamanders are totally aquatic for their entire lives, and some woodland species rarely enter water. One species of Mexican salamander remains in the larval form its entire life. These larval salamanders are

Examples of Salamanders

Marbled Salamander

Pacific Giant Salamander

Red Back Salamander

| **Figure 14-22.**

CHAPTER 14: **VERTEBRATES: PART 1**

Figure 14-23. Axolotl

called **axolotls**. The axolotl – seen in Figure 14-23 – is the larval stage of the tiger salamander. The pond where the axolotls are found lacks iodine, and most amphibians need iodine in their diet to undergo metamorphosis to the adult form. Since the axolotls lack iodine in their diet, they never change form. However, the adult remains aquatic, has gills, and is capable of breeding. Axolotls are found in only one lake near Mexico City, Mexico.

2. Order Apoda

The apods (caecilians) are very unusual amphibians. These legless amphibians live only in the tropical regions of Asia, Africa, Central America, and South America. These animals can be found burrowing under moist soils. The apods are carnivorous and eat earthworms, insects, and insect larvae.

Figure 14-24. A member of the Order Apoda. It is an amphibian that resembles a snake or earthworm.

Section Review — 14.3

1. What are the three orders of amphibians?
2. The term amphibian means "_____."
3. Why do frogs not need to drink water?
4. What are the three structures that frogs use for gas exchange when breathing?
5. Frogs have a protective covering over their eyes known as a _____ membrane.
6. What is aestivation?
7. What is an axolotl?

Biology For Life

Chapter 14 Supplemental Questions

Answer the following questions.

1. Know and be able to label Figures 14-5, 14-10, and 14-20.
2. List an example of a cyclostome and describe its characteristics.
3. Describe the characteristics of the Chondrichthyes.
4. Explain the replacement of teeth in the shark.
5. What is unique about a shark's teeth?
6. Contrast the shark's gills with those of bony fish (specifically relating to how the gills have access to water).
7. State how the internal temperature of fish is determined.
8. Differentiate between shark reproduction and reproduction in other types of fish.
9. List other members of the Chondrichthyes in addition to sharks.
10. What purpose does mucus serve for the perch?
11. List the types of fins of osteichthyes (bony fish) and state their functions.
12. Describe the functions of the perch's digestive and excretory organs.
13. Discuss the pathway of blood flow in a bony fish.
14. Explain the function of the gills and swim bladder in a fish.
15. Name the major portions of the perch brain and spinal cord, and state the function of each part.
16. Describe the five sense organs of fish.
17. Detail the process of reproduction in a bony fish.
18. Describe several adaptations of bony fish.

Chapter 14 Supplemental Questions

19. What is unique about the bony fish known as the salmon?

20. List the three amphibian orders.

21. Describe the major identifying characteristics of the amphibians.

22. Identify the external structures of a frog.

23. Describe the functions of each part of the frog's digestive system, excretory system, and reproductive system.

24. Contrast the frog's circulatory system with the fish's circulatory system.

25. Name the three structures used by the frog for gas exchange.

26. List the major parts of the frog's nervous system.

27. Identify the major senses of the frog.

28. Distinguish the differences between tadpoles and frogs.

29. Explain the features of salamanders and apods.

30. Study all diagrams in this chapter. Be able to draw and label all diagrams.

CHAPTER 15
Vertebrates: Part 2

Chapter 15

Vertebrates: Part 2

Chapter Outline

15.1 Introduction

15.2 Reptiles
  A. Characteristics of Reptiles
    1. Life on Land
    2. Reptile Eggs
  B. Types of Reptiles
    1. Order Squamata: Snakes
      a. External Anatomy
      b. Skeleton
      c. Digestion and Excretion
      d. Circulation and Respiration
      e. The Nervous System and Sense Organs
      f. Reproduction
      g. The Variety of Snakes
    2. Order Squamata: Lizards
    3. Order Chelonia: Turtles
    4. Order Crocodilia
    5. Order Rhynchocephalia: Tuataras

15.3 Birds (Class Aves)
  A. Archaeopteryx
  B. What are Birds?
    1. Characteristics of Birds
      a. Feathers
      b. Structures of the Head
      c. Other External Features
    2. Internal Features of Birds
      a. Skeleton
      b. Feeding and Digestion
      c. Respiratory System
      d. Circulatory System
      e. Excretory System
      f. Nervous System
      g. Reproduction
      h. Embryo Development

15.4 Mammals
  A. Characteristics of Mammals
    1. Common Features of Mammals
    2. Mammalian Development
  B. Monotremes and Marsupials
    1. Characteristics of Monotremes
    2. Characteristics of Marsupials
    3. Representative Marsupials
  C. Placental Mammals
    1. Characteristics of Placental Mammals
    2. Some Types of Placental Mammals

15.1 Introduction

In this chapter, you will learn about the characteristics and features of vertebrates that live mostly on land and in the air. These animals are the reptiles, birds, and mammals.

15.2 Reptiles

Section Objectives

- Describe the characteristics that differentiate reptiles from amphibians.
- List the membranes of the reptile egg and describe the function of each membrane.
- Name the four orders of living reptiles.
- Describe some of the unique adaptations of snake anatomy.
- List two ways that snakes kill their prey.
- Contrast the lizard and the snake.
- Describe the features of the order Chelonia.
- List some of the animals in the order Crocodilia and explain how they differ from other reptiles.
- Describe the uniqueness of the tuatara.

A. Characteristics of Reptiles

1. Life on Land

Reptiles have skin that is thick, tough, dry, and covered with overlapping scales made of a protein called **keratin**. Human fingernails and hair are also composed of keratin. The skin of reptiles is impermeable to water (will not let water penetrate it), which prevents the body from drying out.

Reptiles are better suited to movement on land than are the amphibians. The legs of reptiles are directed downward, rather than straight out from the sides like the amphibians. This downward positioning of the limbs allows the reptile to keep its body off the ground, which allows for easy and rapid movement over land. Another distinguishing feature of the reptilian legs is the presence of keratin claws on each toe. These claws are used for digging and climbing on trees and rocks. Four different species of reptiles are shown in Figure 15-1.

Biology For Life

CHAPTER 15: **VERTEBRATES: PART 2**

Types of Reptiles
Figure 15-1

Caiman Crocodile

Eastern Diamondback Rattlesnake

Green Sea Turtle

Tuatara

Reptiles are cold-blooded, like fish and amphibians. The temperature of these organisms depends upon the temperature of their surrounding environment. Because the temperatures on land tend to fluctuate more quickly than those in the water, reptiles must be equipped to handle these changes. Accordingly, they adapt to these changes by specialized behaviors. One such behavior is known as **basking**. It involves the reptile lying in a prone position and aligning its dorsal surface with rays of the sun. Lizards and snakes often bask in the morning sun. By doing so, the reptiles use the sun's energy to increase their body temperature. On very hot days, reptiles are active but avoid direct sun and appear mostly during twilight hours. In addition, many desert snakes have been designed to avoid heat by being active at night.

Reptile reproduction is characterized by **internal fertilization**. During mating, the sperm cells of the male are deposited inside the body of the female, where fertilization takes place.

2. Reptile Eggs

Reptiles lay large eggs that have leathery shells, which protect the contents of their eggs. These shells function as a barrier against desiccation, or drying out. Several reptilian eggs are shown in Figure 15-2.

There are four membranes within the shell that aid the developing embryo – see Figure 15-3. A thin membrane called the **chorion** (kôr-ē-än) lies directly beneath the shell. It holds the contents of the egg and

Figure 15-2. Turtle eggs in a nest built by a female common snapping turtle. Notice the leathery shell protecting the contents of the egg.

Biology For Life

is permeable to gases. Oxygen and carbon dioxide are exchanged between the egg and its surroundings. The **amnion** is the sac that surrounds the embryo. Fluid contained within the amnion acts like a shock absorber to protect the embryo.

The **allantois** (a-lan-tō-ĭs) is the membrane that contains many blood vessels. This membrane is a respiratory surface for the embryo. Oxygen and carbon dioxide are exchanged between the blood vessels and the surrounding fluid. As the embryo grows larger, nitrogen wastes accumulate within the allantois. The fourth membrane is the **yolk sac**. The yolk sac surrounds the yellow yolk, which serves as food for the developing embryo. Blood vessels in the yolk sac carry the food from the yolk to the growing embryo.

When reptiles hatch, they are able to survive on land. Reptiles do not have an aquatic larval stage like the amphibians.

Section Review — 15.2 A

1. What is keratin?
2. Reptiles are cold-blooded and depend on the temperature of their surrounding environment. To increase body temperature when they need to, reptiles use a behavior known as _____.
3. What are the four membrane layers of a reptile egg?

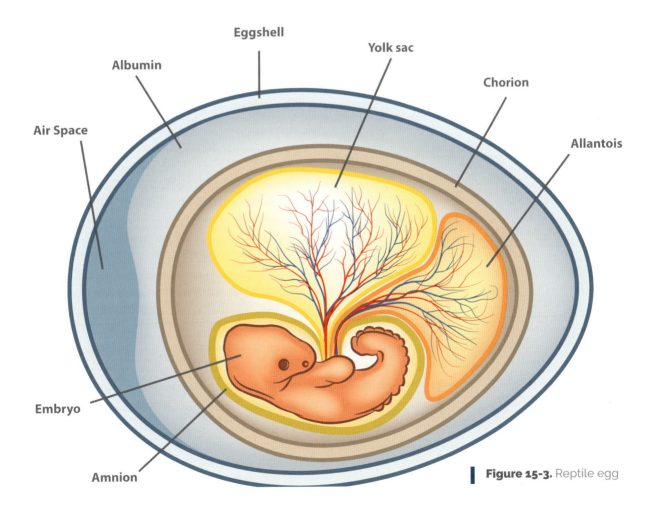

Figure 15-3. Reptile egg

CHAPTER 15: **VERTEBRATES: PART 2**

B. Types of Reptiles

There are four orders of reptiles: the **Squamata** (lizards and snakes), the **Chelonia** (tortoises and turtles), the **Crocodilia** (alligators and crocodiles), and the **Rhynchocephalia** (tuatara). There are about 6,000 species within the four orders of reptiles. In comparison, recall that estimates for the number of living insect species ranges between two and thirty million.

1. Order Squamata: Snakes

The anatomy of the snake shows many features that are characteristic of all reptiles, as well as adaptations that are unique to this particular reptile group – see Figure 15-5.

a. External Anatomy

The most obvious feature of snakes is the absence of limbs. The entire body of the snake is covered with scales. These scales are dry, not moist and slimy as the amphibians' skin. The scales on the ventral (belly) surface of the snake help the animal to gain a hold against the ground and push itself forward.

Figure 15-4. Common Garter Snakes

The snake breathes through nostrils that are found near the front of its head and above the mouth. Snakes do not have eyelids, but instead have

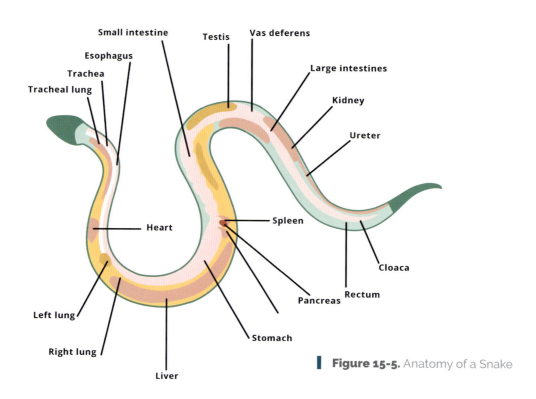

Figure 15-5. Anatomy of a Snake

Biology For Life

a transparent, scale-like membrane that covers their eyes. Snakes do not have external ears but are able to process sound in two ways. First, vibrations that are conducted along the ground are picked up in the bones and muscles of the lower jaw. Second, these vibrations are passed to the bones of an inner ear. The sound is then transmitted along a cochlear mechanism to the auditory nerve of the brain. Recent studies have shown that each side of the snake's jaw can receive sound independently of the other side. This means that snakes can hear in stereo. This is pretty incredible, especially when considering that it was once believed that snakes were deaf! Snake hearing is sufficient for finding prey and avoiding predators. Our Creator God has designed all animals to survive and reproduce.

The cloaca of the snake is a small opening on the underside of the snake. It marks the boundary between the abdomen and the tail. The cloaca is involved with the transfer of sperm from the male to the female, egg laying, and expulsion of wastes

Figure 15-6. The Fangs of a Snake

from the body. In many species, such as the garter snake, a musk gland is located nearby. When these snakes are threatened or handled roughly, the musk gland discharges a foul-smelling substance that repels would-be attackers.

b. Skeleton

The most unusual feature of the snake's skull is the manner in which the jaws are hinged to the head. The snake can open its mouth very wide to swallow large prey, as shown in Figure 15-6.

The skeleton of a snake consists mostly of vertebrae and ribs. Often there are more than 100 vertebrae and ribs attached to the skeleton. Most snakes have no bone structures that resemble a pectoral or **pelvic girdle**. The skeleton of the Indian Python is shown in Figure 15-7.

c. Digestion and Excretion

Snakes have unusual feeding behaviors. All snakes are carnivorous and eat infrequently. Some snakes can live as long as one year between meals. When a snake eats, it eats extremely large meals

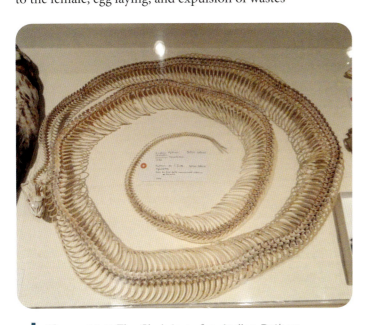

Figure 15-7. The Skeleton of an Indian Python

and swallows its prey whole. The snake's jaws can unhinge to swallow large organisms, and its ribs are also capable of separating. Large prey can easily pass along the digestive tract without obstructions such as pectoral or pelvic girdles. One can easily see this in Figure 15-8, where the snake has swallowed an egg whole.

The teeth of the snake are designed for grasping and holding prey, rather than for chewing or biting. The snake's teeth point backward, which makes it difficult for prey to escape.

In mammals, the esophagus is very muscular and moves food to the stomach. In the snake, however, the esophagus has very little muscle, and food is moved to the stomach by movements of the entire body.

As Figure 15-5 shows, the liver of the snake is elongated. It is the largest internal organ in a snake and fills the space between the heart and stomach. One of the many functions of the liver is to produce bile, which contains digestive enzymes. The digestive organs of the snake have similar functions to those that are found in other vertebrates. A pair of long kidneys makes up the excretory system of the snake. These kidneys are described as staggered; that is, one is in front of the other, rather than on opposite sides of the body. The ureter drains urine from the kidneys into the cloaca.

d. Circulation and Respiration

Reptiles have a four-chambered heart. The oxygenated and deoxygenated blood does not mix together in the ventricle, as it does with amphibians. The ventricle of the reptile is separated into right and left chambers, so that oxygenated blood from the lungs is not mixed with deoxygenated blood from the body. Reptile skin is not a respiratory surface, and therefore reptile lungs are more efficient at gas exchange than the amphibians. The interior surfaces of reptile lungs are highly folded and have a very large surface area for the exchange of gases. Snakes have a single, elongated lung on the right side of the body. The left side of the body contains a very tiny lung or no lung at all. Air enters the lungs through a long tube called the **trachea**.

e. The Nervous System and Sense Organs

The structure of the brain and nervous system of the snake is similar to that of the frog; however, the snake has several types of sense organs. Among the most important of their sense organs is that of smell. Most snakes depend on their sense of smell to locate prey. As an aid to smell, many snakes will continually sample the air with flicks of their forked tongues. The tongue then brings air samples into the mouth. If these samples contain chemicals or odors that have been emitted from potential prey, these are detected by the **Jacobson's organ** that is located in the roof of the mouth (see Figure 15-9). Finally, because the snake has a tongue that is forked, it

Figure 15-8. African Egg-Eating Snake

CHAPTER 15: **VERTEBRATES: PART 2**

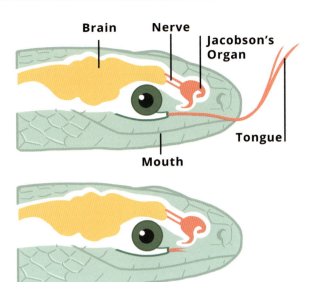

Figure 15-9. The Jacobson's organ in snakes is used with the tongue to detect odors. The snake flicks out its tongue and picks up scent particles. When the snake pulls in its tongue, the particles are transferred to the odor-sensitive Jacobson's organ.

is able to locate its prey "in stereo." That is, as the snake crawls, the forked tongue will instantly detect differing concentrations of chemicals on each side of its tongue. It processes this information and moves in the direction of higher chemical concentrations.

f. Reproduction

Fertilization occurs internally in the reptile. Male reptiles have a structure that transfers the sperm cells into the cloaca of the female during mating. Many snakes lay eggs and then leave these to hatch on their own. The term **oviparous** (ō-vĭ-pă-rus) is used to describe reptiles that lay eggs. Other types of snakes hold the eggs internally until the eggs hatch. These snakes give birth to living young and are described as **ovoviviparous** (ō-vō-vī-vi-pă-rus). In some species, such as the boas and anaconda, the snake embryos develop in the absence of eggshells. Rather, the mother nourishes the developing embryos through a placenta and a yolk sac. As these embryos grow and mature, they are eventually released from the mother as living young. This type of birth is described as **viviparous**.

g. The Variety of Snakes

Living snakes are found on every continent except Antarctica. Fifteen families are currently recognized, comprising 456 genera and over 2,900 species. Snakes range in size from the tiny thread snakes, which are only a few inches long, to pythons and anacondas that are more than 20 feet in length.

Most species of snake are venomous, including many of the so-called harmless snakes. Venoms are secretions of the salivary glands that contain digestive enzymes. The main function of venom is to help digest prey. In some species, the venom consists of powerful chemicals that have been designed to immobilize and even kill prey.

Some of the common snakes that you might find near wooded areas in the eastern United States include the garter snake and black snake. These are relatively harmless snakes and are sometimes called rear-fanged snakes. Although these snakes are considered harmless by some, if they are roughly handled they are capable of inflicting deep bites that could require medical attention. When these species capture their prey, they release a mixture

Figure 15-10. Many snakes lay eggs

Biology For Life

CHAPTER 15: VERTEBRATES: PART 2

Rattlesnake   **Green Tree Python**   **Barbados Threadsnake**   **Corn Snake**

Varieties of Snakes
Figure 15-11

of digestive enzymes into their prey as it passes through the mouth and into the esophagus. The prey remains in the digestive system until the enzymes have turned it into a soupy mixture from which all nutrients have been extracted.

The venomous snakes that most people fear are the front-fanged snakes, such as rattlesnakes and copperheads. These snakes contain powerful nerve poisons called neurotoxins. As strange as it may seem, snakes tend to be shy. In fact, they are more afraid of you then you are of them. Some species of snake will crawl away quickly when they detect vibrations. Others, when cornered, will deliver a warning bite – usually without venom. Others, such as the copperhead, respond aggressively and repeatedly bite their attacker. Although many accidental bites are reported annually in the United States, many bites result from those individuals who choose to harass or pursue fleeing snakes. Herpetologists who have been bitten by snakes can assure you that most people do not have a sense of how quickly snakes can bite! When you see a snake in the wild, it is best to leave it alone. There are many accounts of people having been bitten by snakes when the people attacked the snake first, even though the snake wasn't bothering them. The best advice is to "let sleeping dogs (and snakes) lie."

The best-known poisonous snakes found in North America are thirteen species of pit vipers known as rattlesnakes. Pit vipers possess a small pair of pit-like structures, which are located between the nostrils and the eyes. The pits contain heat sensors that are used to detect and locate small prey, such as birds and small mammals. The fangs of a rattlesnake are hinged and quickly swing forward when the snake is about to strike. These fangs inject poison into a prey organism, which dies quickly and is swallowed whole. The capture and digestion of prey by snakes is truly a unique method of feeding.

Many snakes are constrictors and are fully capable of placing lethal force on their prey. They rapidly grab their prey with their mouth and at the same time quickly coil their body around it. The animal or bird might be able to exhale, but the constrictions prevent them from inhaling. Accordingly, the oxygen supply is cut off and the prey dies from asphyxiation. The snake slowly swallows and digests the dead animal. The large boa constrictors and pythons of South America and Africa are well known for this method of killing very large prey, including pigs and similar-sized animals.

2. Order Squamata: Lizards

The lizard is a four-legged terrestrial animal, and its limbs are supported by a pectoral and pelvic

CHAPTER 15: **VERTEBRATES: PART 2**

girdle. Lizards can move rapidly because their legs are positioned high off the ground. Lizards have two well-developed lungs. Unlike snakes, lizards have external ears and eyelids. Note the iguana in Figure 15-12.

Geckos are small lizards that have special pads on their feet. These pads are somewhat "sticky" and allow them to climb on walls and even run across ceilings. Scientists discovered that the pads on one foot of the gecko contain about 500,000 tiny hairs. Each of these hairs splits into hundreds more. When a gecko attaches to something, it attaches its palm first, then uncurls its toes, all of which form temporary electrical bonds with the surface. (Figure 15-13 shows a gecko.)

Most lizards are small; however, the monitor lizards known as the Komodo dragons are capable of growing over three meters in length. These are found on several islands of Indonesia.

The Galapagos iguana is a marine lizard and herbivore that feeds exclusively on marine

Figure 15-12. Green iguanas are lizards that are popular as pets

Figure 15-13. The Gecko. Notice the spread of its feet, which provide an excellent hold due to their pads. Biologists have recently uncovered the mechanism by which their feet adhere to surfaces.

Figure 15-14. The heaviest living lizard, the Komodo dragon that inhabits several islands in Indonesia and can grow to an average length of 2 to 3 meters (6.6 to 9.8 ft) and weigh around 70 kilograms (150 lb).

Biology For Life

algae. The Gila monster is one of two species of poisonous lizard that are found in North America. It is found in the southwestern deserts of the United States and northwestern Mexico. It is an aggressive lizard that is easily recognized by its reddish-orange and black colors. It eats small squirrels, birds, rodents, other lizards, and insects.

Figure 15-15. The green turtle is found throughout tropical and subtropical seas around the world, with two distinct populations in the Atlantic and Pacific Oceans. Their common name derives from the green fat that is located underneath their shell.

3. Order Chelonia: Turtles

The Chelonia (testudines) contains all of the world's **turtles** and **tortoises**. Turtles live in water, and tortoises live on land. These animals are the only vertebrates that have their bodies fully enclosed in a protective shell. Their skeletons have a unique design. The vertebrae are fused to the shell, and the ribs are widened and flattened near the interior surface of the shell. There are two parts to the turtle or tortoise shell. The

Figure 15-16. A Galapagos Giant Tortoise

CHAPTER 15: **VERTEBRATES: PART 2**

Figure 15-17. A Diamondback Terrapin

dorsal surface of the shell is called the **carapace**, and the ventral surface is called the **plastron**. When the turtle withdraws its head into the shell, the vertebrae bend the neck into an S-shape.

Turtles may occupy freshwater, marine, or brackish habitats. Figure 15-15 shows the green turtle, one species of marine (sea) turtle.

Sea turtles often grow to a very large size. They spend their adult life in the ocean and only come on land to lay their eggs.

Three species of tortoise are found in the United States. Two of these are found in the southern and southwestern deserts, and one species is found along the Gulf Coast, especially along dunes and other upland habitats. Most tortoises are small; however, the Galapagos tortoise can weigh almost 250 kilograms, or nearly 550 pounds – that is, a little over ¼ ton!

The chelonians have no teeth, and most species are herbivores. Freshwater turtles are called **terrapins**; one species is seen in Figure 15-17.

Snapping turtles are terrapins that have very powerful jaws and are carnivorous. These are very common throughout much of the eastern United States. If you spend time near ponds of any size in the early spring, snapping turtles are a common sight as they bask in the sun following hibernation. They often grow to very large sizes, with some reaching the size of a small trash can lid.

4. Order Crocodilia

All members of this order live in shallow water. They have powerful tails, which they use to propel themselves through the water.

The eyes and nostrils of these reptiles are positioned on the top of their head. Most of the time, these are the only parts of the body that are visible above the water. You can see this unique feature in Figure 15-18.

Figure 15-18. The Nile Crocodile

Biology For Life

Figure 15-19. An American crocodile. Notice that its eyes and nostrils are the only parts visible above the water.

The crocodilian heart has four chambers, with the right and left ventricles fully separated. All crocodilians lay their eggs on land.

There are about twenty-five species in the order Crocodilia. The alligators have a much broader and more rounded snout than the crocodiles. The teeth of the alligator are hidden when the mouth is closed, whereas the teeth of the crocodile are visible.

Two other members of this order are the Asian **gavial** and the tropical American **caiman**.

5. Order Rhynchocephalia: Tuataras

The tuatara is the only known species within this order (see Figure 15-21). It can be found on a few remote islands near New Zealand. This reptile has bony skull plates that are different from those of any other living group.

The tuatara has a third eye in the middle of its forehead, which is called a **pineal eye**. This small

Figure 15-20. American Alligator

eye is covered by scales and cannot form images. Scientists are uncertain as to the function of this eye; however, they speculate that it is involved in its diurnal, or daily, behavioral rhythms.

Figure 15-21. The Tuatara belongs to the order Rhynchocephalia

> **Section Review — 15.2 B**
>
> 1. List the four orders of reptiles.
> 2. What is the largest internal organ in a snake?
> 3. What is the function of the Jacobson's organ?
> 4. What is the difference between oviparous and ovoviviparous reptiles?
> 5. What features do lizards have that snakes do not?
> 6. There are two parts of a turtle or tortoise shell. The dorsal surface of the shell is called the _____, and the ventral surface is called the _____.
> 7. How are alligators and crocodiles different?
> 8. The tuatara has a third eye in the middle of its forehead, which is called a _____.

15.3 Birds (Class Aves)

Section Objectives

- Describe the importance of Archaeopteryx.
- List the three types of feathers and explain their function.
- Describe the features of the bird's head, including the nares, beak, eyes, ears, and neck.
- Describe the wings, legs, and feet of a bird.
- Explain the adaptations for flight in terms of the skeletal, digestive, respiratory, and circulatory systems of the bird.
- Describe the function of a bird's excretory and nervous systems.
- Describe the structure of bird eggs.
- Describe the development of the bird embryo.

A. Archaeopteryx

Birds are an important group of animals that are often described as having developed or evolved from reptile ancestors. Indeed, many textbooks in use today point to Archaeopteryx, a fossil bird, as a transitional organism between dinosaurs and birds. In these textbooks, Archaeopteryx is believed to have more in common with a group of dinosaurs known as theropod dinosaurs. As evidence for this transitional form, one group of scientists points to at least twenty-one specialized characteristics, which include sharp teeth, three toes with claws, bony tail, certain skeletal features, and feathers. It is worth keeping in mind that Archaeopteryx was first described in 1861, two years following the publication of Charles Darwin's *Origin of Species*.

CHAPTER 15: VERTEBRATES: PART 2

Figure 15-22. Emperor Penguin and baby

Other scientists insist that Archaeopteryx was a bird and not a feathered reptile. These scientists suggest that Archaeopteryx had the feathers and wing structures of modern-day birds and a well-developed furcula (wishbone). Based on this evidence, they believe that nothing could have prevented it from flying. In addition, when the head of Archaeopteryx was removed from the limestone in which it was held, its cranial structure was found to be bird-like. There exist a number of other findings about Archaeopteryx that have challenged many previous assumptions.

B. What are Birds?

Birds (class Aves) are winged vertebrates that are bipedal (have two legs), endothermic (warm-blooded), and lay eggs. There exist about 10,000 living species of birds. Birds are widely distributed throughout the Earth. They live in remote areas and in a great variety of locations, including the polar regions, deserts, mountains, and lowlands. They exhibit a wide range of behaviors and clearly show the unique designs of the Creator. For example, the Emperor Penguins live and breed

CHAPTER 15: **VERTEBRATES: PART 2**

ability to fly. Flight enables birds to escape predators and avoid harsh climates. Flight also provides birds with the means to locate new food sources when a current source runs out. They can fly or hop around until they find new food. You can see the graceful flight of a bird in Figure 15-24.

1. Characteristics of Birds

There are many structural and functional characteristics of birds, which are not found in other vertebrates. These characteristics have been designed by a Creator and make flight possible. All of the parts of a bird work together to produce flight.

a. Feathers

The most important and distinguishing characteristic of birds are their **feathers**. Feathers cover the entire body of most birds, with the

Figure 15-23. Gambel's Quail

in the extreme cold of the Antarctic. They can withstand temperatures far below -40° F and can dive to 1/3 mile below the surface of the water for up to 20 minutes. Birds that live in the desert, such as the Gambel's Quail, can withstand high temperatures to a point at which they release heat by dilating the blood vessels in their legs. Other birds that occasionally live in the desert are the black vulture and turkey vulture. To cool themselves during extremely hot conditions, they excrete urine, which runs down their legs. As the urine evaporates, it causes a response known as evaporative cooling. The cooling is extended to the blood vessels in the legs, which, in turn, causes cooling of the blood that is transported to the body. Perhaps the most remarkable behavior is found in the barheaded goose that is found in the mountain lakes of central Asia. It can fly at altitudes of 33,000 feet, or over 6 miles high! At these altitudes, the air is very cold and low in oxygen. Accordingly, the barheaded goose has a special type of hemoglobin, which functions well at high elevations and low temperatures.

The birds are the only animals that have feathers. All birds lay eggs, and no birds bring forth live young, as do some fish, amphibians, and reptiles. One of the most distinctive features of birds is their

Figure 15-24. A Common Tern Flying

Biology For Life

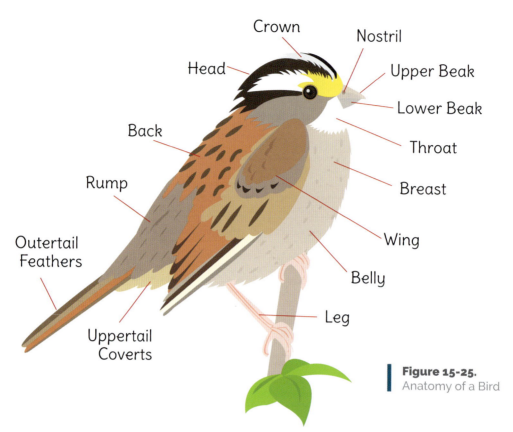

Figure 15-25.
Anatomy of a Bird

exception of the beak, legs, and feet. Feathers are composed of keratin. Feathers are light, but they are also very strong. The strength of the feather is due to its structure. The keratin of the feather forms a hollow tube, which is called a shaft. There are one hundred or more barbs, which are attached to each side of the shaft. Each barb contains a fringe of barbules. There are tiny hooks (hooklets) which connect the barbules of one barb to those of another barb. The structure of the feather is strengthened because there are about four million hooks on the feather.

Figure 15-27 shows the three main types of feathers: down feathers, contour feathers, and quill (flight) feathers. The down feathers are those closest to the bird's body. These feathers insulate the body against heat loss. Throughout history, down feathers have been used in sleeping bags and comforters because of the warmth that they retain. The contour feathers cover the down feathers. These feathers give shape to the bird's body, making it streamlined for flight. These feathers also give the bird its coloration.

Flight feathers are the large feathers of the wing and tail.

Flight feathers of the wing are collectively known as the **remiges**

Figure 15-26.
Parts of a feather

Biology For Life

and are separated into three groups: **primaries**, **secondaries**, and **tertiaries**.

The primaries attach to the metacarpal (wrist) and phalangeal (finger) bones at the far end of the wing and are responsible for forward thrust. There are usually 10 primaries, and they are numbered from the inside out.

The secondaries attach to the **ulna**, a bone in the middle of the wing, and are necessary to supply "lift." They are also used in courtship displays. There are usually 10-14 secondaries, and they are numbered from the outside in. The flight feathers closest to the body are sometimes called tertiaries.

The tail feathers, called retrices, act as brakes and a rudder, controlling the orientation of the flight. Most birds have 12 tail feathers. The bases of the flight feathers are covered with smaller contour feathers called coverts. There are several layers of coverts on the

Figure 15-27.

wing. Coverts also cover the ear. A bird's wing comprised of its remiges and coverts is shown in Figure 15-28.

Most birds have oil glands located at the base of their tails. A bird uses its beak to take oil from these

Figure 15-28. Structure of a Bird's Wing

Biology For Life

Molting

Figure 15-29. A Loggerhead Shrike in mid-molt (top) compared to one with regular plumage (bottom).

glands, then spreads the oil over its feathers. This behavior waterproofs the feathers so that any water is repelled and rolls off the bird as small beads.

Many birds periodically lose their feathers. The feathers are lost gradually, and new feathers continuously replace the lost ones. This replacement of feathers is known as molting. You can see molting taking place in Figure 15-29.

b. Structures of the Head

The beak of a bird is also made up of keratin. There are two nares, or nostrils, on the beak, which function in breathing. The beak is toothless. Birds use their beaks for scratching, cleaning, repairing

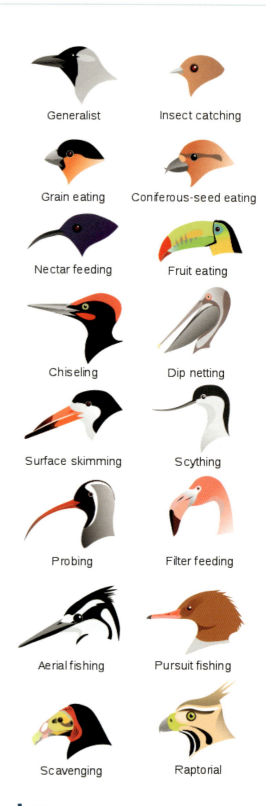

Figure 15-30. Variations in Bird Heads.

Figure 15-31. Barn Owl

feathers, collecting nesting materials, and defending themselves against enemies. The most important function of the beak is to obtain food. A wide selection of bird heads is shown in Figure 15-30.

The size and shape of a bird's beak is related to the type of food the bird eats. For example, the beak of the sword-billed hummingbird is four times longer than its body, because the food source for this bird is nectar, which lies deep within a flower. If the hummingbird had a shorter beak, it could not reach the nectar. A flamingo's beak contains a sieve that strains tiny crustaceans out of the water.

Birds have very keen vision. They are able to see colors and discern objects at great distances. Keen vision is useful to predatory birds, such as hawks, when searching for mice or other small prey such as insects. Most birds have eyes on the sides of their head. As a result, both eyes cannot look in the same direction at the same time. A few birds, such as owls, have both eyes located at the front of the head. The owl can focus both eyes on a single object and accurately judge distances.

The bird's ears are located directly behind the eyes. These are not easily seen, because they are covered and protected by feathers. Birds have an acute sense of hearing. The canals of the outer ear lead to the eardrums, or tympanic membranes, within the head. The bird has Eustachian tubes that connect the mouth cavity with the eardrums and equalize the pressure on both sides of the bird's eardrum.

Birds do not have a well-developed sense of smell or taste, with the exception of vultures. Vultures are attracted from long distances to the odor of decaying animals. For the most part, birds rely on their senses of hearing and sight. The bird's tongue is used to pick up food.

A bird has a very flexible neck and can turn its head 180 degrees. It can look directly behind itself. The bird is capable of this movement

CHAPTER 15: VERTEBRATES: PART 2

because it has many more vertebrae in its head that do other vertebrates.

c. Other External Features

Unlike most vertebrates, birds are capable of standing on two legs. A bird has wings for forelimbs, rather than arms or fins. The tail of a bird consists only of feathers, and not of bones, as in reptiles. Efficient flight would not be possible if a bird had bones in the tail area.

A bird's foot is clawed and covered with scales made of keratin. The feet of a bird vary according to the particular lifestyle of the bird. A typical bird foot has four toes, three pointing forward and one pointing backward. This arrangement facilitates perching, since it allows the feet to wrap around the perch. Some birds, such as ducks, have webbed feet, which aid the bird in swimming. Hawks, eagles, and owls have pointed talons, which are used to kill prey. Birds such as the great blue heron have feet that are used for both wading and walking near shore.

The wings of a bird vary greatly in size and shape. Birds with long wings, such as swifts, swallows, falcons, and many sea birds, are long distance fliers. These birds need wide-open spaces in order to accelerate, turn, glide, and brake. Birds that are short distance fliers live in woods, bushes, and reeds. Their short, wide wings allow them to turn easily in crowded areas. A blackbird is an example of a short distance flier. Some birds, such as ostriches, penguins, and cassowaries, are not able to fly. Flightless birds have other means with which to protect themselves from predators. The cassowary, for example, can kick a large animal hard enough to rip open the animal's body.

2. Internal Features of Birds

The anatomy of a bird differs markedly from that of a reptile or amphibian. Most of the differences serve to enhance the bird's

Figure 15-32. Great Blue Heron

Biology For Life

ability to fly. For example, the internal organs of a bird are packed closely together around the **center of gravity**. The center of gravity refers to the point in the body where the mass is concentrated. It is the balance point between the wings in one direction and the head and tail in the other. This arrangement of the organs ensures stability during flight.

a. Skeleton

The bird skeleton is very light compared to that of the other vertebrates. Although it is the heaviest part of the bird's body, the bones are hollow. The strength of the bones is provided by cross struts, which add very little weight. These are small structures on the inside of bones that attach one side of the bone to the other and can be likened to the struts that are found in old-fashioned steel girder bridges. The skeleton of a bird is made up of the pectoral and pelvic girdles, cranium, backbone, and limbs.

The **pectoral girdle** is made up of the **sternum** (breastbone), **clavicle**, **coracoid**, and **scapula**. The clavicles come together to form the furcula, or "wishbone." The furcula provides a flexible attachment site for the breast muscles, and the coracoid functions as a strut. During flight the coracoid resists pressure that is created by the wing stroke. Simultaneously, flight muscles that connect the sternum to the humerus work to alternately elevate and depress the wing.

The pelvic girdle is a group of fused bones that provides rigid support for the legs. It functions to reduce mechanical stress that occurs from both take-offs and landings. The pelvic girdle can be divided into two groups of bones, the synsacrum and the pygostyle. The synsacrum is a fusion of the pelvic and vertebrae of the tail. The pygostyle is located at the end of the spinal column and represents a fusion of the terminal vertebrae of the tail. It supports the tail feathers and musculature.

Most of the skeletal features just described are shown in Figure 15-33.

The breastbone is very large and serves as one point of attachment for the powerful flight muscles.

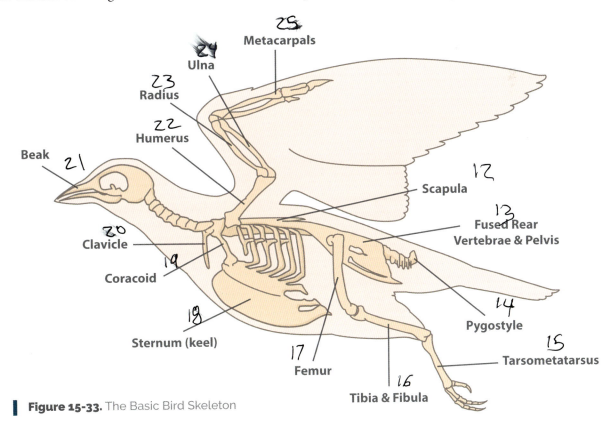

Figure 15-33. The Basic Bird Skeleton

The other ends of the flight muscles attach to the bones in the upper part of the wing. As a bird flies, its wings go downward and backward, then upward and forward to the original position. This series of motions, along with the shapes of the wings, results in flight. It is worth summarizing that bird flight is made possible by several things working together: (1) wing motions, (2) wing structure, (3) hollow bones, (4) light skeleton, and (5) organs positioned in a small, fixed center of gravity.

b. Feeding and Digestion

An enormous amount of energy is required for flight. Birds must obtain this energy by eating food that is high in calories. Depending on the species, birds eat fruit, nectar, seeds and nuts, insects, fish, captured prey (for example, hawks snatching mice, etc.), and carrion. As birds typically have higher body temperature requirements than mammals, they require larger portions of food or frequent feedings or both. These requirements can easily be observed in the wintertime, especially if you have a birdfeeder in your yard. The colder the air temperature becomes, the more time birds will spend at the feeder.

Birds have no teeth and are unable to chew their food. Although they cannot chew, most seed-eating birds are very good at breaking the seeds they eat into smaller pieces. Similarly, birds that eat insects may crush their prey or use their beaks to tear it into tiny pieces. This ability to tear is also found in the owls, hawks, eagles, and vultures.

Birds swallow their food whole, and it passes through the pharynx and down into the esophagus. The food is then stored in the crop, where it is moistened and prepared for digestion. These parts of the digestive system are shown in Figure 15-34.

The bird's stomach is divided into two parts, the proventriculus and the gizzard. The walls of the proventriculus secrete gastric fluid, which contains digestive enzymes. The gastric fluid mixes with food and the mixture is passed to the gizzard. The gizzard contains opposing hard plates that rub against each other and crush the food between them. This process of breaking the food up into smaller pieces aids in digestion. You should note that the function of the gizzard is similar to the teeth of many vertebrates. In both cases, the grinding reduces the surface area of the food, thus facilitating digestion. Some birds, such as pigeons, ingest small pebbles that make their way to the gizzard. These help break apart tough seed coats. Turkeys have very strong gizzards and are capable of breaking apart acorns and some walnuts.

After spending time in the gizzard, the partially digested food passes into the intestine, which absorbs nutrients from the food and allows indigestible materials to pass into the rectum. These wastes are then expelled through the cloaca.

The digestive process in birds occurs quickly. For example, a bird can eat a berry and evacuate

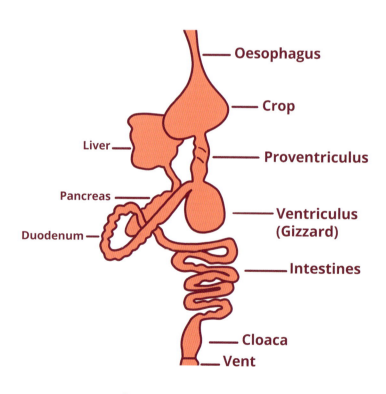

Figure 15-34. The Bird's Digestive System

Figure 15-35. Bird Migration

the seed from its body in about twelve minutes. There are two advantages to this efficient digestive system. First, the bird can process a large amount of food in a short time, and, second, it quickly eliminates weight-bearing wastes that would hinder flight.

Because a bird takes in large quantities of food and has feathers for insulation, the bird is able to regulate its internal body temperature. This characteristic, called warm-bloodedness, is not found in reptiles or amphibians. Warm-blooded animals have higher metabolic rates and therefore require more food. Since birds must have a constant supply of food, many birds migrate to locations where food is abundant. The birds return in the spring when food is more plentiful. Although seasonal migrations of birds are predictable and routine, scientists know very little about how these are conducted by birds.

c. Respiratory System

The respiratory system of a bird contains air sacs, as well as lungs and airways. The air sacs fill up many spaces in the body cavity, and some extend into the hollow bones. Air sacs allow the bird to inhale a large volume of air at one time.

Breathing begins with air that enters the respiratory system through nares (nostrils) in the beak. The air travels through the pharynx, larynx, trachea, and syrinx (sir-inks), or voice box. Below the syrinx the trachea branch off into two tubes called bronchi. Each bronchus passes through a lung and then enters a posterior air sac. Even smaller bronchi branch off the main bronchi and lead to anterior air sacs. Gas exchange occurs through these small bronchi. Recall that gas exchange refers to the absorption of oxygen by the body and its release of carbon dioxide.

d. Circulatory System

The heart pumps blood throughout the body of the bird. The heart contains four distinct chambers, including two atria and two ventricles. As in humans, the atria have thin walls and are located at the upper end of the heart. Conversely, the ventricles are thick-walled and muscular and situated at the lower part of the heart. The characteristics of the ventricles make them well-suited for pumping blood. The right atrium pumps deoxygenated blood into the right ventricle, which, in turn, pumps it to the lungs. The blood releases carbon dioxide and acquires oxygen. This newly oxygenated blood is transported to the left atrium and then to the left ventricle. With each contraction, oxygenated blood is transported through the aorta and to the entire body of the bird.

In comparison with humans that have a resting heartbeat rate (heart rate) of between 60-100 beats per minute, birds have a much higher heart rate. On average, the heart rate will range from about 250-500 beats per minute. However, the hummingbird has one of the highest heart rates among birds. When at rest its heart rate is about 250 beats per minute. When it is feeding it uses significantly more energy and oxygen, and the heart rate increases upwards of 1,200 beats per minute!

e. Excretory System

Birds have a simple excretory system. The breakdown of proteins results in the formation of ammonia, which is the principal waste product in most living organisms. Most animals convert ammonia to uric acid and then to urea. The urea is eventually excreted from the body as urine.

The formation of urine requires relatively large amounts of water. However, because of their size and light weight requirements that are needed for flight, birds can neither hold nor lose a large volume of water. Accordingly, our Creator solved this problem by having birds produce and then excrete uric acid. The production of uric acid requires little water but also requires the expenditure of additional energy. Do you see the trade-off? That is, birds need to conserve space in order to keep their bodies both small and light in weight. The formation of uric acid does not require great amounts of water but does require energy. If birds passed liquid waste, they would need a larger body size by which to process it. The price of having lightweight bodies as well as

Figure 15-36. A Robin

conserving space requires the expenditure of additional energy.

The two kidneys of the bird filter uric acid. The uric acid then enters a pair of ureters that empty into the cloaca. Uric acid and undigested wastes are expelled from the body.

f. Nervous System

A bird's nervous system is highly developed. In comparison with a reptile of the same body weight, a bird's brain is six to eleven times larger. The olfactory lobes of birds are very small as birds have a poor sense of smell. Birds have very good vision and have large optic lobes. The cerebrum of the bird is relatively large. It is the control center of the brain and directs the bird to hop, run, fly, and swim. The cerebellum, which controls muscular coordination, is also well-developed in the bird. The medulla lies at the base of the brain and joins to the upper part of the spinal cord. The medulla controls functions such as breathing.

g. Reproduction

There is only one ovary in the female bird, through which eggs develop and leave via the oviduct. The testes of the male bird lie above its kidneys. The sperm leave the testes by small tubes called the **vas deferens**, which empty into the cloaca.

When breeding season arrives, the birds begin to build nests and, shortly thereafter, they mate. Sperm is released from the cloaca of the male directly into the cloaca of the female. Fertilization takes place internally in the upper part of the oviduct.

As can be seen in Figure 15-37, the newly fertilized egg consists of a tiny embryo resting on the top of the yolk. The yolk contains fats and proteins. These are the nutrients for the developing embryo.

Following fertilization, glands in the oviduct secrete substances that surround the zygote and yolk. The first material to be laid down is the albumen, or the "egg white." The albumen is an additional protein food source for the developing bird. Strands of a stringy material called **chalaza** suspend the embryo from the ends of the egg. The purpose of the chalaza is to keep the embryo from pressing against the wall of the egg (two chalaza are shown in Figure 15-37). The gland in the oviduct also secretes an additional membrane and shell around the albumen, which protects the developing embryo. The shell also has pores through which oxygen and carbon dioxide can pass.

The embryo must remain at a high temperature; therefore a bird must incubate the eggs to keep them warm. The nest materials function to insulate the bottom of the eggs to keep them warm. To prepare for incubation, a parent will loosen or pull off a patch of feathers from part of its chest. This area is called a brood patch. Beneath the brood patch lie enlarged blood vessels that transfer heat from the surface of the skin to the surface of the eggs. The incubation period varies with the size of the bird. The larger the bird, the longer the time that is required for incubation. The incubation times range from about two weeks to almost two months.

h. Embryo Development

After fertilization has occurred and the egg has been warmed, the zygote begins to develop. The zygote divides until the cells that result are

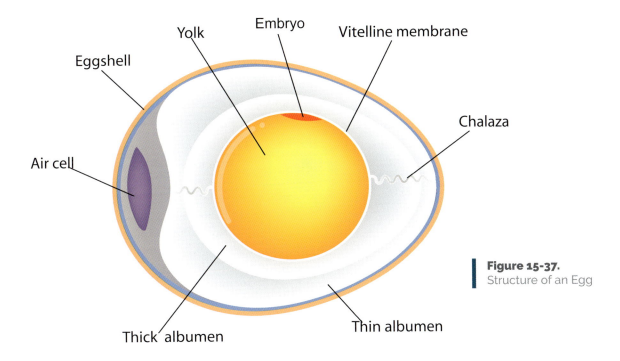

Figure 15-37. Structure of an Egg

plentiful enough to cover the yolk surface. The cells develop into the different types of tissues and organs. The heart is the first organ to develop and is visible and functioning after three days within the chick embryo.

A membrane grows out of the digestive system of the embryo and envelopes the yolk. This membrane, called the yolk sac, produces digestive enzymes. The enzymes digest the food in the yolk, and blood vessels in the yolk sac deliver the digested food to the embryo. A bird's egg contains three membranes in addition to the yolk sac. These include the amnion, allantois, and chorion.

When the embryo has completed its growth within the shell, it uses a special structure called an egg tooth to peck its way out of the egg. The egg tooth is lost after the chick breaks out of the shell (Figure 15-38 shows a one-day-old chicken with its egg tooth). The number of eggs that a species of bird will lay depends upon the amount of care the brood requires. If the young require a great deal of care, then fewer eggs will be laid.

Some birds, which are called **precocial** birds, are very advanced in their development when they hatch. They have down feathers, can feed themselves, and can hop and swim. Some examples of precocial birds are ducks, chickens, turkeys, and swans. Some of these birds can run within a short time following hatching, especially turkeys and quail. Most birds belong to a group called **altricial**, which means they are completely dependent upon their parents. These birds are quite helpless upon hatching, are mostly naked, and cannot see. Blackbirds, nuthatches, robins, and pigeons are examples of altricial birds.

Figure 15-38. A one-day-old chicken with an egg tooth at the tip of its beak on the upper mandible

Section Review — 15.3

1. What is a distinctive feature that birds have?
2. Name the three types of feathers and the function of each.
3. The process of the replacement of feathers is known as _____.
4. To what does the "center of gravity" in birds refer?
5. How is a bird's flight made possible?
6. What is the average heart rate of most birds? Which bird has the highest heart rate?
7. How long does the incubation period last?
8. What is the difference between precocial and altricial birds?

CHAPTER 15: **VERTEBRATES: PART 2**

15.4 Mammals

Section Objectives

- List some of the common features of the Class Mammalia.
- Explain the necessity of the great length of a mammal's developmental period.
- List several characteristics of monotreme and marsupial mammals.
- Describe the several types of marsupial mammals.
- List the functions of the placenta in placental mammals.
- Describe the characteristics of the orders of placental mammals and name a representative animal from each group.

A. Characteristics of Mammals

Mammals are the most specialized group of vertebrates. The features of mammals, including the structure and function of the brain, heart, body covering, and development of the young, are considered in the next section of the text. Four different mammals are shown in Figure 15-39.

1. Common Features of Mammals

Among the vertebrates, the mammalian brain is the most complex. It consists of folded tissues that appear as ridges and grooves, called **convolutions**. These increase the surface area of the brain and result in additional brain capacity. The brain consists of three lobes, the **cerebrum**, **cerebellum**, and **medulla oblongata**, or medulla. The cerebrum is the largest structure of the brain and is responsible for complex behaviors. The cerebellum is also a very large and convoluted region of the brain. It is responsible for motor control, or the coordination of muscle movement. Finally, the medulla is the smallest part of the brain. It has coordinating functions that are called autonomic. That is, these functions are conducted without our consciously thinking about them. These include the heart rate and breathing.

The Mammalia, or mammals, are a class of vertebrates that are warm-blooded and have hair, and the females give birth to living young. The mothers also provide the young with milk and some level of parental care. In fact, all mammals provide care for their young for many years following their birth. Humans, in particular, often care for their children for twenty years.

Varieties of Mammals
Figure 15-39

Giraffe

Fruit Bat

Lion

Hedgehog

Biology For Life

CHAPTER 15: VERTEBRATES: PART 2

Mammals have four-chambered hearts. This arrangement ensures the separation of deoxygenated and oxygenated blood. Mammals breathe with well-developed lungs. A strong muscle, called the **diaphragm** (dī-ə-fram), is located between the lungs and the abdomen and aids in the intake of air. When a mammal inhales, the chest cavity expands, the diaphragm moves downward, and the lungs fill the air. Conversely, when a mammal exhales, the chest cavity contracts, and the diaphragm moves upward, helping to push air out of the lungs.

All mammals have some hair or fur at some point in their life. The purpose of the hair is to help to reduce the loss of body heat through the skin. Many aquatic mammals do not have hair, for example, whales, dolphins, sea lions, etc. As a consequence, they are more streamlined and can easily swim through water. These mammals have a layer of blubber, or fat, beneath their skin that works to prevent heat loss.

Mammalian skin contains four types of glands. The most important of these glands are the mammary glands, from which milk is secreted to nourish the young. The sweat glands help to regulate body temperature and rid the body of wastes. The sebaceous glands produce oil, which helps to lubricate the hair and skin. Scent glands produce chemical substances that help mammals communicate with each other.

Mammalian teeth are very unique. There is a great deal of diversity among the shapes and sizes of mammalian teeth, which reflects the variety of the mammalian diet. Incisors are used for biting and cutting food. The canines are pointed teeth and used for both piercing and tearing. Finally, molars and bicuspids are used for grinding food (see Chapter 8 for a review of the teeth and tooth functions).

2. Mammalian Development

Mammals reproduce by internal fertilization and give birth to living young. However, God has placed a number of exceptions to our classification rules, perhaps to keep us from developing a system that explains everything – and to keep us humble. He gave us egg-laying mammals, such as the duck-billed platypus and the spiny anteater. However, in the other mammals, the developing embryo is protected by its mother's body and is attached to a muscular organ known as the uterus.

After birth, the young receive maternal care. They obtain nutrition from the milk of the mother that is secreted by mammary glands. The mother also protects them from natural enemies and teaches them to search for their own food. In general, the more complex the behaviors that the mammal is capable of performing the longer it will be associated with its mother after birth.

B. Monotremes and Marsupials

Scientists have divided mammals into three groups, based on their methods of reproduction. These three groups are the monotremes, marsupials, and placentals. The monotremes lay eggs and incubate them in a birdlike manner. The marsupials are pouched mammals. Marsupials bear tiny, premature young, called joeys, which leave the mother's uterus and crawl up into the pouch. They attach to nipples and complete their development in the pouch of the mother. The monotremes and marsupials are found mainly in Australia, New Guinea, and New Zealand.

Figure 15-40. A North Pacific Sea Lion

CHAPTER 15: **VERTEBRATES: PART 2**

1. Characteristics of Monotremes

Monotremes are covered with hair, are warm-blooded, and produce milk that nourishes their young. Their cloaca has excretory and reproductive functions. Monotremes have internal fertilization, but the embryos develop externally in eggs. Like birds, monotremes must incubate their eggs.

The heads of monotremes are birdlike in appearance (this is an obvious feature in the three species seen in Figure 15-41). They have long, leathery extensions on their faces that seem somewhat beak-like. The adults have no teeth and lack external ears. There are only three species of monotremes: the duck-billed platypus and two species of spiny anteaters. These are located in New Guinea and Australia.

The duck-billed platypus is a small mammal, about the size of a large squirrel. The platypus spends some of its time in water. Its feet are webbed, and it has claws. When the platypus swims, it paddles with its forefeet and steers with its hind feet. The platypus uses its leathery bill to probe for crustaceans and plants that live at the bottom of the water. The claws allow the platypus to burrow for food when on land.

The platypus is warm-blooded but not as consistently warm-blooded as other mammals. The body temperature of a platypus fluctuates, but averages about 30° C (86° F). The brain of the platypus is small, and it has a smooth surface.

A female platypus digs a long burrow after she mates. She lays two eggs on a bed of moist leaves at the end of the burrow. The female platypus curls her body around the eggs during the incubation period, which lasts for ten days. The mother platypus has no nipples, and milk is produced by sweat glands on the underside of the female. The young are nourished by licking milk from the mother's fur.

The two species of spiny anteater have coarse hair and sturdy spines that cover their bodies. They range in weight from two to ten kilograms (about four to twenty pounds). The spiny anteater has a large, convoluted brain. Its limbs have claws that it uses for digging. The anteater escapes predators by

Monotremes

The Duck-billed Platypus

The Short-beaked Anteater also known as the short-beaked Echidna

The Long-beaked Anteater also known as the Long-beaked Echidna

Figure 15-41. The Three Species of Monotremes

Biology For Life

CHAPTER 15: VERTEBRATES: PART 2

Figure 15-42. The Virginia Opossum, the only North American marsupial

The young are helpless when they hatch and will remain in the pouch until their spines form.

2. Characteristics of Marsupials

The marsupials have internal fertilization, and development of the embryo occurs in the uterus of the mother. The mother develops a tiny yolk sac around the embryo, which furnishes it with nutrients. Marsupials give birth after a very short gestation period, perhaps four to five weeks. At birth, the baby marsupial crawls from the uterus of the mother and onto a nipple, which is located inside a pouch on the underside of mother, or is exposed to the outside environment. The pouch protects the young animal, and the milk provides nourishment until the young can forage for food on their own. The length of time spent in the pouch varies depending upon the species of marsupial.

3. Representative Marsupials

The opossum (see Figure 15-42) is the only North American marsupial. Most marsupials come from Australia. Opossums live mostly in rural

burrowing rapidly into the ground. They have long slender snouts that are used for foraging. Anteaters turn over stones and break into termite and ant nests. They capture their prey with their long, sticky tongues.

The female spiny anteater lays one leathery egg, which she incubates in a pouch on her belly.

Figure 15-43. The Koala. Note the claws that firmly attach the koala to the tree

Biology For Life

CHAPTER 15: **VERTEBRATES: PART 2**

Figure 15-44. Female Eastern Grey Kangaroo with a joey in her pouch

Figure 15-45. A young Eastern Grey Kangaroo in motion

areas and are active at night. Their diet consists of small birds and mammals, eggs, and insects.

The opossum has the shortest period of gestation among marsupials. The young are born only 12.5 days after fertilization. The number of young ranges from six to fourteen, and they are very small at birth. It has been estimated that a litter of fourteen baby opossums can fit in one teaspoon!

The marsupial mole is a rodent-like marsupial that lives in Australia. It is a burrowing animal that has a pouch, which opens at the posterior end of its body. The young are protected in the pouch when the mother burrows through the dirt.

The koala is a large, slow-moving marsupial, which lives in the eucalyptus tree in Australia. Its diet consists of the leaves from the eucalyptus tree. The mother koala gives birth to only one young at a time. The young koala spends six months in the mother's pouch and another six months on her back.

Kangaroos are the best-known marsupials that come from Australia.

The diet of the kangaroo consists mainly of grass. Smaller kangaroos are called wallabies. These animals have large and powerful hind legs. They also have a large tail, which helps them maintain their balance when they hop. The powerful hind legs, as well as the large tail, are clearly evident in the mother kangaroo shown in Figure 15-44.

The kangaroo can hop as fast as sixty kilometers per hour (about 36 mph) and can jump over fences about three meters high (9 feet) – they should be easily able to dunk a basketball!

Section Review — 15.4 A & B

1. What are the three groups into which mammals are divided into?
2. What characteristics distinguish monotremes and marsupials?
3. Give some examples of monotremes.
4. Give some examples of marsupials.

Biology For Life

C. Placental Mammals

The placental mammals are the most abundant group of mammals. Marsupials and placentals give birth to live young; however, the two groups differ in the way that the young are nourished before birth. The developing marsupial leaves the uterus so early that sometimes it does not survive the journey into the pouch. The young of placental mammals remain inside the uterus until they are more fully developed.

1. Characteristics of Placental Mammals

A special structure called the placenta allows these mammals to remain in the mother's uterus until they can develop enough to withstand the hazards of the outside environment following birth. The placenta is a flattened organ that is rich in blood vessels. It is connected to the developing baby animal by a structure called the **umbilical cord**.

The placenta and the wall of the uterus have convoluted surfaces. These convolutions greatly increase the surface areas of the placenta and the uterine wall that contact each other. Materials are exchanged between the mother and the young through these points of contact. Although the blood of the mother does not cross the placenta, the oxygen and food are transported through the placenta to nourish the young. Carbon dioxide and nitrogen wastes from the embryo travel through the umbilical cord, then through the placenta, and finally enter the bloodstream of the mother. Her kidneys excrete her own wastes along with those that have been produced by her young.

The gestation period varies with the size of the animal. The mouse has a gestation period of only twenty-one days, whereas the elephant has a gestation period of almost two years. Humans have a gestation period of nine months.

2. Some Types of Placental Mammals

Scientists recognize seventeen orders of placental mammals. A description of the most well known of these orders follows.

Rodentia: Rodents

The rodents are the largest order of mammals, are found worldwide, and contain about 1,700 species. Mice, rats, gerbils, and squirrels are typical rodents. Most rodents are plant eaters, but some eat insects, fish, reptiles, small mammals, and birds. These animals may live in trees, on the ground, in burrows, or in water. No doubt you have seen all of the rodents in Figure 15-46.

All rodents have a pair of incisor teeth on each jaw. These incisors grow continuously. When the upper teeth move against the lower teeth in the

Members of the Order Rodentia
Figure 15-46

Eastern Gray Squirrel

House Mouse

Common Brown Rat

American Beaver

CHAPTER 15: **VERTEBRATES: PART 2**

Members of the Order Lagomorpha
Figure 15-47

Pika

Eastern Cottontail

Jackrabbit

European Hare

process of gnawing, part of the back surface of the upper teeth wears off, creating sharp, chisel-shaped teeth.

The beaver is an outstanding example of how an animal can change its environment to suit its needs. Beavers use their sharp teeth to cut down trees and build dams across streams. These dams help turn streams into lakes, and, in the process, the dams become their homes. They have large hind feet that are webbed to aid in swimming. The openings to their eyes and ears can close when they are submerged in the water.

Lagomorpha: Rodent-like Animals

The lagomorphs consist of rabbits, hares and jackrabbits, and pikas – see Figure 15-47. These animals are found throughout much of the world. They have been able to adapt to environments such as the Arctic and the desert. They live in forests as well as treeless mountains.

Pikas are small hamster-like animals, with short limbs, rounded ears, and short tails. They are native to cold climates, mostly in Asia, North America, and parts of Eastern Europe. Many persons mistakenly think that rabbits and hares are rodents. They have incisors that grow continuously; however, they have additional, small, peg-like incisors that are located behind the two larger pair of incisor teeth. Rodents do not have these.

All rabbits (except the cottontail rabbits) live in underground burrows or warrens, while hares and jackrabbits (and cottontail rabbits) live in simple nests above the ground and usually do not live in groups. Hares and jackrabbits are generally larger than rabbits and have longer ears. They also have

Biology For Life

black markings on their fur. Hares and jackrabbits have not been domesticated, while rabbits are often kept as house pets.

Chiroptera: Bats

The Chiroptera represent the second largest order of class Mammalia. This order contains about 850 bat species (one such species is shown in Figure 15-48). Bats are the only group of mammals that are able to fly.

Bats live everywhere except the polar regions and a few islands. The fruit bat is the largest of all bats, with a wingspan of 1.5 meters (about 4.5 feet; see Figure 15-49).

Figure 15-48. Bats are the only group of mammals that can fly

Most bats fly at night and use echolocation to guide them. Echolocation is a process that involves the production of high-pitched (high frequency) sounds that make echoes when they are reflected off of objects. The reflected high-frequency sounds are "heard" by the bat that produced the sound. The bat uses the echo, or returned sound, to locate the object. The bat uses echolocation to avoid flying into obstacles in its paths and to capture flying insects such as flies, mosquitoes, and moths.

The diet of the bat varies, depending upon the species. Sources of food include insects, nectar, pollen, frogs, lizards, birds, and other bats. Vampire bats drink blood. Their saliva contains a protein that prevents the blood of their prey from clotting. After a bat has bitten an animal, it licks the free-flowing blood from the wound.

The young develop for a long time within the female bat. Following birth, the young leave the uterus and climb on the front side of her body to acquire milk. Most bats live for a very long time, with some species living up to twenty years. This is a relatively long time for such a small mammal, when considering that the average mouse may live up to two and a half years in captivity (but only several months in the wild).

Figure 15-49. Fruit Bat

Biology For Life

CHAPTER 15: **VERTEBRATES: PART 2**

Members of the Order Cetacean
Figure 15-50

Humpback Whale

Bottlenose Dolphin

Harbor Porpoise

Mammals That Live in Water

Many species from among several orders of mammals live in water. Aquatic mammals tend to be hairless and have a thick layer of blubber, which insulates them from the cold. Their forelimbs are usually paddle-shaped and clawless. They often have no visible hind limbs.

The order Cetacean (si-tā-shen) is a large group of aquatic mammals that includes whales and dolphins.

Whales breathe air like all other mammals. These animals can alternate breathing with periods of non-breathing. Whales may remain submerged for as long as two hours. They accomplish this by having twice as many oxygen-carrying red blood cells in relation to body weight as do the land-dwelling mammals. Whales can also store oxygen in their muscles.

Porpoises tend to be smaller but stouter than dolphins. They have small, rounded heads and blunt jaws instead of beaks. Both porpoises and dolphins have a structure called a "melon," which scientists believe is used for echolocation. The melon found on dolphins is round and bulbous, whereas on porpoises it is less obvious. The teeth of porpoises are spade-shaped, whereas dolphins have conical teeth. In addition, a porpoise's dorsal fin is generally triangular, rather than curved, like that of many dolphins and large whales. Some species have small

Biology For Life   423

Members of the Order Carnivora
Figure 15-51

Australian Shepherd Dog

Siamese Cat

Raccoon

Brown Bear

Short-Tailed Weasel

bumps, known as tubercles, on the leading edge of the dorsal fin. You can see the shapes of the two types of dorsal fins in Figure 15-50.

Carnivora: Flesh-eating Mammals

The order Carnivora contains about 280 species, most of which live on land. The carnivores include familiar animals such as cats, dogs, bears, raccoons, and weasels (see Figure 15-51). These animals have long pointed canine teeth that are used for shredding flesh.

Examples of Aquatic Carnivores
Figure 15-52

Sea Lion

Pacific Walrus

Common Seal

Biology For Life

Members of the Order Perissodactyla
Figure 15-53

Zebras

Horse

Black Rhinoceros

Members of the Order Artiodactyla
Figure 15-54

Goat

Hippopotamus

Giraffes

Llama

CHAPTER 15: **VERTEBRATES: PART 2**

Hooves of Perissodactyla
Figure 15-55

Horse

Rhinoceros

Some mammals in this order are aquatic and include seals, sea lions, and walruses (see Figure 15-52).

Ungulates: Hoofed Mammals

There are two orders of mammals in this group that include the Perissodactyla, or odd-toed hoofed mammals, and the Artiodactyla, or even-toed hoofed mammals. You can see three members of the Perissodactyla in Figure 15-53 and four members of the Artiodactyla in Figure 15-54.

The odd-toed mammals include the zebras, horses, tapirs, and rhinoceroses. The even-toed mammals include the goats, sheep, pigs, hippopotamuses, camels, llamas, deer, giraffes, and various types of cattle.

Ungulates have no claws. Instead, they walk around on the tips of their toes. Their hooves are actually modified toenails.

The ungulates have wide teeth with large surfaces that are used for grinding their food. Most of these animals are very large, which gives them an advantage in fighting or escaping their predators.

The cells of plants are surrounded by a cell wall made of cellulose; therefore, plant material is much harder to digest than animal matter. Most mammals do not have the digestive enzymes needed to break down the cellulose in plants. Ungulates, however, have microorganisms that live in their digestive tract that aid the animals in cellulose digestion. Some ungulates, called ruminants, have several compartments to their stomachs. Food is partially digested and then regurgitated and re-chewed. The food then passes on to another compartment of the stomach. When a cow "chews its cud," this means that it is re-chewing food that has been regurgitated so that the food can be more fully digested.

Hooves of Artiodactyla
Figure 15-56

Giraffe

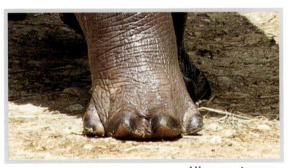
Hippopotamus

Biology For Life

CHAPTER 15: **VERTEBRATES: PART 2**

Insectivora: Insect-eating Animals

There are about 400 species of insectivores. These mammals live on all the continents except Australia and Antarctica. These animals are generally small with longish snouts. Two of these insect-eating animals can be seen in Figure 15-57.

Most insectivores have a well-developed sense of smell. Members of this order of mammals include European hedgehogs, moles, and shrews.

Edentata: Toothless Mammals

This order includes about thirty species that live in North and South America. One of these mammals is the anteater. It has a long tongue with sticky saliva that it uses to catch termites.

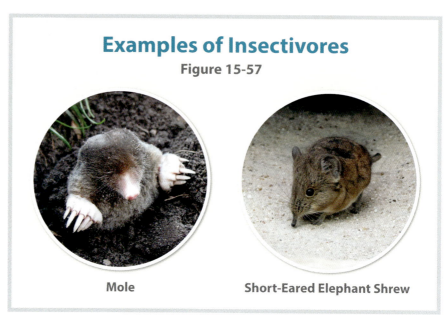

Examples of Insectivores
Figure 15-57

Mole   Short-Eared Elephant Shrew

Figure 15-58. Some Edentata: Anteater, Armadillo, and Sloth

Biology For Life

Figure 15-59. Black-and-White Ruffed Lemur

The tree sloth lives in the rain forests of South America. It is very slow-moving and eats plants. The three-toed sloth moves so slowly that it accumulates green algae that live in its fur.

The armadillo is also a toothless mammal that is covered with bony plates. They burrow into the ground or curl up for protection.

Primates

The primate order contains about 170 species of mammals. Recall that we are Primates, and we can live anywhere. The non-human primates live in the tropical and subtropical regions of the world. These include the monkeys and apes as well as the prosimians (lemurs, lorises, and tarsiers); one of these is shown in Figure 15-59.

The four types of apes include the gorillas, orangutans, chimpanzees, and gibbons. Gorillas are the largest of the living primates. They are ground-

Figure 15-60. A young male white-fronted capuchin monkey

Biology For Life

Chimpanzees

dwelling herbivores that inhabit the forests of Africa. The orangutans are a species of great ape. Known for their intelligence, they live in trees and are the largest living arboreal (tree-living) animal. In comparison with the other great apes, they have longer arms, and their hair is reddish-brown, instead of brown or black.

The feet of the chimpanzee are better suited for walking than those of the orangutan. This is because they have broad soles and short toes. Both the common chimpanzee and bonobo (another species of chimpanzee) can walk upright on two legs when carrying objects with their hands and arms. Their face, fingers, palms of the hands, and soles of the feet are hairless, and they have a dark coat. They do not have tails.

Gibbons are referred to as the lesser apes and differ from great apes (chimpanzees, gorillas, and orangutans) in being smaller and pair-bonded and by not making nests. In addition, there exist minor anatomical details by which they are more closely aligned with monkeys than great apes.

Section Review — 15.4 C

1. How do marsupial and placental mammals differ?
2. What is echolocation?
3. The order Cetacean is a large group of aquatic mammals that includes _____ and _____.
4. Perissodactyla are _____ hoofed mammals and Artiodactyla are _____ hoofed mammals.
5. Most insectivores have a well-developed sense of _____ and include mammals such as the _____, _____, and _____.
6. What are the four types of apes?

CHAPTER 15: **VERTEBRATES: PART 2**

Orangutans

Gibbon

Silverback Gorilla

Figure 15-61. The Apes

Chapter 15 Supplemental Questions

Answer the following questions.

1. Know and be able to label Figures 15.3, 15.25, and 15.33.
2. Describe the characteristics that differentiate reptiles from amphibians.
3. Name the four orders of living reptiles.
4. List two ways that snakes kill their prey.
5. What is unique about the birth of snakes?
6. Explain the following terms: oviparous, ovoviviparous, viviparous.
7. Describe the features of chelonians.
8. Contrast the lizard and the snake.
9. List some of the animals in the order Crocodilia and explain how they are different from other reptiles.
10. Describe the uniqueness of the tuatara.
11. Explain the functions of the yolk, albumen, and chalaza in a bird egg.
12. List the three types of feathers and explain their function.
13. Describe the features of the bird's head, including the nares, beak, eyes, ears, and neck.
14. Describe the wings, legs, and feet of a bird.
15. Explain the adaptations for flight in terms of the skeletal, digestive, respiratory, and circulatory systems of the bird.
16. What is "warm bloodedness," and which type of vertebrates have this characteristic?
17. Describe the function of a bird's excretory and nervous systems.
18. Describe the structure of bird eggs.
19. Explain the function of nest building and incubation of bird eggs.

Chapter 15 Supplemental Questions

20. Describe the development of the bird embryo.

21. List some of the common features of the class Mammalia.

22. How do the terms "altricial" and "precocial" apply to birds?

23. What are the three groups into which mammals have been divided, and what is the basis of this differentiation?

24. List several characteristics of monotreme and marsupial mammals and give three members of each group.

25. Describe the several types of marsupial mammals.

26. Which marsupial is unique to North America? Where do most marsupials come from?

27. List the functions of the placenta in placental mammals.

28. Describe the characteristics of the orders of placental mammals and name a representative animal from each group.

29. Study all diagrams in this chapter. Be able to draw and label all diagrams.

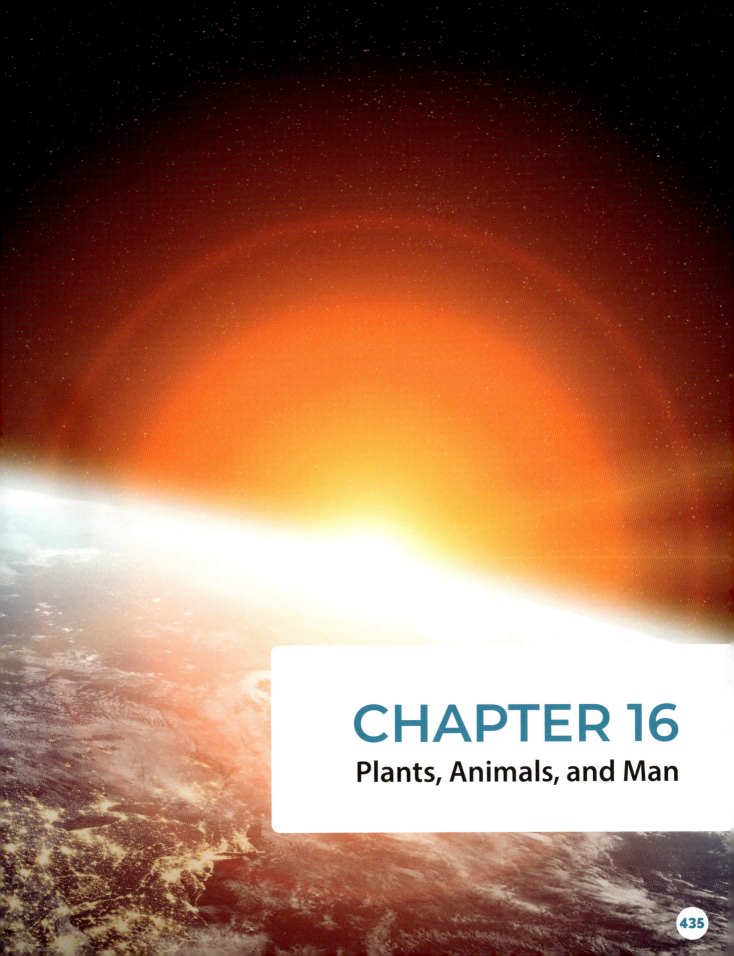

CHAPTER 16
Plants, Animals, and Man

Chapter 16

Plants, Animals, and Man

Chapter Outline

16.1 Introduction

16.2 Life and the Soul
　　The Concept of Life
　　The Soul

16.3 The Vegetative Soul

16.4 The Sensitive Soul
　　Plants and Animals
　　Animal Behavior
　　Are Animals Intelligent?

16.5 The Rational Soul
　　Human Beings and Creation
　　The Purpose of the Human Life
　　Natural Moral Law

16.1 Introduction

So far we have looked at living organisms beginning with the basic building blocks from which these have been created. That is, all living things have been created from organic molecules and consist of carbon, hydrogen, oxygen, and nitrogen. These molecules are organized into cells, the cells into tissues, tissues into organs, and organs into systems. The similarities and differences that are found among the constituent parts of organisms have been used to construct the biological classification system.

In this chapter we intend to study organisms as living things. We will address such questions as "Why are living things alive?" "What can organisms do as living things?" and "Why are humans unique among all living things?" In our time these questions are treated in such branches of biology as Animal Behavior (Ethology) and Animal Psychology; in previous times they were treated in the science of Philosophical Psychology.

16.2 Life and the Soul

Section Objectives

- Define what it means for a thing to be alive.
- Explain what a soul is.
- Explain why there are three types of soul.
- List the characteristic activities of the vegetative life, animal life, and human life.

The Concept of Life

Life can best be described as the capacity for self-change, a concept that is best understood in the processes of nutrition, growth, and reproduction. For a thing to be considered living, it must be able to conduct all of these processes. Some organisms, such as those found in the Kingdom Animalia, can change themselves in more complex ways – for example, moving to new, suitable locations. However, all living things acquire nutrients, grow, and reproduce. They are always changing until the moment they die. For humans, death is where the physical body separates from the soul, and we enter the spiritual world. At the time of death, we are no longer capable of changing in a physical manner.

1. Nutrition ranks among the most basic of functions found in living things. Its purpose is to maintain the organism and allow it to carry out its other activities. It involves obtaining or making

Biology For Life

one's own food. For organisms that cannot make their own food, nutrition involves the ingestion and breakdown of complex substances and converting these into new molecules that become part of the organism. When Billy eats an apple, he does not become apple-like. Rather, Billy's digestive system breaks the apple down into its various molecules that are used by his body. So the apple becomes part of Billy.

2. Growth refers to the capacity for living things to change in both size and shape. It also refers to the differentiation of cells, tissues, organs, and changes in systems that result in new functions. Hair, teeth, nails, bones, fat, blood, skin, and muscle all arise from the tiny, undifferentiated fertilized ovum produced at the moment of conception. The newly created zygote has all of the information that it will ever need to grow, mature, and become an adult that is capable of reproducing itself.

3. Reproduction provides for the continuity of the same species and involves the production of another being by one or more parents. Typically, we are most familiar with the kind of reproduction where an egg is fertilized to produce an embryo that develops within a seed, an egg, or the mother. However, reproduction is often of the simplest kind, such as that which is observed in a paramecium; for example, one organism divides into two new organisms.

The Soul

Scientists are typically not trained to ask "why?" but "how?" Nevertheless, when we ponder the existence of organisms, we are presented with creations that are all capable of self-change. This remarkable ability permeates the entire organism, from its individual cellular processes to its complex systems that work together in unison.

When we ask "why?" we embark on a philosophical journey that is best understood as the pursuit of Wisdom. We see unity in creatures and are led to the concept of the soul of the organism. The Ancient Egyptians believed in a

Figure 16-1. Saint Thomas Aquinas

soul, which consisted of five parts and was called a "Ka." Much later, the Ancient Greeks advanced the concept of the soul, and their understanding was accepted up until the time of St. Thomas Aquinas. St. Thomas gave us the modern concept of the soul that was developed in his famous work, the *Summa Theologica*, especially in Question 75.

All the changes in molecules that make up nonliving things display all the features of being moved by something else. For example, the changing of glucose into water and carbon dioxide is accompanied by a release of energy. This process can only occur when something external to the glucose changes its environment (heats it up). On a natural level, this change results from the atoms proceeding to a state of low energy. In nature, matter always proceeds to its lowest level. However, the change we see in a collection of molecules that makes up a living thing does not display this type of feature. The molecules act together in a way that is different from if they were simply placed in a container. For example, the energy stored in ATP results from the conversion of glucose into carbon dioxide and water. This change cannot be explained in terms of energy and must involve some other principle.

In living systems, molecules are subservient to higher principles that proceed from information. This is what we mean by "life" – the ability to organize non-living matter into a system that is capable of nutrition, growth, and reproduction. We say that the collection of molecules known as a living thing has a soul and has unity. The soul is the animating principle of a living being that organizes the chemical and mineral components to be what that being is. In other words, the soul specifies what a living thing is, the "what-ness" of the being. The same proportionate amount of particular chemicals and minerals will be organized differently, for example, by an asparagus soul and by the soul of a rose. The soul animates the matter to perform the fundamental tasks of living beings, which are nutrition, growth, and reproduction. The souls of higher beings will have higher level tasks to control, such as self-movement and sensing the environment. The soul of man is also responsible for intellect and will, which permit man to know and understand both material and spiritual realities.

Living things have been given unity and all of their capabilities by some external agent that is typically known as the parent or parents of the organism. This manifestation is known as the Law of Biogenesis, and was scientifically proved by Louis Pasteur in 1864. It states that "organisms do not spontaneously arise in nature from nonliving things." In other words "Life begets life." When biologists experiment with living things, they must start with material taken from a living thing and not with simple chemical compounds. Even a simple cell that is killed cannot be restored to life. On a larger scale, once a living organism has died, it quickly reverts to nonliving matter.

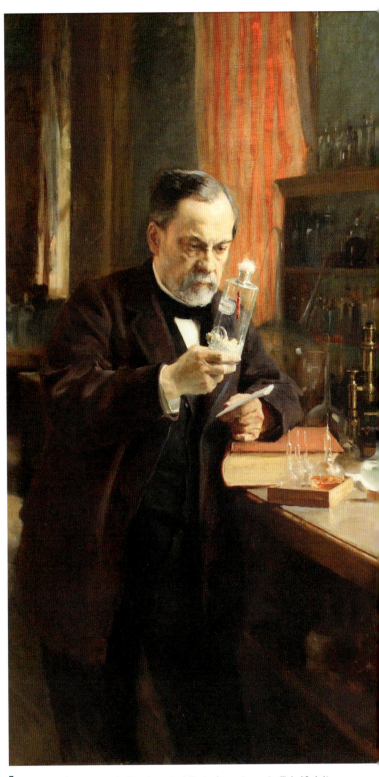

Figure 16-2. *Louis Pasteur In His Laboratory*, A. Edelfeldt

CHAPTER 16: PLANTS, ANIMALS, AND MAN

All living creatures have souls. The souls of plants and animals are considered to be material souls. Human souls are immaterial. A plant or animal soul is called material because it is dependent upon matter for its existence and for its purpose. The material soul in a plant or animal provides the principle of life of that creature, and this soul will cease to exist with the death of the plant or the brute animal. While the plant or animal depends for its life upon its soul, the material soul depends upon a suitable organization of the body for it to be able to maintain its own existence and its life-giving influence. For the material soul is not created or made out of nothing, but is generated from the creature's parents, along with all of the material parts of the creature. The human soul, on the other hand, is immaterial and eternal, and is created by God at the moment of conception. It is not generated from the parents as are plant and animal souls. The human or rational soul continues to infuse all parts of the living body until the death of the body, but this soul does not cease to exist, then or ever.

Based on the types of activities carried out by various living things, we are able to distinguish three distinct types of souls.

1. The plant is the simplest type of living organism. We find in plants the self-directed activities of nutrition, growth, and reproduction. All the processes of the plant are explainable in terms of the laws of physics and chemistry, since every chemical reaction that goes on inside a plant can be duplicated in a test tube. These chemical reactions are carried out so that the plant can obtain food to renew itself, grow, and reproduce. Aristotle and St. Thomas say that the chemical and physical processes are ordered to these fundamental activities by the vegetative soul, which infuses the entire plant.

2. Animals are capable of all the activities of a plant (nutrition, growth, and reproduction), but manifest two additional capabilities:

    a. The animal can sense the world in a manner that is not explained in terms of chemical or physical processes. The sense organs of animals produce only electrical signals, yet animals clearly experience an awareness of a "thing" that is exterior to itself. These signals are received and produced by sense organs,

Biology For Life

CHAPTER 16: **PLANTS, ANIMALS, AND MAN**

and the animal is "conscious," or aware of them. Although plants can react to certain physical stimuli, such as turning their leaves towards light (phototropism), they do not exhibit awareness.

   b. The animal performs an action in response to its awareness of the world around it. For example, consider a dog that is sitting in a backyard resting and becomes aware of rustling leaves. Aristotle and St. Thomas attribute the cause of this higher level of activity to the animal (sensitive) soul. It infuses all the material parts of the animal and gives it sense awareness and self-direction.

3. Man stands higher than the animals in that his capacity to understand the nature and "why" of things is unique in the animal kingdom. In this understanding man is aware of himself as a living, thinking organism – he is self-conscious. This is a difference **in kind** and not simply **of degree**. The difference between apes and fish is simply one of degree, since they all possess the same capabilities of awareness and self-direction to a greater and lesser extent. A greater gap separates man from the apes than that which separates the animals from each other.

Aristotle and St. Thomas attribute this highest level of activity in man to the presence of the human (rational) soul. This soul infuses the entire human and provides it with the vegetative activities of nutrition, growth, and reproduction. It allows the animal activities of sense awareness and self-direction but also adds a sense of understanding. This sense seeks the what and why of things and includes a self-awareness that is not present in plants and animals. Since the beginning of time, people seek to know who they are, what is their purpose, and what is their destiny. No individual escapes these thoughts, and it is clear that God made our minds and souls to seek Him.

Section Review — 16.2

1. What are the three processes by which one can best understand self-change in living things?
2. What are the three distinct types of souls?
3. Are all types of souls immaterial?

Biology For Life

CHAPTER 16: PLANTS, ANIMALS, AND MAN

16.3 The Vegetative Soul

> **Section Objectives**
>
> - List the characteristic activities and purpose of vegetative life.
> - List ways plants react to the world around them.

A plant is the simplest kind of living thing. All processes that occur in a plant can be explained in terms of the laws of physics and chemistry, because they can be reproduced individually in a test tube. However, the profound ordering of these processes towards the basic activities of the plant – nutrition, growth, and reproduction – leads to the concept of the vegetative (plant) soul.

There are various species of vegetative soul corresponding to the different species of plant. It is the parent plants that are the source of the vegetative soul in their offspring. They provide the sperm and ovum as a part of their own activity. The seeds produced by fertilization are dispersed and carry out the activities proper to a plant.

It is the vegetative soul that organizes and unites the billions of molecules that are found within a plant. How else is it possible to account for the cellular structure of the plant, in light of the fact that a typical cell carries out hundreds of chemical reactions every second? Membranes separate these reactions, but the cells work together to produce the proper reactions that are necessary for self-preservation, growth, and reproduction.

Many of the reactions of a plant involve nutrition. The plant takes carbon dioxide from the air and water and minerals from the soil, then combines them in new ways to form new molecules. These become parts of the plant's own structure through growth and self-repair or self-regeneration.

Nutrition is also essential to the growth of the plant. As the plant grows in size, it manifests an ever-increasing differentiation and specialization of new parts. An astonishing variety of new structures arise from the tiny, undifferentiated fertilized seed of a plant, including roots, stems, and leaves. Each of these is made with different types of cells, all of which work in unison and perform unique tasks. The purpose of this growth and differentiation is to help the plant attain the optimal framework to carry out its basic activities.

The growth of a plant or other living organism is different from what we refer to as growth in a crystal or other piece of nonliving matter. First, the process of "growth" in nonliving matter usually takes place by addition. That is, something from the environment is added to it, rather than the taking in and change of material (nutrition) that we find in plants and other organism. Secondly, the additions to nonliving matter resemble the older or former parts. This is entirely different from the makeup and function of the new "parts" of a plant.

Figure 16-3. Phototropism. The seedlings are growing toward the light on one side.

CHAPTER 16: **PLANTS, ANIMALS, AND MAN**

Figure 16-4. Gravitropism. After the tree fell down, it continued to grow and gradually curved upward again.

The growth of a plant is essential to its ability to reproduce. It is during reproduction that the plant passes on "life" to its offspring. The newly reproduced organism is always the same species (kind) as the parent.

Although plants can react to various stimuli, plants have no genuine awareness of the world. When plants react to stimuli, the response is called a tropism. The plant alters its growth to make possible its optimum growth. One of the most commonly observed tropisms is phototropism, in which plant stems grow toward light. If the plant were growing in the presence of sufficient light, it would not bend. However, in the process of bending, it maximizes its surface area that is exposed to light. In this way, it is able to optimize its ability to conduct photosynthesis. You may have seen phototropism firsthand if you have ever spent time watching plants growing near a window.

Another commonly observed tropism is called geotropism or gravitropism. All organisms respond to gravity, and plants are no exception. As a plant grows, its main stem or trunk will grow in an upward direction. Conversely, its roots will grow in the opposite direction, downward through the soil. Occasionally, when traveling along mountain roads, you may encounter trees that seem to have grown awkwardly. Sometime during their lives, they began to fall down slopes. Their roots did not completely come out of the soil, and the tree ended up growing at some angle, rather than pointing straight up. If the tree was a sapling at the time, its new growth would have been evident as a curved main trunk. If it happened while it was an older tree, it would continue growing at an angle other than straight up, but its new branches would point upwards. These trees that grow on unstable hillsides are often interesting to look at and display the power of the response to gravity.

Section Review — 16.3

1. All process that occur in a plant can be explained in terms of the laws of _____ and _____.
2. What is phototropism?
3. What is geotropism?

Biology For Life

443

CHAPTER 16: PLANTS, ANIMALS, AND MAN

16.4 The Sensitive Soul

> **Section Objectives**
>
> - List the characteristic activities and purpose of animal life.
> - Describe the way an animal is aware of the world around it.

Plants and Animals

Animals have capabilities beyond those that are exhibited by plants. We refer to these when we say that an animal is conscious.

1. Animals have a sense of the world around them.
2. Animals respond to this sense – they become impelled by a desire to do something in response to this awareness of the world.
3. Animals are able to move themselves from place to place as required to survive.

Each species of animal has its unique kind of soul. Similar to plants, the parent organism(s) are the source of the animal soul in their offspring. The cells that develop from a fertilized egg act in a unified manner that is characteristic of a new organism. At some point in its life, or upon separation from its parent, the new organism carries out the activities proper to its species.

We can compare the animal to a computer or even a robot if we so choose, assuming the computer and robot have sensors. These may also include mechanical devices that function like various organs or even arms and legs. All of these structures are connected to electrical circuits that connect them to a central processing unit and together function with a computer program. The program has been added to the simple machine, giving it capabilities that a simple machine does not have. The activity of the robot requires electrical signals that it receives from its sensors to perform some action, in conjunction with its program. In the same way, the animal is more than a plant; it has something similar to the computer program that enables it to receive signals from its sense organs and undertake some action. This is why we say the animal soul differs in kind from the plant soul. Scientists refer to this principle as continuous environmental tracking. This mechanism allows animals to continuously track changes in the environment and respond accordingly. It has an added level of "depth" and "complexity."

Note that mechanical, physical, and chemical processes are necessary for sense perception to occur within an animal. Sight would not be possible were it not for the eye, optic nerve, and visual cortex. The electrical and chemical activities in these organs are not sensations, but they make sensations possible. The animal has the capability of relaying these electrical signals, produced by sense organs, to the brain. In turn, the brain relates these to the thing or object that produced them. This connection of the brain, senses, and objects is what we know as sensation. For example, your father is cooking hamburgers on an outdoor grill. The aroma of the hamburger wafts over to your pet dog resting on the other side of the house. The olfactory receptors transmit electrical signals to

CHAPTER 16: **PLANTS, ANIMALS, AND MAN**

the brain. The brain responds by stimulating the dog's appetite and making him salivate. The hamburger is the object, and hunger is the sensation.

Sensations enable an awareness of the world and the state of an animal. Here there exist two worlds: the **subjective world** of self (this instinctive-awareness is a form of self-consciousness) and the **objective world** outside the self. An animal couldn't function without having awareness of these two worlds, and it must be able to form a sense image, or representation, of these two worlds. Since an animal is able to remember sense images, it must have a memory that enables it to make composite images from previous occasions. We know this as learning. However, this subjective awareness in the sensitive soul is not a form of self-awareness, or awareness of the "self" as a special being. It is only an awareness of the condition at the time, for example, the animal is hungry and is aware that it is hungry.

When we speak of the animal's awareness of the world around it, nothing obliges us to give the same degree of awareness to all species. Different species of animals have different kinds of souls. Some species possess fewer sense capabilities than others. Furthermore, the capabilities of a sense organ will be related to the complexity of the organism. For example, the awareness of a barnacle is minimal when compared to a chimpanzee, because it possesses fewer senses. Similarly, while many worms and shellfish are able to detect light that passes through their skin, they cannot produce visual images. Their awareness of the world is less complex than that of fishes, which have eyes that are fully capable of forming sharp images.

It is "awareness" that causes animals to act in ways that best serve their health and survival. The animal has a capacity to size up any situation with emotion, or motivating force. These emotions, or "forces to act," are also called instincts, or appetites, of the animal. The term **estimative sense** is the name given to the capacity of the animal to size up a given situation.

As in the case of awareness, nothing dictates that we apply the same range of instincts or

appetites to all species. In the case of the single cell amoeba, it would seem that the only sense power present is the most fundamental – the sense of touch – and the amoeba has two basic instincts or appetites:

1. an inclination to move toward that which is beneficial – attraction
2. an inclination to avoid that which is harmful to it – avoidance

However, more complex animals such as the chimpanzee manifest a host of appetites ranging from desire to pleasure.

Through the impulses of its appetites, the animal is directed to act in a way that is appropriate to its awareness of its world. In many cases this involves moving from one place to another. We call this activity locomotion. Two good examples are the bird flying to its nest or a horse trotting back to its barn.

Biology For Life

CHAPTER 16: PLANTS, ANIMALS, AND MAN

In both cases, these activities are responses that have arisen from an awareness of the world.

Animal Behavior

Animals experience the world around them and behave according to their instincts. Furthermore, comprehending the world and responding to it imply awareness or consciousness. But what exactly do we mean when we say consciousness?

An animal with a pair of eyes that are structurally identical to ours clearly experiences some sort of vision that is comparable to our own. Through experimentation we can deduce the sharpness of the animal's vision, its color perception, depth perception, etc. The problem of understanding the nature of the experience is more difficult when the animal's eyes or sense organ is very different from our own; however, experiments can yield clues.

A great deal of research has shown that memory and the ability to discriminate occur in animals. Experiments have shown that digger wasps memorize landmarks around their nest, allowing them to easily return to their nest after hunting flights. During one investigation, researchers deliberately moved the landmarks by one foot. The wasp returned to the landmarks but had difficulty finding its nest. Similar evidence has been found for such discrimination in octopuses, fishes, and birds.

Chimpanzees are complex animals that clearly experience emotion. The range of emotions indicated through gesturing and expressing is easily comprehensible to us in terms of contentment, happiness, rage, terror, despair, grief, etc. Some animals even show evidence of imagination. Occasionally, sleeping dogs and cats move and vocalize in ways that suggest they are

Figure 16-5. Evidence of shape perception in insects. The digger wasp always memorizes significant features around its nest so as to find it easily when it returns. The wasp learns to recognize the circle of pine cones around its nest.

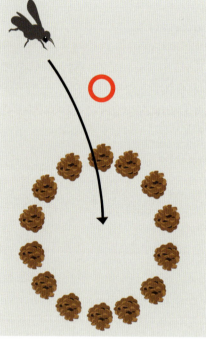

Circle of pine cones moved, and the wasp, searching in center of circle, cannot find its nest only a foot away.

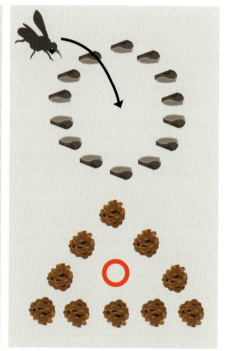

Wasp still seeks its nest in the center of the circle, ignoring its nest that is now surrounded by a triangle of cones, so it is focusing on the shape and not the cones.

Biology For Life

CHAPTER 16: **PLANTS, ANIMALS, AND MAN**

dreaming. These behaviors are acts of the imagination.

However, granted that an animal sees, hears, remembers, imagines, and has emotions, the most important questions to ask are as follows: What does it perceive in its world? What does an animal look at? What does it look for and how does it look for it? Experiments indicate the peculiar fact that an animal does not react to all features of its environment but only a small portion of these. This is best understood in the light of our own reactions to stimuli. Can you imagine having to be conscious of the clothing on your body, the air passing in and out of your nose with each breath, and all of the background sounds as you sit and read this book? Just as we are able to filter out background information, so too are other animals.

The remarkable specificity of animal awareness is illustrated in the jackdaw. It is incapable of perceiving a grasshopper that is motionless, but it becomes animated when the grasshopper begins to move. In a similar vein, a frog does not perceive details of the unmoving parts of the world around it; it will starve to death if it is surrounded by motionless insects (food). The frog will leap to capture objects the size of a worm or insect, provided the object moves like one. It can easily be fooled by a dangling piece of meat or any small "wriggling" object in front of it. Such behavior of animals is a basic property of instinctive behavior. That is, the animal responds "blindly" to only part of its present situation, even though its sense organs are perfectly capable of responding to other situations. So the animal truly has some kind of awareness of the world around it. This awareness is both appropriate and sufficient for its way of life. It is worth mentioning that the range of behaviors (also known as repertoires) an animal exhibits is related to its size or intelligence, or both.

Figure 16-6. Animals respond to stimuli that are important for their everyday functions. The cat sitting calmly hears the ticking metronome. Its perception can be recorded on a graph if an electrode is inserted into its brain and connected to a meter. When the cat sees a mouse, it concentrates its attention solely on the mouse and tunes out the irrelevant ticking.

Biology For Life

Are Animals Intelligent?

Animals are clearly conscious self-moving things, acting on the basis of sense perception, emotion, and instinct, but do they possess intellectual understanding? Are they able to go beyond the sense perception to the nature of the thing itself?

When we observe animal behavior we note that they respond without comprehending, or *internalizing*, the situation. Their response is related to external stimuli that are associated with specific objects, but which can be produced by similar, related objects, for instance:

1. A deaf turkey hen pecks all her own chicks to death as soon as they are hatched; the chicks cheeping – which she can't hear – are the only stimulus that can inhibit her natural aggression in defense of her nest.

2. A hen with normal hearing will attack an imitation chick if it emits no sound and is pulled towards the nest by a string. Conversely, she will respond maternally to a stuffed weasel if it has a built-in speaker that produces the cheeping of a turkey chick.

3. Jackdaws will mob anyone who holds in their hands something that is black and moving, thus acting as they were an "eater of jackdaws." Yet if the person held in their hands a featherless nestling jackdaw, they would not respond.

Animals then deal with phenomenon associated with objects and not the object itself; that is, the animal responds to the appearance of the object and its sounds, colors, and movements. They do not distinguish between the sense qualities of an object and what an object is. The ability to do so is characteristic of intellectual understanding.

> **Section Review — 16.4**
>
> 1. What are three characteristics that distinguish animals from plants?
> 2. Sensations enable an awareness of the world and the state of an animal; these two worlds are the _____ world and the _____ world.
> 3. What does the term "estimative sense" mean?

16.5 The Rational Soul

Section Objectives

- List the characteristic activities and purpose of human life.

We reason the existence of the human (rational) soul because humans have the ability to respond to objects and situations beyond those shown by plants and animals.

1. Humans learn about the nature of objects from experience with those objects. So humans know the concepts of "dogness" and "rockness" which is found in the many dogs and rocks in the world. They also have knowledge of things not found concretely in this world like truth, goodness, and beauty.

2. Based on this knowledge, humans can choose for themselves a particular goal. They are able to substitute another if they wish and then freely choose an appropriate means to attain that goal.

We refer to the above capabilities when we say that the human being is rational.

Moreover, in this understanding of the nature of objects a human becomes aware of himself as a living, thinking organism. The human has self-awareness and sees himself/herself as a rational being in a world of objects and space. This is a type of consciousness known as self-consciousness. It is the result of a human's rationality, because of the human's ability to reflect on his/her nature. For example, if you ponder the possibility of a vocation as a concert violinist, it is possible for you to attain that goal. In time, you practice, apply yourself, and receive input from others about your progress and talent. You might even assess your own level of happiness. On the other hand, our most intelligent animals such as the pig, chimpanzee, or dolphin can have no such inclinations and dreams. They are limited by the designs of their Creator and by their soul.

If we liken man to a computer or a robot, we will discover that man is something more than a computer program. In 1930, Kurt Godel, a German mathematician, showed that man has the ability to consider the universal idea of a computer program. Man can discover obvious facts about any computer program. These facts cannot be discovered with the set of rules and instructions which form the basis of a program. So man is more than a computer, and his mind does more than process physical signals. The Creator imbued man with a neural network and a brain. We will never be able to understand our own rules and instructions, just as man can neither understand himself, nor angels, nor the Infinite Spiritual Being.

It is this ability of the human to deal with universal concepts, such as the nature of things, that leads to the idea that the human soul must be immaterial (spiritual). For the human soul is not limited to dealing with things on an individual basis. Such a limitation is characteristic of material things.

The human soul is not just a special capacity added to the material parts of a human, like the animal and vegetable souls. It is immaterial and eternal and is created by God at the moment of conception. The soul continues to infuse all parts of the living body until death. The human body then reverts to a collection of molecules referred to as dust in Sacred Scripture. The parents are the source of the material part of their child, but God is responsible

Figure 16-7. Kurt Godel, a German mathematician

for the human soul that infuses it. This is why the act proper to marriage is so special in the eyes of God. He is involved in the act of creation. There is only one species of human soul, but this soul is uniquely suited to each person.

Since the soul is immaterial, it has no parts and therefore is immortal; it cannot decompose. God could put an end to the soul at the death of the person, but His wisdom and goodness demand that He does not defeat the desire of the soul for truth and happiness – just as His justice demands that He reward the good and punish the wicked.

Like the animals, humans have a "sense knowledge" of the material world. Humans are completely aware of all the sensible qualities of objects, like color, shape, texture, hardness, and sound. The qualities of objects produce a composite mirror image in our brains, which is very rich and detailed. However, beyond the sense image, we are able to abstract, or understand, the nature of the object creating the image. This involves going beyond the material world of the physical senses. We begin to enter the realm of the immaterial. These are the notions or concepts of "rockness" or "dogness" that we previously discussed.

The power of the human soul that enables humans to form ideas (concepts) about objects and reason with them to attain other true ideas about reality is called the intellect. The soul uses the composite sense image of things to draw out their underlying natures. Because of their rational souls, humans can rise above what is true for the present. They are capable of attaining what they were designed for, the pursuit of Truth.

In the human there are two levels of knowledge. One kind is called sensitive and is associated with images. The other is known as immaterial and is associated with ideas regarding the images. Yet humans are a union between the body and soul, so there is a close connection between the two levels of knowledge. While ideas are not composite sense images, they are always accompanied by images in the brain's imagination. So when little Billy looks out the window and sees a squirrel playing in a tree, he "sees" in two ways: he has a visual image of the squirrel in his brain and has in his mind the idea "squirrel." It is for this reason that we cannot think clearly when we are tired. Our immaterial soul does not tire, but the brain does. If the brain is tired and needs rest, it cannot properly form images. Similarly, one can observe brain activity with a device known as the electroencephalograph. It shows brain activity even when the subject is engaged in abstract thought about numbers or love.

As with the animals, our knowledge of the world as humans gives rise to emotions regarding what is known. Some of these emotions are on the sense level, as in animals, and are instinctive. Unlike the animals, though, we are able to combine knowledge with emotions, which leads to inclinations and behavior that are not instinctive. Furthermore, our inclinations influence the emotions and, in turn, are influenced by them. These inclinations arise from the other fundamental activity of the human soul, which we refer to as the intellectual appetite or the will. It refers to the source of the human's desire

for things as perceived by the power of the intellect.

The will controls all human behavior – even behavior motivated primarily by the sense emotions. The sight of a chocolate cake might arouse the person's emotions because food is pleasant to eat, but he wills to eat only if he perceives, by the power of his intellect, that it would be good for him to do so. Hence a man can also starve himself in spite of a contrary urging from the sense appetites, if he judges that it is reasonable to do so. Since the intellect can often see the alternatives associated with a given choice or rejection of a good, the will has a freedom to choose among them. The term "free will" is used to describe the ability of choosing a limited good when this choice is presented.

Because of their rational souls, humans can rise above what is good for only the here and now and attain that which is the purpose of a human's intellectual appetite, the attainment of The Good. The Good is that which is good for humans for all times and for all places.

Human Beings and Creation

According to divine design, it is in the human being that we find creation at its finest. In humans, we find the fullest expression of the capacities for self-change: nutrition, growth, reproduction, and consciousness. Moreover, we can also distinguish in a human being the various means of acting found in material things.

1. We propose for ourselves a particular goal, knowing that we could substitute another if we wished. Then we freely choose the appropriate means to attain that goal. It is because of our ability to understand that we are fully aware of ourselves as living beings. Understanding comes from reflecting on one's own acts and therefore

learning about oneself. Our choosing power, the will, is the prime agent that sets other capabilities into operation: setting the intellect to think of this thing or that thing, directing the sense powers to this or that object, moving the voluntary muscles. Our will can even override our emotions when necessary, but it moves itself to choose.

2. We have some idea what it is like for an animal to perceive without understanding intellectually, since we can experience in ourselves nonrational associations that are found in animals. For example, the smell of pencil shavings evokes in us an unexpected memory of grammar school.

   We have insight into the way an animal is inclined toward or away from an object by appetites. Like higher animals, we also possess appetites, although we have an intellect that helps us override feelings.

   We have a notion of what an animal's instinctive actions may be like when we respond to something impulsively, without verbal thought or deliberation, such as when we jump at a loud sound. This operation, when reason and will are suspended, is reflected in our everyday language when we refer to an angry person as "acting like an animal."

   Our nonverbal thinking and involuntary memory associations give us some understanding of animal consciousness. We experience nonverbal thinking when we search for a matching piece of a jigsaw puzzle or when, without words, we rearrange a room's furniture in our imagination. Such thinking employs images, not definitions.

3. We have some idea of the unconscious operations of a plant when we observe the healing of a bruised arm. This activity proceeds without our choice and mostly without our sense awareness. The activity is clearly our own, but it proceeds from us unconsciously and ends with a goal not determined by us.

4. Finally, we have some idea of the operations of inanimate matter. If we fall off a cliff, we fall like a rock. The falling motion and its limits are completely determined by the environment – characteristic of the change associated with a nonliving thing being moved by something else.

CHAPTER 16: PLANTS, ANIMALS, AND MAN

The Purpose of the Human Life

We have noted that the ultimate purpose of plants and animals is to grow and reproduce. All their activities are ordered in the most efficient way possible. However, when we consider what is the ultimate purpose of humans, we are struck by the fact that material things in this world do not satisfy any of the purposes of human knowledge or the human intellectual appetite.

There must be some idea and some good, which, when possessed by humans, completely satisfies the **human intellect's need for attaining truth and the human will's need for the good**. It is in this sense that we can reason to the idea that the possession of God is the ultimate purpose of man's existence.

Natural Moral Law

The will controls all human behavior. A human chooses to do an action if, by the power of his intellect, he "sees" that it would be good for him to do so. Because humans have this capacity of freely choosing to do an action, there arises the notion of the moral life, the way of life a man should follow. This deals with the choices a human should make based on his nature, his membership in society, and the fact that possession of God (The Truth and The Good) is his purpose in life. These choices are known as the natural moral law or the law of human nature. Thus humans should avoid gluttony (it damages the body); they should contribute to the good of society (it helps them fulfill their desire to live with others); and they should love God (for the possession of God is what humans are made for).

Section Review — 16.5

1. What distinguishes humans from animals?
2. What defines the rational capability of a human being?
3. What is meant by the natural moral law or the law of human nature?

Chapter 16 Supplemental Questions

Answer the following questions.

1. Explain what it means for a thing to be alive.
2. Explain what a soul is.
3. Explain why there are three types of soul.
4. List the characteristic activities of the vegetative, animal, and human life.
5. List the characteristic activities and purpose of vegetative life.
6. List the ways that plants react to the world around them.
7. List the characteristic activities and purpose of animal life.
8. Describe the way an animal becomes aware of the world around it.
9. List the characteristic activities and purpose of human life.

Image Attributions

Chapter 1

Unborn Background: Copyright: Zffoto, Adobe Stock
Tertullian: Public Domain: Wikimedia
St. Jerome, El Greco: Public Domain
Conceptual Motherhood Image: Copyright: Zffoto, Adobe Stock
Structure of DNA and RNA: Copyright: Designua, Shutterstock
Fertilized Egg: Nathan Puray, Seton Home Study School
Human Fertilization: Copyright: Marochkina Anastasiia, Shutterstock
Sperm and Egg: Public Domain: Wikimedia, www.PDImages.com
Jerome Lejeune and a Child with Down Syndrome: Wikimedia: CC-BY-SA-3.0, Denis-Soto
Jerome Lejeune: Wikimedia: CC-BY-SA-3.0, Jerome Lejeune Foundation
Development of the Embryo: Copyright: Designua, Adobe Stock
Implanted Embryo: Copyright: sakurra, Adobe Stock
Early Human Development in Oviduct: Copyright: olando, Adobe Stock
Placenta and Umbilical Cord: Copyright: Alila Medical Media, Adobe Stock
Endoderm and Mesoderm: Copyright: 7activestudio, Adobe Stock
Four-Week-Old Human Embryo: Copyright: stihii, Shutterstock
Six-Week-Old Human Embryo Nathan Puray, Seton Home Study School
Medical Accurate 3D Illustration of a Fetus Week 8: Copyright: Sebastian Kaulitzki, Adobe Stock
3D ultrasound of baby in mother's womb. Copyright: Valentina Razumova. Shutterstock.
Fetal Growth: Copyright: pablofdezr, Adobe Stock
Unborn Human Fetus: Copyright: mrallen, Adobe Stock
20-28 Week Development: Nathan Puray, Seton Home Study School
6 month fetus: Copyright: Sebastian Kaulitzki: Shutterstock
7 month fetus: Copyright: Sebastian Kaulitzki: Shutterstock
32-36 Week Development: Nathan Puray, Seton Home Study School
40 Week Development: Nathan Puray, Seton Home Study School
Newborn Baby: Copyright: demphoto, Adobe Stock
Suffer the Children: Public Domain: Wikimedia, Carl Heinrich Bloch
Candle: Myriams Fotos, CC0 Public Domain, pixabay

Chapter 2

Cells chapter image: Copyright: rost9: Adobe Stock
Cells header background: qimono, CC0 Public Domain, pixabay
Waterweed: Vierschilling, CC0 Public Domain, pixabay
Atom: Wikimedia: CC-BY-SA-3.0, Shizhao
E Coli Bacteria: WikiImages, CC0 Public Domain, pixabay
Robert Hooke Micrographia 1665: Public Domain: Wikimedia, Robert Hooke
Hooke Microscope: Public Domain: Wikimedia
Felix Dujardin: Public Domain: Wikimedia, Louis Joubin
Matthias Jacob Schleiden: Public Domain: Wikimedia, Popular Science Monthly Volume 22
Theodor Schwann: Public Domain: Wikimedia, Rudolf Hoffmann
Rudolf Verchow: Public Domain: Wikimedia
Bacteria: Public Domain Pictures, CC0 Public Domain, pixabay
Cell Structure: Copyright: Tefi, Shutterstock
Cell Membrane: Copyright: Soleil Nordic, Shutterstock
Cell-Nucleus-Chromosome-DNA, Medical Vector Diagram: Copyright: VectorMine, Adobe Stock
Endoplasmic Reticulum: Wikimedia: CC-BY-3.0, BruceBlaus
Plant Cell: Copyright: Snapgalleria: dollarphotoclub
Prokaryotic Cell (3D): Copyright: Decade3D: dollarphotoclub
Eukaryotic Cell (3D): Copyright: Naeblys: Shutterstock
Bacteria: qimono, CC0 Public Domain, pixabay
Human Body Anatomy: Copyright: Terriana, Adobe Stock
Structure of Carbon: Copyright: Iraidka, Adobe Stock
Abstract Particles Connection: Copyright: Belkin & Co, Adobe Stock
Mix Fruits: Copyright: episterra, Adobe Stock
Sugar Cubes: Humusak, CC0 Public Domain, pixabay
Triglyceride Molecular Structure: Copyright: Leonid Andronov, Adobe Stock
Butter and Oil: Copyright: Multiart: dollarphotoclub
Molecular Structural Formula of Amino Acid: Copyright: MariLee, Adobe Stock
Proteins: Copyright: M.studia, Adobe Stock
Peptide Bond Formation: Public Domain: Wikimedia, YassineMrabet
Structure of the Protein Molecule: Copyright: Sergunt, Adobe Stock
Lock and Key Model of Enzymes: Copyright: Designua, Shutterstock
Friedrich Miescher: Public Domain: Wikimedia
DNA and RNA: Copyright: udaix, Shutterstock
RNA: Copyright: nobeastsofierce, Shutterstock
Structural Chemical Model of ATP: Copyright: chromatos, Shutterstock
ATP-ADP Cycle: Copyright: Designua, Adobe Stock

Chapter 3

Heart background: Copyright: Sergey Nivens, dollarphotoclub
Red Blood Cells: Copyright: Alexandr Mitiuc, Adobe Stock
Sacred Heart of Jesus: Public Domain: Wikimedia
Anatomy of the Human Heart: Copyright: blueringmedia, dollarphotoclub
Systemic circulation: Copyright: Matthew Cole, dollarphotoclub
Arteries and Veins: Copyright: Matthew Cole, dollarphotoclub
Heart: Copyright: CLIPAREA.com, dollarphotoclub
Blood Cells in Vein: Copyright: adimas, dollarphotoclub
Karl Landsteiner: Public Domain: Wikimedia, Albert Hilscher, ONB-1074462
RH Factors During Pregnancy: Nathan Puray, Seton Home Study School

Biology For Life

Eucharistic Miracle of Lanciano: Wikimedia: CC-BY-SA-3.0, AFC photo
White Blood Cell: Copyright: Sebastian Kaulitzki, Shutterstock
Lymphatic System: Copyright, Alila Medical Media, Shutterstock
Scanning Electron Microscope Image of Blood Cells: National Cancer Institute: Public Domain, Bruce Wetzel and Harry Schaefer
Neutrophil Engulfing Anthrax Bacteria: Wikimedia: CC-BY-2.5, Volker Brinkmann
Hand Holding a Thermometer: Copyright: grib_nick, Adobe Stock
Clogged Arteries: Copyright: Lightspring, Shutterstock
Blood Pressure: Copyright: wutzkoh, Adobe Stock
Fetus Circulation: Copyright: ellepigrafica, Shutterstock
Newborn: Kimpton_house, CC0 Public Domain, pixabay
Eucharist: kisistvan77, CC0 Public Domain, pixabay

Chapter 4

Chapter Background: Copyright: magicmine, Adobe Stock
Border Collie Dog: 825545, CC0 Public Domain, pixabay
Human Endocrine System: Copyright: Designua, Shutterstock
Adrenal Gland: Copyright: Designua, Shutterstock
Parathyroid Gland: Copyright: zuzanaa, dollarphotoclub
Pancreas: Copyright: freshidea, Adobe Stock
Thymus Gland: Copyright: nerthuz, Adobe Stock
Brain: Copyright: Tefi, Shutterstock
Negative-Feedback System: Copyright: Designua, Adobe Stock
Sleeping Baby: isaiasbartolomeu, CC0 Public Domain, pixabay
Man: RitaE, CC0 Public Domain, pixabay

Chapter 5

Lungs Background: Copyright: Sergey Nivens, dollarphotoclub
Urinary System / Kidney: Copyright: lightspring, Shutterstock
Woman Praying: StockSnap, CC0 Public Domain, pixabay
Chemical Model of Oxygen: Copyright: LoopAll, Adobe Stock
Respiratory System: copyright: snapgalleria, dollarphotoclub
Lungs: Copyright: Designua, Shutterstock
Cellular Respiration, Copyright: VectorMine, Adobe Stock
ATP-ADP Cycle: Copyright: Designua, Adobe Stock
Liver Anatomy: Copyright: stockshoppe, dollarphotoclub
Human Urinary System: Copyright: Mr High Sky, Shutterstock
The Kidney: Copyright: GunitaR, Shutterstock
Nephron: Copyright: Aldona Griskeviciene, Shutterstock
Male Drinking Water: Copyright: Kaew6566, Adobe Stock
Newborn: sathyatripodi, CC0 Public Domain, pixabay

Chapter 6

Nerve Cells Background: Copyright: Giovanni Cancemi, Shutterstock
Brain: Copyright: Alexandr Mitiuc, dollarphotoclub
Neuron: Copyright: kateryna zakorko, dollarphotoclub
Synaptic Transmission: Copyright: nobeastsofierce, Adobe Stock
Nervous System: Copyright: Stockshoppe, Shutterstock
Brain: Copyright: stockshoppe, dollarphotoclub
Spinal Cord: Copyright: joshya, Adobe Stock
Skin Diagram: Copyright: Neokryuger, Shutterstock
Basic Taste: Copyright: Designua, Shutterstock
Olfactory System: Copyright: p6m5, dollarphotoclub
Structure of the Cochlea: Nathan Puray, Seton Home Study School
Anatomy of the Ear: Copyright: Alila Medical Media, Shutterstock

Woman's Eyes: minanfotos, CC0 Public Domain, pixabay
Human Eye Anatomy: Copyright: Alila Medical Media, Shutterstock
Brain: Copyright: stockshoppe, dollarphotoclub
Neuron: Copyright: kateryna zakorko, dollarphotoclub
Baby Eyes: Piipaaa, CC0 Public Domain, pixabay
Baby's Hand: Pexels, CC0 Public Domain, pixabay
Woman and Grandmother: Copyright: Alexander Raths, Adobe Stock

Chapter 7

Blue Skeleton X-ray Background: Copyright: angkhan, Adobe Stock
Skeleton Hand: Copyright: Alexandr Mitiuc, Adobe Stock
Bone Anatomy: Copyright: 7activestudio, dollarphotoclub
Bone Structure: Copyright, studiovin, Shutterstock
Medical Education Chart of Biology for Human Skeleton Diagram. Copyright: Vecton, Shutterstock
Joints: OpenStax CNX. Located at: http://cnx.org/contents/185cbf87-c72e-48f5-b51e-f14f21b5eabd@10.8. License: CC BY: Attribution. License Terms: Download for free at http://cnx.org/contents/185cbf87-c72e-48f5-b51e-f14f21b5eabd@10.8
Skull / Brain Diagram: Copyright: 7activestudio, dollarphotoclub
Backbone: Copyright: HANK GREBE, Adobe Stock
Rib Cage: Copyright: solar22, Shutterstock
Pectoral Girdle: Copyright: 3drenderings, Adobe Stock
Pelvic Girdle and Leg: Copyright: 7activestudio, Adobe Stock
Articular Cartilage: Copyright: Artemida-Psy, Adobe Stock
Rheumatoid Arthritis: Copyright: designua, Adobe Stock
Types of Muscles: Copyright: blueringmedia, dollarphotoclub
Skeletal Muscle: Copyright: Designua, Shutterstock
Myofibril: Copyright: Alila Medical Media, Adobe Stock
Anatomy of the Elbow: Copyright: blueringmedia, dollarphotoclub
Cardiac Muscle: Copyright: Kateryna_Kon, Adobe Stock
Smooth Muscle: Copyright: Kateryna_Kon, Adobe Stock
Skin Anatomy: Copyright: 7activestudio, dollarphotoclub

Chapter 8

Digestive System Background: Copyright: Crystal light, Adobe Stock
Digestive System: Copyright: blueringmedia, dollarphotoclub
Human Digestive System: Copyright: snapgalleria, dollarphotoclub
Mouth Anatomy: Copyright: stockshoppe, dollarphotoclub
Tooth Structure: Copyright: sakurra, Adobe Stock
Stomach Diagram: Copyright: Stockshoppe, Shutterstock
Small Intestines: Copyright: snapgalleria, dollarphotoclub
Intestines: Copyright: 7activestudio, dollarphotoclub
Pancreas & Liver: Copyright: peterjunaidy, Adobe Stock
Close-up of Intestinal Villi: Copyright: rost9, Adobe Stock
Large Intestines: Copyright: switchpipi, Adobe Stock
Healthy Food: Copyright: Dusan Zidar, Adobe Stock
General Carbohydrates Molecular Structures: Copyright: logos2012, Adobe Stock
Oranges: Hans, CC0 Public Domain, pixabay
Fresh fruit and Vegetables: Copyright: jarik2405, Adobe Stock
Food Pyramid Pie Chart: Copyright: Okea, Adobe Stock
Harvesting Tomatoes: Copyright: goodmoments, Adobe Stock
Mother and Newborn: Copyright: kieferpix, Adobe Stock
Gastroesophageal Reflux: Copyright: designua, Adobe Stock

Chapter 9

Cell Division Background: Copyright: Yurchanka Siarhei, Shutterstock
DNA: PublicDomainPictures, CC0 Public Domain, pixabay
Tree: OpenClipart-Vectors, CC0 Public Domain, pixabay
DNA and Nitrogenous Bases: Copyright: udaix, Shutterstock
DNA-Double Helix Molecules and Chromosomes: Copyright: nobeastsofierce, Adobe Stock
DNA & RNA: Copyright: Designua, Shutterstock
t-RNA: Copyright: designua, Adobe Stock
DNA Duplications: Copyright: ellepigrafica, Shutterstock
Human Chromosomes: Wikimedia: CC-BY-SA-3.0, Steffen Dietzel
Dyad: Nathan Puray, Seton Home Study School
Mitosis: Copyright: achiichiii, Adobe Stock
Kinetochore: Public Domain: Wikimedia, Afunguy at English Wikipedia
Cell Division: Copyright: kiss, Shutterstock
Chromosomes: Copyright: Giovanni Cancemi, Adobe Stock
Gregor Mendel: Public Domain: Wikimedia
Peas: PublicDomainPictures, CC0 Public Domain, pixabay
Meiosis: Copyright: Designua, Shutterstock

Chapter 10

Viruses Background: Copyright: Kristy Pargeter, Adobe Stock
Parameciums: Copyright: wire_man, Adobe stock
Scientist with Microscope: fernandozhiminaicela, CC0 Public Domain, pixabay
Prokaryotic Cell: Copyright: Mark Rasmussen: dollarphotoclub
Eukaryotic Cell: Copyright: Mark Rasmussen: dollarphotoclub
Halobacteria: Public Domain: Wikimedia, NASA
Klebsiella Pneumoniae Bacteria: Public Domain: Wikipedia, David Dorward, Ph.D.; National Institute of Allergy and Infectious Diseases
Parameciums: Copyright: wire_man, Adobe stock
Fungi: lindaharrison64, CC0 Public Domain, pixabay
Rose: suju, CC0 Public Domain, pixabay
Koala: winampdevil, CC0 Public Domain, pixabay
Classifications: Copyright: alinabel, Adobe Stock
Tiger: Heidelbergerin, CC0 Public Domain, pixabay
Carl von Linnaeus: Public Domain: Wikimedia, Nationalmuseum (Stockholm)
Average Prokaryote Cell: Public Domain: Wikimedia, LadyofHats
Curvularia geniculata: Public Domain: CDC / Dr. Ajello (PHIL #4512)
Nitrogen Atom: Copyright: generalfmv, Adobe Stock
Prokaryotic Fission: Copyright: Andrea Danti, Adobe Stock
Bacterial Conjugation: Wikimedia: CC-BY-SA-2.5, Mike Jones
E Coli Bacteria: WikiImages, CC0 Public Domain, pixabay
Fruiting Bodies: Wikiimages, Trance Gemini, CC-BY-SA-3.0
Treponemapallidum: Public Domain: CDC / Dr. David Cox (PHIL # 1977)
Rickettsia: Public Domain: Wikimedia, Rovery, Rrouqui, and Raoult
Mycoplasma gallisepticum: Wikimedia: CC-BY-SA-3.0, M.H. B. Catroxo and A.M.C.R.P.F. Martins
Anabeana sphaerica: Wikimedia: CC-BY-SA-3.0
Herpes Zoster Virus: Public Domain: CDC / Dr. Erskine Palmer / B.G. Partin (PHIL #1878)
Bacteriophage: Copyright: nobeastsofierce, Shutterstock
Lytic Cycle-Nathan Puray, Seton Home Study School
Viroid-Flickr, AJC1, CC BY-NC-SA-2.0
Euglena: Public Domain: Wikimedia, Environmental Protection Agency
Euglena Structure: Copyright: snapgalleria, dollarphotoclub
Ceratium Hirundinella: Copyright Rattiya Thongdumhyu, Shutterstock
Diatoms: Public Domain: National Oceanic and Atmospheric Administration / Department of Commerce, Prof. Gordon T. Taylor
Trypanosoma: CDC/Dr. Myron G. Schultz - This media comes from the Centers for Disease Control and Prevention's Public Health Image Library (PHIL), with identification number #613.
Amoeba Proteus: Copyright: micro_photo, Adobe Stock
Paramecium: Wikimedia: CC-BY-SA-3.0, Barfooz
Paramecium Diagram: Copyright: Snapgalleria, dollarphotoclub
Plasmodium vivax: Public Domain: CCDC / Steven Glenn, Laboratory and Conultation Division (PHIL #5863)
Malaria Life Cycle: Public Domain: National Institute of Allergy and Infectious Diseases
Mushrooms: adege, CC0 Public Domain, pixabay
Amanita muscaria: PublicDomainPictures, CC0 Public Domain, pixabay
Sarcoscypha coccinea: Wikimedia: CC-BY-SA-3.0, Ryane Snow
Bread Mold Closeup: Copyright: Armando Frazao, dollarphotoclub
Chytrid: Wikimedia: CC-BY-SA-2.5, Dr. David Midgley
Penicillin Conidiophore: Public Domain: CDC / Luccille K. Georg (PHIL #8397)
Basic Structure of Fungi: Copyright: Mari-Lea, Shutterstock
Slim Mold: Copyright: sleepyhobbit, Adobe Stock
Watermold: Wikimedia: CC-BY-SA-3.0, Keisotyo
Potato - Late Blight: Public Domain: Wikimedia
Bread Mold: Njoy Harmony, CC0 Public Domain, pixabay
Truffles: Copyright: kab-vision: Adobe Stock
Morels: PublicDomainImages, CC0 Public Domain, pixabay
Button Mushrooms: Copyright: dasuwan, Adobe Stock
Bracket Fungi: Wikimedia: CC-BY-SA-3.0, Eric Guinther
Lichen: Hans, CC0 Public Domain, pixabay

Chapter 11

Large Tree (Chapter Opening): Copyright: vovan, Adobe Stock
Pine (Chapter Header): LUM3N, CC0 Public Domain, pixabay
Leaves: Copyright: Rostislav Sedlacek, dollarphotoclub
Flowers: manfredrichter, CC0 Public Domain, pixabay
Bamboo: PublicDomainPictures, CC0 Public Domain, pixabay
Rose: suju, CC0 Public Domain, pixabay
Tree: giani, CC0 Public Domain, pixabay
Pine cones: Kapa65, CC0 Public Domain, pixabay
Hawortia: Foto-Rabe, CC0 Public Domain, pixabay
Ginkgo leaves: Erko, CC0 Public Domain, pixabay
Flower: HolgersFotografie, CC0 Public Domain, pixabay
Ferns: Copyright: Scisetti Alfino, dollarphotoclub
Micrasterias: Public Domain: Wikimedia, Ajburk
Blueberries: Heidelbergerin, CC0 Public Domain, pixabay
Strawberry runner: Copyright: Kazakova Maryia, Adobe Stock
Fern rhizomes: Copyright: Kazakova Maryia, Adobe Stock
Potato sprout: Copyright: Kazakova Maryia, Adobe Stock
Life Cycle of Ulva: Nathan Puray, Seton Home Study School
Life Cycle of Conifer: Copyright: Kazakova Maryia, Adobe Stock
Flower Anatomy: Copyright: 7activestudio, dollarphotoclub
Life Cycle of a Flower: Public Domain: Wikimedia, Lady of Hats
Chlorella: Wikimedia: CC-BY-SA-4.0, Vladi Damian
Algae: B.Ash: CC0 Public Domain, pixabay
Chlamydomonas: Public Domain: Wikimedia, Dartmouth Electron Microscope Facility, Dartmouth College
Volvox Colony of Freshwater Green Algae: Copyright: Pawel Burgiel, Adobe Stock
Spirogyra: Public Domain: Wikimedia, spicywalnut
Oedogonium: Nathan Puray, Seton Home Study School
Laurencia: Wikimedia: CC-BY-SA-3.0, Eric Guinther
Brown Algae: Copyright: kichigin19, Adobe Stock
Moss: Nacymac, CC0 Public Domain, pixabay
Life Cycle of Moss: Public Domain: Wikimedia, Lady of Hats
Liverwort: Copyright: Henrik Larsson, dollarphotoclub
Life Cycle of Liverwort: Public Domain: Wikimedia, Lady of Hats

Biology For Life

Vascular Plants: shogun, CC0 Public Domain, pixabay
Water molecule: Clker-Free-Vector-Images, CC0 Public Domain, pixabay
Root Tip: Wikipedia: CC-BY-SA-2.5, SuperManu
Celery Stalks: Copyright: Tanya Rusanova, Adobe Stock
Anatomy of Vascular Plants: Copyright: udaix, Shutterstock
Whisk Ferns: Copyright: sinhyu, Adobe Stock
Lycopodium Plant: Wikimedia: CC-BY-SA-3.0, Eric Guinther
Northern Giant Horsetail: Wikimedia: CC-BY-AS-3.0, Rror
Fern Rhizomes: Copyright: Kazakova Maryia, Adobe Stock
Life Cycle of a Fern: Copyright: magemasher, Adobe Stock
Berry: MabelAmber, CC0 Public Domain, pixabay
Seeds with Wings: pasja1000, CC0 Public Domain, pixabay
Pine Cone: JensEnemark, CC0 Public Domain, pixabay
Bean Pod: 4639459, CC0 Public Domain, pixabay
Catkins: MabelAmber, CC0 Public Domain, pixabay
Acorns: MabelAmber, CC0 Public Domain, pixabay
Ginkgo tree: punch_ra, CC0 Public Domain, pixabay
Cycas circinalis: Wikimedia: CC-BY-SA-3.0, Raul654
Young Norfolk Island Pine: Public Domain: Wikimedia, Kahuroa
Life Cycle of Conifer: Copyright: Kazakova Maryia, Adobe Stock
Life Cycle of a Flower: Public Domain: Wikimedia, Lady of Hats
Magnolia Flowers: steinchen, CC0 Public Domain, pixabay
Life Cycle of a Bean: Copyright: GraphicsRF, Adobe Stock
Day-lilies: philsimaging, CC0 Public Domain, pixabay

Chapter 12

Close up on live octopus in aquarium: Copyright: ND700, Shutterstock
Sponges (chapter header): manfredrichter, CC0 Public Domain, pixabay
Snail: Alexas_Fotos, CC0 Public Domain, pixabay
Sponges: manfredrichter, CC0 Public Domain, pixabay
Spherical Symmetry - Volvox aureus im Dunkelfeld mit Tochterkolonien: Copyright: micro_photo, Adobe Stock
Radial Symmetry - Stony Coral: Copyright: aquapix, Shutterstock
Bilateral Symmetry - Crab: xu_ming-xm, CC0 Public Domain, pixabay
Marine Sponges: Copyright: Kolevski.V, Adobe Stock
Hydra: Copyright: sakurra, Adobe Stock
Body Structure of a Hydra: Copyright: Yaroslava, Adobe Stock
Jellyfish: klawson, CC0 Public Domain, pixabay
Sea Anemone: Free-Photos, CC0 Public Domain, pixabay
Polyps of Great Star Coral: Copyright: John Anderson, Adobe Stock
Pillar Corals: WikiImages, CC0 Public Domain, pixabay
Flatworm Diagram: Copyright: Desihnua, Shutterstock
Planarian Dugesia: Copyright: sinhyu, Adobe Stock
Fasciola hepatica: Wikimedia: CC-BY-SA-3.0, Flukeman
Life Cycle of the Sheep Liver Fluke: Copyright: Aldona Griskeviciene, Shutterstock
Schistosoma: Public Domain: Wikimedia, David Williams, Illinois State University
Taenia solium: Public Domain: Wikimedia, Centers for Disease Control and Prevention
Hookworms: Public Domain: CDC (PHIL #5205)
Rotifers: Copyright: sinhyu, Adobe Stock
Anatomy of an Earthworm: Copyright: Aldona Griskeviciene, Shutterstock
Earthworm: Natfot, CC0 Public Domain, pixabay
Head Anatomy of an Earthworm: Copyright: Borbely Edit, Shutterstock
Haemadipsa zeylanica: Public Domain: Wikimedia, Pieria
Mollusk Anatomy: Copyright: nicolasprimola, Adobe Stock
Caribbean Reef Squid: Wikimedia: CC-BY-SA-3.0, Nick Hobgood
Helix pomatia: claude05alleva, CC0 Public Domain, pixabay
Marisa comuarietis: Public Domain: Wikimedia, Katrin-die-Rauberbraut
Limax maximus: ariesa66: CC0 Public Domain, pixabay

Giant Clam: Copyright: Sebastien Burel, Adobe Stock
Octopus Tentacles: Copyright: cloud7day, Adobe Stock
Spear Squid: Copyright: feathercollector, Adobe Stock
Octopus: edmondlafoto, CC0 Public Domain, pixabay
Starfish: financesafterforty, CC0 Public Domain, pixabay
Sea Urchins: charzz1913, CC0 Public Domain, pixabay
Dorsal Surface of a Starfish: ©DaveCarlson/CarlsonStockArt.com
Starfish is regenerating from an arm Raja Ampat Indonesia: Copyright: zaferkizilkaya, Shutterstock
Starfish: Pexels, CC0 Public Domain, pixabay

Chapter 13

Butterfly (chapter opening): Copyright: oliverroberts, Adobe Stock
Butterfly (chapter header): PublicDomainPictures, CC0 Public Domain, pixabay
Grasshopper: ulleo, CC0 Public Domain, pixabay
Trilobite: PublicDomainPictures, CC0 Public Domain, pixabay
Millipede: Josch13, CC0 Public Domain, pixabay
Crab: 12019, CC0 Public Domain, pixabay
Scorpion, Patrizia08, CC0 Public Domain, pixabay
Blue Butterfly, Paulbr75, CC0 Public Domain, pixabay
Open and Closed Circulation: Nathan Puray, Seton Home Study School
Damselfly molting: CC-BY-SA-4.0: 2016 Jee & Rani Nature Photography
Tarantula: Copyright: Danny, Adobe Stock
Orb Weaver Spider: Brett_Hondow, CC0 Public Domain, pixabay
Spider Anatomy: Wikimedia: CC-BY-SA-3.0, John Henry Comstock & Pbroks13
Phidippus audax: Public Domain: Wikimedia, Kaldari
Jumping spider: 12019, CC0 Public Domain, pixabay
Golden orb weaver spiders: Wikimedia: CC-BY-SA-3.0, DirkvdM
Scorpion with Young: Public Domain: Wikimedia, Fusion121
Daddy Long-legs: Copyright: Henrik Larsson, Adobe Stock
House Dust Mite: Public Domain: Wikimedia, FDA
Crayfish: Public Domain: Wikimedia, Gusmonkeyboy
Crayfish Anatomy: Nathan Puray, Seton Home Study School
Rusty Millipede: Wikimedia: CC-BY-4.0, Zoltan Korsos
Centipedes: Alexas_Fotos, CC0 Public Domain, pixabay
Dragonfly: MacroMan, CC0 Public Domain, pixabay
Yellow-jacket Wasp: Public Domain: Wikimedia, Bernie
Honey Bee: chezbeate, CC0 Public Domain, pixabay
Completed & Incomplete Metamorphosis: Copyright: Panaiotidi, Shutterstock
Grasshopper: bogdanchr, CC0 Public Domain, pixabay
Grasshopper Mouth Anatomy: Public Domain: Wikimedia, Westeros91
Internal Anatomy of a Grasshopper: ©DaveCarlson/CarlsonStockArt.com
Brimstone Butterfly: Wikimedia: CC-BY-SA-4.0, Zeynel Cebeci
Geometer Moth: Copyright: Henrik Larsson, Adobe Stock
Grasshopper: makamuki0, CC0 Public Domain, pixabay
Yellow-jacket Wasp: Public Domain: Wikimedia, Bernie
Monarch Butterfly: gyulche1, CC0 Public Domain, pixabay
Palm Beetle; Angeleses, CC0 Public Domain, pixabay
Viceroy Butterfly: skeeze, CC0 Public Domain, pixabay
Flower-fly: Public Domain: Wikimedia, subhrajyoti07
Sesiid Moth: Wikimedia: CC-BY-2.0, Patrick Clement
Honey Bee: Capri23auto, CC0 Public Domain, pixabay
Honey Bee Anatomy: Wikimedia: CC-BY-SA-3.0, Wikipedian Prolific
Honey Bee Swarm: Wikimedia: CC-BY-3.0, Bidgee

Chapter 14

Sea corals (chapter opening): Copyright: vlad61_61, shutterstock
Clown-fish (chapter header): LauraRinke, CC0 Public Domain, pixabay
Red-eye frog: Capri23auto, CC0 Public Domain, pixabay
Clown-fish and Sea Anemone Background: Pepril, CC0 Public Domain, pixabay
Pacific Hagfish: Public Domain: Wikimedia, Linda Snook, NOAA/CBNMS
Sea Lamprey: Wikimedia: CC-BY-SA-3.0, Tiit Hunt
Lamprey Mouth: Copyright: Gena, Adobe Stock
Lampreys and Lake Trout: Public Domain: Wikimedia, USGS
Grey Reef Shark: Copyright: Richard Carey, Adobe Stock
Parts of a Shark: Wikimedia: Public Domain, Chris_huh
Tiger Shark Teeth: Copyright: Matthew R. McClure, Shutterstock
Egg case of Port Jackson Shark: Wikimedia: Public Domain, AYArktos
Skate: Wikimedia: CC-BY-2.0, expl6380
Ray: Wikimedia: CC-BY-2.0: Martin Holst Friborg Pedersen
Yellow Perch: Copyright: Andrey Burmakin, Adobe Stock
Fish Anatomy: Copyright: anton_novik, Adobe Stock
School of fish: Copyright: Leonardo Gonzalez, Shutterstock
Salmon Eggs: Copyright: ronniechua, Adobe Stock
Atlantic Salmon: Copyright: Jakub, Adobe Stock
Great Barracuda: Public Domain: Wikimedia, Florida Keys National Marine Sanctuary
Lion fish: Hans, CC0 Public Domain, pixabay
Lungfish: Wikimedia: CC-BY-SA-4.0: Mitch Ames
Western Spadefoot Toad: Public Domain: Wikimedia, Chris Brown, USGS
Dumpy Tree Frog: WikiImages, CC0 Public Domain, pixabay
Green Frog: Copyright: Eastman Arts, Adobe Stock
Frog Anatomy: Copyright: snapgalleria, Adobe Stock
Frog Face: joncressey, CC0 Public Domain, pixabay
Three Green Frogs: Copyright: kuritafsheen, Adobe Stock
Life cycle of a frog: Copyright: designua, Adobe Stock
Marbled Salamander: Wikimedia: CC-BY-2.0, Peter Paplanus
Pacific Giant Salamander: Public Domain: Wikimedia, Jeffrey Marsten
Red Back Salamander: Copyright: Jason Patrick Ross, Shutterstock
Axolotl: Tinwe, CC0 Public Domain, pixabay
Member of the Order Apoda: Wikimedia: CC-BY-SA-2.5, Dawson

Chapter 15

Lion (chapter opening): Copyright: Photocreo Michal Bednarek, Shutterstock
Parrot (chapter header): Couleur, CC0 Public Domain, pixabay
Chameleon: Pixel-mixer, CC0 Public Domain, pixabay
Caimen Crocodile: manfredrichter, CC0 Public Domain, pixabay
Eastern Diamondback Rattlesnake: skeeze, CC0 Public Domain, pixabay
Green Sea Turtle: skeeze, CC0 Public Domain, pixabay
Tuatara: Successful4, CC0 Public Domain, pixabay
Turtle eggs: Wikimedia: CC-BY-2.0, Rachel Kromrey/USFWS
Reptile egg: Copyright: NoPainNoGain, Shutterstock
Garter snakes: skeeze, CC0 Public Domain, pixabay
Anatomy of a Snake: Nathan Puray, Seton home Study School
Snake Fangs: Copyright: mgkuijpers, Adobe Stock
Python Skeleton: Public Domain: Wikimedia, Daderot
African Egg-Eating Snake: Public Domain: Wikimedia, Mond76
Jacobson's Organ: Nathan Puray, Seton Home Study School
Snake eggs: Copyright: PIXATERRA, Adobe Stock
Rattlesnake: skeeze, CC0 Public Domain, pixabay
Green Tree Python: Pony Paparazzi, CC0 Public Domain, pixabay
Barbados Thread Snake: Wikimedia: Attribution: Blair Hedges, Penn State
Corn Snake: Kapa65, CC0 Public Domain, pixabay
Iguana: Salao, CC0 Public Domain, pixabay
Gecko lizard: Kapa65, CC0 Public Domain, pixabay
Komodo dragon: TLSPAMG, CC0 Public Domain, pixabay
Green Sea Turtle: skeeze, CC0 Public Domain, pixabay
Galapagos Giant Tortoise: Public Domain: Wikimedia, Bernard Spragg
Diamondback Terrapin: Public Domain: Wikimedia, Ryan Hagerty
Nile Crocodile: Copyright: Lefteris Papauakis, Adobe Stock
Crocodile: Sponchia, CC0 Public Domain, pixabay
American Alligator: James DeMers, CC0 Public Domain, pixabay
Tuatara: Public Domain: Wikimedia, knutschie
Emperor Penguins: MemoryCatcher, CC0 Public Domain, pixabay
Gambel's Quail: Public Domain: Wikimedia, Alan Schmierer
Common Tern Bird: kanenori, CC0 Public Domain, pixabay
Bird Anatomy: Copyright: Jakinnboaz, Shutterstock
Feather Diagram: Copyright: crspix, Adobe Stock
Types of Feathers: Copyright: designua, Adobe Stock
Bird Wing Diagram: satyatiwari, CC0 Public Domain, pixabay
Loggerhead Shrike molting: Wikimedia: CC-BY-2.0, Kevin Cole
Loggerhead Shrike: Public Domain: Wikimedia, David Menke
Variety of Bird Heads: Wikimedia: CC-BY-SA-2.5: L. Shyamal
Barn Owl: LubosHouska, CC0 Public Domain, pixabay
Great Blue Heron: paulbr75, CC0 Public Domain, pixabay
Bird Skeleton diagram: Copyright: roadrunner, dollarphotoclub
Bird's Digestive System: Nathan Puray, Seton Home Study School
Bird Migration: 7435465, CC0 Public Domain, pixabay
Robin: Capri23auto, CC0 Pubic Domain, pixabay
Egg Structure: Copyright: kawin302, Adobe Stock
Chick with Egg Tooth: Wikimedia: CC-BY-SA-3.0: uberprutser
Giraffe: Alexas_Fotos, CC0 Public Domain, pixabay
Fruit Bat: Copyright: subphoto, Adobe Stock
Lion: Copyright: PHOTOCREO Michal Bednarek, Shutterstock
Hedgehog: Alexas_Fotos, CC0 Public Domain, pixabay
North Pacific Sea Lion: gkgegk, CC0 Public Domain, pixabay
Platypus: pen_ash, CC0 Public Domain, pixabay
Short-beaked Anteater: PublicDomainImages, CC0 Public Domain, pixabay
Long-beaked Anteater: Wikimedia: CC-BY-SA-3.0: Jaganath
Virginia Opossum: daynaw3990, CC0 Public Domain, pixabay
Koala: skeeze, CC0 Public Domain, pixabay
Eastern Grey Kangaroo with Joey: Copyright: Harlz, Adobe Stock
Young Eastern Grey Kangaroo: Wikimedia: CC-BY-SA-3.0: J.J. Harrison
Eastern Grey Squirrel: skeeze, CC0 Public Domain, pixabay
House Mouse: Georg_Wietschorke, CC0 Public Domain, pixabay
Common Brown Rat: Public Domain: Wikimedia, National Park Service
American Beaver: Public Domain: Wikimedia, Groucho M
Pika: Public Domain: Wikimedia, Alan Schmierer
Eastern Cottontail: Public Domain: Wikimedia, William R. James
Jackrabbit: skeeze, CC0 Public Domain, pixabay
European Hare: Wikimedia: CC-BY-2.0: Bengt Nyman
Flying Bat: Copyright: Chamnan phanthong, Adobe Stock
Fruit Bat: Copyright: Andrea Izzotti, Adobe Stock
Humpback Whale: Wikimedia: CC-BY-2.0, Christopher Michel
Bottle-nose Dolphin: PublicDomainImages, CC0 Public Domain, pixabay
Harbor Porpoise: WIkimedia: CC-BY-SA-4.0, Ecomare/Salko de Wolf Den Hoorn Texel
Australian Shepherd Dog: Couleur, CC0 Public Domain, pixabay
Siamese Cat: webandi, CC0 Public Domain, pixabay
Raccoon: GrammarCop, CC0 Public Domain, pixabay
Brown Bear: Bergadder, CC0 Public Domain, pixabay
Short-Tailed Weasel: 12019, CC0 Public Domain, pixabay
Sea lion: PublicDomainPictures, CC0 Public Domain, pixabay
Pacific Walrus: skeeze, CC0 Public Domain, pixabay
Common Seal: Wikimedia: CC-BY-4.0, Vejlenser
Zebras: 12019, CC0 Public Domain, pixabay

Biology For Life

Horse: Joachim_Marian_Winkler, CC0 Public Domain, pixabay
Black Rhinoceros: skeeze, CC0 Public Domain, pixabay
Goat: violetta, CC0 Public Domain pixabay
Hippopotamus: onkelramirez1, CC0 Public Domain, pixabay
Giraffes: hbieser, CC0 Public Domain, pixabay
Llama: Alexas_Fotos, CC0 Public Domain, pixabay
Horse Hoof: Kaz, CC0 Public Domain, pixabay
Rhinoceros Hoof: Marabu, CC0 Public Domain, pixabay
Giraffe Hoof: zoosnow, CC0 Public Domain, pixabay
Hippopotamus Hoof: adege, CC0 Public Domain, pixabay
Mole: Beeki, CC0 Public Domain, pixabay
Short-Eared Elephant Shrew: 7854, CC0 Public Domain, pixabay
Sloth: WikiImages, CC0 Public Domain, pixabay
Anteater: ddouk, CC0 Public Domain, pixabay
Armadillo: PublicDomainPictures, CC0 Public Domain, pixabay
Lemur: skeeze, CC0 Public Domain, pixabay
Capuchin, Wikimedia: CC-BY-SA-3.0, Whaldener Endo
Chimpanzees: sasint, CC0 Public Domain, pixabay
Orangutans: Public Domain: Wikimedia, Denis & Chris Luyten-De Hauwere
Gibbon Monkey: Wildfaces, CC0 Public Domain, pixabay
Silver-back Gorilla: herbert 2512, CC0 Public Domain, pixabay
Lion: Copyright: PUTSADA, Adobe Stock

Chapter 16

Earth (chapter opening): Copyright: sdecoret, Adobe Stock
Earth (chapter header): qimono, CC0 Public Domain, pixabay
DNA: PublicDomainPictures, CC0 Public Domain, pixabay
St. Thomas Aquinas: Public Domain: Wikimedia, Francisco de Zurbaran
Louis Pasteur in His Laboratory: Public Domain: Wikimedia, Albert Edelfelt
Plant Sprout: TillVoigt, CC0 Public Domain, pixabay
Cat: Sbringser, CC0 Public Domain, pixabay
Man: Pexels, CC0 Public Domain, pixabay
Phototropism: Copyright: adastra, Shutterstock
Gravitropism: Wikimedia: CC-BY-SA-40, Rufus22181496
Dog: Jasch77, CC0 Public Domain, pixabay
Chimpanzee: Rabenspiegel, CC0 Public Domain, pixabay
Shape Perception: Nathan Puray, Seton Home Study School
Metronome: Copyright: ThomasLENNE, Adobe Stock
Cat: Copyright: Sergii Figurnyi, Adobe Stock
Cat and Mouse: Copyright: Sergii Figurnyi, Adobe Stock
Kurt Godel: Public Domain: Wikimedia
Small chocolate cake: stux, CC0 Public Domain, pixabay
Searching puzzle pieces: Copyright: OGI75, Shutterstock

Glossary

acetylcholine (Ach)
a neurotransmitter that causes a contraction in a muscle

acetylcholinesterase (AChE)
an enzyme that inactivates acetylcholine by degrading it into its precursors acetic acid and choline molecules. AChE works extremely quickly and can process 5,000 acetylcholine molecules per second.

actin
a thin protein filament within muscle fiber

active transport
movement of materials across the cell membrane that requires the use of energy to get substances to cross the cell membrane, either because their chemical properties do not permit diffusion, or because they must be transported from an area of lesser concentration to an area of greater concentration.

ADP (adenosine diphosphate)
a nucleotide that contains less energy than ATP and is both a building block and a product of ATP construction and use

adrenal cortex
the outer part of the adrenal gland. It produces corticoid hormones, which regulate metabolism, salt, and water balance, control production of certain blood cells, and influence the structure of connective tissue.

adrenal glands
paired structures that sit on top of the kidneys. Each adrenal gland is made of an outer cortex and an inner medulla. The adrenal cortex controls salt and water balance in the body. The adrenal medulla controls the body's response to stressful events, controls metabolism, and influences sexual development.

adrenaline
a hormone secreted by the adrenal medulla. It is also known as epinehrine. Adrenaline increases blood pressure by dilating or widening the blood vessels in the liver, heart, and skeletal muscles. At the same time, adrenaline will constrict blood vessels in the skin and other internal organs to allow more blood to flow to the muscles, heart, and liver. Adrenaline helps convert glycogen to glucose, so that more energy is made available to the body.

adrenocorticotropic hormone (ACTH)
a hormone produced by the anterior pituitary gland that stimulates hormone secretion by the adrenal cortex

agnatha
the class of jawless fish, also called the cyclostomata, and sometimes referred to as the cyclostomes. The jawless fish, the hagfish and the lamprey, are parasites and scavengers.

algae
a plantlike organism that lives in water, contains chlorophyll

alimentary canal
a long, hollow, twisted, and coiled tube extending from the mouth to the anus. The alimentary canal is sometimes referred to as the gastrointestinal tract, or GI tract.

alveoli
clusters of balloon-like microscopic air sacs at the end of each bronchiole

amino acids
organic molecules that are the building blocks of proteins. There are twenty acids that are found in nature.

amphibians
cold-blooded, vertebrate animals, that live part of their life cycle in the water, and part on the land. Amphibians include frogs, toads, salamanders, and caecilians.

anaphase
part of the process of mitosis. In this stage, the two chromosomes that made up each chromatid pair begin to move away from each other and go to the opposite poles of the cell. One part of the pair moves to one pole of the cell, and the other part moves to the opposite pole.

androgens
male sex hormones produced in the testes. One of the androgens is known as testosterone, which during puberty causes the development of the male secondary sex characteristics.

angiosperms
flowering plants which produce seeds protected by a cover

animal-like protists
known as protozoa. Protozoa are single-celled eukaryotes that share some traits with animals. Animal-like protists can move, and they obtain nutrition from outside of themselves instead of producing their own food.

antibiotics
powerful medicines that fight bacterial infections. Antibiotics either stop the growth of bacteria or directly kill bacteria.

antibodies
proteins that attack and neutralize foreign substances that may enter the body, such as bacteria and viruses

anticodon
a triplet of bases that are complementary to the codon of the m-RNA to allow the t-RNA to "recognize" the m-RNA and to allow placement of the proper amino acid

antigen
a substance that causes an immune response and triggers the production of antibodies. (In red blood cells, the presence or absence of certain antigens determines the four different types of red blood cells: A, B, O, and AB.)

aorta
the largest artery in the body, where the blood enters into smaller arteries and is transported to the whole body

apoda
a group of worm-like, legless amphibians that live in tropical regions. These animals can be found burrowing under moist soils. The apods are carnivorous, and eat earthworms, insects, and insect larvae.

appendix
a small projection located near the point where the small and large intestines meet. At one time it was believed that the appendix was a useless, or "vestigial" organ. Recent research has shown that the appendix contains many infection-fighting cells and that it helps regulate intestinal bacteria.

arachnids
arthropods that have bodies comprised of two segments and eight legs. Arachnids, including spiders, mites, scorpions, and ticks, have an abdomen and a cephalothorax. A cephalothorax means the head and the thorax are joined. Thus, the arachnids have only two body segments.

archaea
single-celled organisms that lack a membrane-bound nucleus and membrane-bound organelles

arteries
vessels that carry blood away from the heart. Arteries have thick muscular walls with much elastic connective tissue and a smaller central cavity because they carry blood pumped under high pressure from the heart.

arterioles
small blood vessels that branch out from arteries. Blood travels from arterioles into even smaller vessels called capillaries.

asymmetric body symmetry
describes bodies with no orderly repeating parts

ATP (adenosine triphosphate)
a nucleotide with two additional phosphate groups attached to it. It is formed in the mitochondria using energy produced during cellular respiration in which the cells "burn" glucose.

autonomic nervous system
part of the peripheral nervous system which controls involuntary body functions through smooth muscle, cardiac muscle, and glands

axon
the extension from the nerve cell body at the bottom or end of the cell. Each neuron has only one axon. An axon carries impulses away from the cell body and transmits impulses to other cells.

bacteria
the second of the prokaryotic domains. Bacteria are single-celled organisms that lack a membrane-bound nucleus and membrane-bound organelles.

bacteriophage
a virus that infects bacteria

ball-and-socket joint
the most freely movable joint in the body, in which the rounded head of a bone fits into a hollow socket of another bone. The two shoulder joints and the two hip joints are the only ball-and-socket joints in the human body.

basal disk
the structure by which a hydra attaches to a surface

beak
a hard, toothless, external part of the mouth of birds which is used for eating and for grooming, manipulating objects, killing prey, fighting, probing for food, courtship and feeding the young. A beak is also called a bill, or rostrum.

bilateral symmetry
describes a body structure whereby the right half mirrors the left half

bile
a liquid produced by the liver which is used in digestion. Bile does not contain any enzymes; however, bile contains substances that aid in the digestion of fats, or lipids.

binomial nomenclature
a two-part naming system for living things. In this system, organisms are identified according to the genus and species.

biogenesis
a scientific law that "life begets life." The Law of Biogensis was scientifically proved by Louis Pasteur in 1864. It states that, "organisms do not spontaneously arise in nature from nonliving things."

blastocyst
an embryo of about six days old. It has not yet implanted in the walls of the uterus. A blastocyst consists of about 70 to 100 cells which will differentiate into the placenta and the baby when the blastocyst does attach to the uterine wall.

blood pressure
the force needed to keep blood flowing through the vessels to the whole body. Blood pressure must be high enough so that all the tissues are adequately supplied, but also low enough so that the tissues are not damaged.

blood vessels
the "pipes" that carry blood throughout the body. The types of blood vessels include arteries, arterioles, veins, venules, and capillaries. Blood vessels have walls made of smooth muscle cells, elastic connective tissue, and are lined with endothelial cells.

bone marrow
a jelly-like substance within bones where blood cells are produced

bony fish
the largest group of vertebrates with approximately 30,000 different species of fish divided into two major subgroups: the ray-finned fish and the lobe-finned fish. Over 99% of all bony fish are ray-finned; there are only eight living species of lobe-finned fish.

Bowman's capsule
a cup-like structure within a nephron that encloses the glomerulus. Each Bowman's capsule is the beginning of a long, continuous tube, which has sections that differ in structure and function.

brain
the organ that controls most of the functions of the body. Lower functions controlled by the brain include heart rate, respiration, blood pressure, and digestion, and all voluntary movement. Higher functions controlled by the human brain include consciousness, memory, planning, language, creativity, emotion, personality, expression, problem solving, and reasoning.

brain stem
the portion of the brain connecting the brain to the spinal cord. The brain stem is mostly white matter from motor neurons and sensory neurons traveling between the brain and spinal cord. The brain stem coordinates motor signals sent from the brain to the body.

bronchioles
smaller and finer tubes that branch out from the bronchi

bronchus
one of the large air tubes which lead to the lungs from the trachea, or windpipe. (plural: bronchi)

capillaries
the smallest vessels in the body and allow the exchange of oxygen and other materials between blood and tissue cells. Capillaries also provide the connection between large arteries and large veins.

carbon
the basic element of life

carbon dioxide
a waste product produced by cellular respiration. This waste gas must be removed from the cell through the circulatory system and then exhaled from the body.

carpals
wrist bones

cartilage
firm but rubbery tissue that serves as a cushion at the joints

cell
the basic unit of structure and function of all living organisms

cell division
the process by which a cell divides into two or more cells. Cell division is the source or cause of tissue growth and repair in multicellular organisms.

cell membrane
a biological membrane made up of proteins and lipids that separates the interior of all cells from the outside environment and protects the cell from its surroundings.

cell wall
the rigid structure on the outside of plant cells that provides their basic structure

cellular respiration
the process in cells by which the chemical energy of "food" molecules is released and used to make ATP

cementum
a thin layer of bony material that fixes teeth to the jaw

centrioles
organelles in animal cells, they are small paired cylindrical structures that help to pull apart the two halves of a cell during cellular reproduction

ceplahothorax
a fused section of a chelicerate consisting of the head and thorax

cerebral cortex
the outer surface of the two cerebral hemispheres. The cerebral cortex is a convoluted surface made of folds and fissures of gray matter.

cheliped
appendages in crustaceans that bear pincers

chemical digestion
the breakdown of food particles by chemical means, through the action of enzymes, into simpler substances

chitin
a complex, large, sugar molecule that makes up the cell walls of fungi. Chitin is also found in the hard outer shell of insects and mollusks.

chlorenchyma
a type of parenchyma that contains chloroplasts. Chlorenchyma is specialized for photosynthesis.

chlorophyll
a green pigment that is used to capture the energy of the sun

chloroplast
a type of plastid that contains the green pigment, chlorophyll

chordates
animals with bilateral symmetry or duplicate body parts on both sides, a body cavity, and segmented bodies. All chordates are animals with a digestive track that has an intake opening or mouth, and an exit opening or anus.

chromosome
structures in organisms that contain DNA. In humans, chromosomes contain all of the traits or characteristics for the child that he or she will ever have. Chromosomes reside in the nucleus of egg and sperm cells, and provide the chemical instructions for human development.

cilia
short hair-like structures that extend from the surface of some cells

circulatory system
system that includes the heart, blood vessels, and the blood. Another name for the circulatory system is the cardiovascular system. This system brings oxygen and nutrients to the cells in the body, and bring waste products, like carbon dioxide, to the lungs and other organs of waste disposal for elimination from the body.

clavicle
collarbone

cleavage
the process in which the living fertilized egg divides, or cleaves, into two smaller cells

cochlea
the portion of the inner ear that contains hair cells which are the sensory receptors for sound

codons
each group of three nucleotides, or "word" in a strand of DNA, which corresponds to the production of one amino acid

coelom
a cavity within the body of a mollusk

conjugation
the transfer of genetic material between bacteria through direct cell-to-cell contact

conifers
cone-bearing, woody gymnosperms that are found in a wide variety of climates on all continents except Antarctica

contractile vacuole
a vacuole that allows certain organisms to control the water balance of the cell. The water flows into this vacuole, causing expansion of the vacuole. When the vacuole has expanded completely, it contracts, releasing the excess water out through the gullet.

convolutions
ridges and depressions formed into the surface of the cerebrum (cerebral cortex). These convolutions greatly increase the surface area of the cerebral cortex.

coracoid
a short projection from the shoulder blade in mammals, to which part of the biceps is attached

cornea
the front portion of the sclera. The cornea is transparent, which means that all light can easily pass through it. The cornea is more curved than the rest of the eye. Its curved surface causes the incoming light rays to bend, thus helping to form an image of the object.

crown
the chewing surface of a tooth that we can see

crustaceans
mostly aquatic arthropods, that typically have the body covered with a hard shell or crust. Crustaceans include lobsters, crabs, crayfish, shrimp, krill, barnacles, and pill bugs.

cuticle
a type of skin that is made of chitin and proteins

cuticle (plant)
the waxy layer on a leaf that protects the leaf from outside injury and from excessive water loss

cytoplasm
clear, gel-like substance outside the nucleus of the cell of plants and animals

cytoskeleton
a miniature internal support system for the cell, made up of microtubules and other small proteins. It gives the cell its shape, much like wooden beams give shape to a church or your skeleton gives shape to your body.

decomposer
organisms that break down dead organisms and organic matter into simpler organic compounds so they can release carbon and minerals back into the environment

deoxyribose
a five-carbon sugar contained in DNA. The difference between ribose and deoxyribose is that ribose has one more oxygen atom.

determination
also called restriction, means a part of the embryo becomes restricted to a specific role or function

diaphragm
a large muscle at the bottom of the chest cavity. The contraction and relaxation of this muscle controls the movement of air into and out of the lungs.

diastole
the phase of the heartbeat when the ventricles relax and are not pumping blood. During diastole, the ventricles fill with blood to prepare for the next contraction.

dicots
seed-bearing plants that contain two embryonic seed leafs. Dicots produce flowers with petals in multiples of four and five. Dicot leaves have veins that form a branching, or netted, pattern from a central mid-vein.

differentiation
a process whereby cells become specialized and the appearance of a new property emerges such as the development of muscle cells

digestion
the actual breaking down of complex food into the basic building blocks of carbohydrates, proteins, fats, vitamins, and other small nutrients

digestive system
the organ system responsible for breaking down food into carbohydrates, proteins, fats, vitamins, and other small nutrients which are used by the human body for energy to fuel all the life processes

DNA
deoxyribonucleic acid. DNA carries the genetic code of the organism.

domain
classifies organisms based on the complexity of their cell structure. Modern scientists now consider this to be the broadest classification category of living things.

double helix
the spiral shape, more like a spiral staircase, of DNA. This shape is known as a double helix. The rails are linked with covalent bonds and the rungs are held together by hydrogen bonds.

duodenum
the first section of the small intestine. Its length is about 20 centimeters (8 inches).

ear
the organ of hearing

eardrum
a thin membrane that separates the external ear from the middle ear. Its function is to transmit sound vibrations from the air to the ossicles inside the middle ear.

Edentata
an order that includes about thirty species that live in North and South America. The collared anteater, nine-banded aramdillo, and brown-throat three-toed sloth are members of the edentata.

electrolytes
include sodium, potassium, chloride, and other substances. Electrolytes are chemicals that allow cells to maintain the correct electrical voltage to work properly.

embryo (human)
the developing human child during the first eight weeks of life in the womb

endocrine system
the system responsible for maintaining the body's proper functions including all aspects of metabolism, reproduction, growth, and development. The endocrine system is a system of glands.

endoplasmic reticulum
a network of canals that connect the nuclear membrane with the plasma membrane. The job of the endoplasmic reticulum is to prepare proteins for transport through a process called synthesis.

enzymes
large biological molecules responsible for the thousands of metabolic processes that sustain life. Enzymes are catalysts; they make a chemical reaction work faster, or better, without being used up themselves. Enzymes accelerate or increase the rate of metabolic reactions, like the digestion of food.

epidermis (root)
the outermost tissue of a root which protects the root and helps prevent water loss

epidermis (layer of the skin)
the outer layer, or top layer of the skin. It is made up of epithelial cells.

epiglottis
a small flap of tissue at the base of the throat. The epiglottis is a protective feature of the respiratory system. The epiglottis flap automatically closes over the voice box, that is, the opening of the windpipe, when a person swallows, to prevent chocking.

epinephrine
See adrenaline.

erythrocytes
red blood cells

esophagus
a tube that connects the throat to the stomach. Also called the food tube.

estrogens
female sex hormones produced by the ovaries. These cause the beginning of the menstrual cycle. In addition, estrogens cause the development of secondary sex characteristics, which develop during puberty.

eukaryotes
multicellular organisms with cells that have an organized nucleus as well as specialized, membrane-protected tiny organelles (which are a specialized part of a cell)

euglena
single-celled protists that contain chlorophyll and are capable of photosynthesis. Euglena do not have cell walls. Some species of euglena have a red eye spot that helps the euglena detect light and move toward the light.

excretion
an active process through which living organisms remove by products of metabolism, or waste products, from themselves. Excretion is the final step in the process of digestion.

eye
the organ of sight. The eye is responsible for vision.

eyespot
a light-sensitive spot found on some organisms

fallopian tube
a pair of tubes that extend from the ovaries to the uterus

fatty acid
a very long chain of carbon and hydrogen molecules. Fatty acids are the building blocks of the fat in food, and in the body. Fatty acid molecules are usually joined together in groups of three, forming a molecule called a triglyceride.

femur
thigh bone

fertilization
the process that results in the joining of the chromosomes of an egg cell and a sperm cell to produce a new, unique genetic code

fetus
the word used to describe the unborn child beginning with the ninth week; a completely formed baby in miniature

fibrin
a protein that solidifies a clot

fibula
calf bone

flagella
whip-like structures, similar to tails, that extend from the cell body and enable those cells to move through their environment

flatworms
relatively simple bilateral, un-segmented, soft-bodied invertebrates. Unlike other bilaterals, they have no body cavity), and do not have specialized circulatory and respiratory organs.

flowers
those parts of angiosperms that house the reproductive structures

fructose
fruit sugar

fruiting bodies
the part of a fungus that forms on the surface of the soil. The fruiting bodies are the structures most commonly recognized as a fungus and are found in a wide variety of shapes, sizes, and colors. The part of the common mushroom that you can see on your lawn is a fruiting body. The fruiting bodies are specialized to produce and release spores, which are the reproductive cells of fungi.

gallbladder
a pear-shaped muscular-walled organ that stores and releases bile into the duodenum to aid in digestion by breaking down fats. Bile travels from the gallbladder to the small intestine in the common bile duct.

ganglia
masses of tissue which contain nerve cells

gastrovascular cavity
the primary structure for digestion and circulation in jellyfish, hydras, flatworms, and other invertebrates

gene
a specific part of a chromosome in a section of DNA

genetic code
the part of DNA that contains all information needed to build and maintain a complete organism, and to pass on traits from parents to offspring

genetics
the scientific study of how parents pass on their characteristics (heredity traits) to their offspring

genome
the term for the genetic code or the hereditary information that is contained within the DNA

genus
a group of organisms or species that are structurally similar

geotropism
the growth of parts of a plant towards or away from the source of gravity. For example, roots grow down towards gravity, and stems grow up away from gravity.

gill arches
bony or cartilaginous arches located on either side of the pharynx and supporting the gills in fish and amphibians. Fish have four gill arches on each side of the head. Gill arches support the gill rakers, which are located on the front of each arch.

gill filaments
threadlike processes forming the respiratory surface of a gill.

glands
organs responsible for producing hormones which are important for good health and development of the human body.

glomerulus
part of the nephron located inside a cup-like structure known as Bowman's capsule.

glucagon
a hormone that converts glycogen into glucose within the liver and muscles. It does this when the concentration of glucose in the blood is low.

golgi bodies
organelles made of stacks of membrane pouches organized throughout the cytoplasm of a cell. Golgi bodies are the "post office" of the cell. Here, newly made proteins are sorted and packaged according to type.

gonads
the reproductive glands. Male gonads are called testes, and female gonads are called ovaries.

gravitropism
a turning or growth movement by a plant, fungus, or animal in response to gravity. For example, the roots of plants generally grow towards the center of gravity. (also known as geotropism)

gray matter
axons without myelin

Biology For Life

growth hormone
a hormone produced by the pituitary gland that regulates growth of bone and other tissues

gullet
a canal that opens into the reservoir of the euglena

heredity
the transmission of traits from the parent to the child

holdfast
an organ by which some alga attach to rocks or other solid objects

hormones
secretions of endocrine glands that regulate functions such as growth, maturation, and other functions of the body

humerus
the largest bone of the arm or fore limb in mammals. The humerus extends from the shoulder to the elbow.

hydrocarbons
molecules made from carbon and hydrogen. Coal, oil, natural gas, kerosene, and gasoline are types of hydrocarbons.

hypertension
the medical term for high blood pressure

hyperthyroidism
a condition that occurs if too much thyroxine is produced by the thyroid. A rise in the metabolic rate increases the heart rate, blood pressure, and body temperature. People with hyperthyroidism may sweat heavily, become nervous, and develop bulging eyes. This condition may be treated with drugs that will reduce thyroxine secretions. Another treatment for hyperthyroidism consists of surgically removing a part of the thyroid gland.

hyphae
long filaments of cytoplasm surrounded by a cell membrane and cell wall of fungi

hypothalamus
the master switch of the endocrine system that connects directly to the brain and receives feedback about the entire body through the nervous system. Based on this information, the hypothalamus then sends signals to stimulate or inhibit hormone secretion by the pituitary gland.

hypothyroidism
a condition that results from a deficiency of thyroxine. Persons with this condition have a very low metabolic rate, usually lack energy, and may be overweight. If hypothyroidism occurs during childhood, it can cause stunted growth and severe mental retardation. This condition is known as cretinism. Hypothyroidism cannot be cured, but it can be corrected with thyroxine medication.

immune system
system that has the function to protect the body and rid the body of germs. The immune system is composed of the following: leukocytes or white blood cells, the lymph system, the thymus gland, the spleen, and other specialized tissues.

immunity
the body's ability to prevent or resist infection or illness caused by germs

implantation
the stage at the beginning of pregnancy when the blastocyst attaches to the wall of uterus. It occurs around six days following fertilization. Implantation of the blastocyst also causes its cells and the cells of the uterus to divide rapidly, forming the rudiments, or beginnings, of a placenta and various protective membranes.

inferior vena cava
a large vein that returns deoxygenated blood to the heart from the lower part of the body

ingestion
the means by which food enters the digestive system. Ingestion is simply eating and drinking.

insestine (vertebrates)
part of the digestive system, specifically the part of the alimentary canal that extends from the end of the stomach to the anus

insulin
a hormone that lowers the levels of glucose in the blood and promotes the uptake of glucose by the body cells

integumentary system
the system that is the outer layer of the body. The skin is the largest organ of the body which includes the epidermis, dermis, subcutaneous tissue, hair, nails, sweat glands, sebaceous glands, and specialized sensory nerves, along with related muscles and blood vessels.

interneurons (or associative neurons)
neurons that form connections or serve as a link between sensory neurons and motor neurons

interphase
the first of the five stages in mitosis. During interphase, the cell is not yet dividing, but growing and replicating its DNA in preparation for cell division. At this point in the mitotitc cycle the chromosomes are not visible as long spaghetti-like strands but are bunched together and called chromatin.

invertebrates
animals that do not have a backbone. Invertebrates make up over 95% of all species of animals. The invertebrates include: sponges, flatworms, mollusks, roundworms, and many other groups.

iris
the colored portion of the eye. It is located behind the cornea. The iris contains smooth muscle tissue that can dilate (enlarge) or constrict (make smaller) the pupil.

isogamy
reproduction by isogametes, the simplest form of sexual reproduction

joint
the location at which bones connect. Most joints allow movement and provide mechanical support.

keratin
a protein that makes up hair, nails, and parts of the skin

kidney
the most important and most complex organ of the urinary system. The kidneys remove wastes from the blood. Through re-absorption, the kidneys put water and electrolytes back into the blood. Each kidney is divided into three major regions. The kidneys monitor and help control blood pressure.

kilocaleries
the units used to measure the energy content of food. A calorie (with a lower case "c") is a unit of energy and is defined as the amount of energy required to raise the temperature of one gram of water by one degree Celsius. One thousand calories is equivalent to one kilocalorie, or one Calorie (with an upper case "C"), or one food calorie.

large intestine
also called the colon. The large intestine has a wider diameter than the small intestine, and it is much shorter in length. The primary function of the large intestine is to absorb water and salts from the lumen.

larynx
the voice box. It is located just below the epiglottis in the upper part of the trachea. In the larynx, a pair of vocal cords stretches across the trachea. Speech and other sounds we make are made possible by vibrations that are conducted along the vocal cords.

lateral line
a system of sensory organs unique to fish that detects vibrations in the water, and even helps the fish determine the position of their own bodies.

leaves
the organs in which photosynthesis occurs. The leaves are involved also in excretion, and in homeostasis, that is, maintaining healthy functioning, particularly in water balance. Leaves are generally made of flat thin structures called blades that catch the sun's rays.

left hemisphere (of the brain)
the left half of the cerebrum. It is largely responsible for language processing, logic, and mathematical computation.

lens
the eye's light-focusing structure, located behind the pupil. Muscles are attached to the lens and when these muscle contract the lens changes shape. This process allows the lens to bend light and to form sharp images of objects that are at varying distances from the eye.

leukocytes
white blood cells. These white blood cells locate and destroy enemy germs and prevent illness.

lichen
a symbiotic combination of fungus with algae. The algae provide carbohydrates for the fungus, and the fungus protects the algae and helps to collect and retain minerals and water from the surroundings that are needed by the algae for growth and development.

ligaments
tough, fibrous, connective tissues that attach bones in moveable joints

lipids
organic molecules that will not dissolve in water but will dissolve in non-polar substances such as alcohols

liver
a solid organ that assists in digestion. It helps with digestion but the liver is not part of the actual digestive tract. The liver is responsible for many essential functions related to digestion, to metabolism or bodily chemical processes, to immunity from disease, and to the storage of nutrients within the body.

liverworts
plants that anchor to ground with rhizoids and often form a carpet-like layer of small plants. Like mosses, they are capable of both sexual and asexual reproduction.

locomotion
the act or power of moving from place to place

lumen
the hollow interior space of the alimentary canal. This includes the esophagus, the small intestine, and the large intestine.

lungs
part of the respiratory system, the two air-filled, spongy internal organs located on either side of the chest

lymph fluid
a clear-to-white, somewhat milky liquid, that contains extracellular fluid called plasma, that has left the circulatory system by way of capillaries in the tissues

lymphatic system
part of the body's circulatory system. Body fluids are carried in lymphatic system vessels and blood vessels. Together these vessels form the vascular system of the body. The two main functions of the lymphatic system are absorption of fats and combating disease.

lymphocytes
a type of white blood cell. The lymphocytes manufacture antibodies (proteins found in the blood to detect and destroy foreign invaders, like bacteria). These lymphocytes help protect the body against disease, which could be caused by microorganisms.

lysosomes
organelles; a specific type of vesicle formed by the Golgi body. Lysosomes contain enzymes that are used by the cell for digesting almost all biological molecules.

macromolecules
large organic molecules. There are four basic groups of macromolecules: carbohydrates, proteins, lipids, and nucleic acids.

malaria
a very harmful–often fatal–disease that is widespread in tropical and subtropical regions, including parts of the Americas, Asia, and Africa.

mammals
are warm-blooded animals with a four-chambered heart, that breathe exclusively with lungs, have hair, and nourish their young with milk produced in mammary glands. Mammals may be divided into three large groups based on the development of their offspring: monotremes, marsupials, and placental mammals.

mammary glands
a unique trait found in both male and female mammals. However, they develop fully only in females to produce milk to feed the young.

marsupials
mammals that give birth to very undeveloped young, called joeys, that develop in a pouch usually located on the mother's abdomen. The marsupials include opossums, kangaroos, wallabies, wombats, koalas, and some others.

maxillipeds
crustacean appendages found in pairs that serve as sense organs and are used to help pass food to the mouth

mechanical digestion
the physical breaking down of larger pieces of food into smaller particles. Mechanical digestion begins with chewing in the mouth, but continues through the entire digestive tract.

meconium
a greenish black stool that begins collecting in the intestines of the unborn child in the sixth month of development, and is expelled shortly after birth

medulla oblongata
the enlarged portion of the brain stem that extends down from the center of the brain and connects to the spinal cord. This part of the brain controls basic life functions, such as the rate of breathing and the heartbeat.

megasporangia
a plant structure that produces megaspores

melatonin
a hormone that is involved in the sleep-wake cycle and is believed to cause drowsiness and lowering of the body temperature, thus preparing the body for sleep. It is also thought to regulate daily rhythms of the body.

meninges
the three protective membranes that cover the brain and spinal cord. The outermost layer of the meninges is called the dura mater. The middle layer is called the arachnoid mater. The innermost protective layer is called the pia mater.

mesoderm
middle germ layer in the developing embryo

mesoglea (hydra)
a thin jelly-like material found between the gastroderm and epidermis of the hydra. Nerve cells found in the mesoglea coordinate movement of the hydra.

metabolism
the sum of all chemical reactions occurring within the cells of living organisms

metacarpals
the bones of the palm of the hand

metaphase
part of the process of mitosis. In this stage, the chromatid pairs arrive at the middle of the nucleus, and the kinetochores split apart. The chromatid pairs begin to unwind and separate, and, once again, the term "chromosomes" is used to describe these structures. The spindle attaches to each chromosome's kinetochore, and the chromosomes line up at the equator.

metatarsals
the bones of the foot

mieosis
a special type of cell division found in eukaryotes. It refers to germ cells (male and female reproductive cells) that are permanently transformed into their final products, either sperm or eggs.

mitochondria (singular: mitochondrion)
bean-shaped organelles that release the energy stored in food during the process known as cellular respiration

mitosis
a process in the cell cycle where the genome duplicates. That is, during mitosis the chromosomes in the cell nucleus prepare for duplication and distribution of chromosomes. There are five distinct stages to the process of mitosis: interphase, prophase, metaphase, anaphase, and telophase.

molecule
the functional substance of most of creation. A molecule is a combination of two or more atoms of either the same kind, or of two or more kinds.

mollusks
animals that have a soft body without a backbone and usually live in a shell. The mollusk group includes over 80,000 different species of snails, squids, oysters, and octopuses.

monocots
seed-bearing plants that contain one embryonic seed leaf. Monocots produce flowers with petals in multiples of three. Monocot leaves have veins that are parallel to each other and run the length of each leaf.

monotremes
pouched mammals that lay eggs and incubate them in a birdlike manner

monosaccharides
sugars that are the simplest carbohydrates

morphogenesis
the origin of form or shape such as the appearance of fingers and toes

motor nerves
nerves that are made exclusively of axons from motor neurons. Motor nerves transmit information away from the central nervous system to the rest of the body, specifically to muscles and glands.

motor neurons
neurons that relay information from the brain, through the spinal cord, and to the muscles to produce movement

muscle tissue
tissue made of interconnected, elongated cells that have the ability to contract and relax. Muscle tissues are responsible for movement in almost all animals. There are three basic types of muscle tissue: skeletal (or voluntary) muscle tissue, smooth (or involuntary) muscle tissue, and cardiac muscle tissue.

muscles
soft tissues composed of hundreds of thousands of muscle fibers. They support and protect other body organs, but most importantly, muscles are the agents or the cause of movement for the body.

musculoskeletal system
the system of skeletal bones and the muscles that move them

mycelium
the underground structure of fungi. It is a network of hyphae spread out under the soil.

myelin sheath
an insulating material made of lipids and proteins that protects the axons and speeds up the transmission of impulses

myosin
a thick protein filament within muscle fiber

nephron
the basic functional unit of the kidney. The nephron is a tubular or tube structure made of two distinct parts: the renal corpuscle and the renal tubule.

nerves
bundles of axons outside of the central nervous system. These nerves carry impulses between the brain and the spinal cord, and the remainder of the body.

nervous system
part of an animal's body that coordinates its voluntary and involuntary actions and transmits signals between different parts of its body. The human nervous system includes the brain, the spinal cord, the sensory organs, and the nerves.

neurons
or nerve cells, are the functional cells of the nervous system. Neurons have three functions: (1) to transmit sensory information to the central nervous system; (2) to process, integrate, and interpret incoming sensory information; and (3) to transmit motor impulses to muscles and glands to affect a change as a result of the sensory information.

neurotransmitter
chemicals which diffuse, or flow, across synapses and attach to receptor sites on neurons, glands, or muscle cells

nipple
an area of fleshy skin on mammals to allow milk to exit when feeding the young

nitrogen fixation
the process by which some bacteria are able to remove nitrogen gas from air and convert it into ammonia for use in the growth and metabolism of plants and animals.

non-vascular plants
the group of plants which does not have vessels which carry fluid; includes mosses, liverworts, and hornworts. Non-vascular plants have cell walls containing cellulose and contain chlorophyll. These plants also produce sugars through photosynthesis.

nuclear membrane
the double membrane around the nucleus of a cell. The nuclear membrane contains large pores which allow the RNA to pass into the cytoplasm.

nucleic acids
the large organic molecules which store the genetic code

nucleolus
the part of a nucleous of a cell that is involved in the production of ribosomes

nucleus
a spherical structure which is located near the center of the cell. Its job is similar to a factory that makes proteins for the cell.

obligate anaerobe
bacteria that are poisoned by oxygen

olfaction
the sense of smell

olfactory receptors
receptors in the nose that are sensitive to odors

optic nerve
a special nerve that brings sensory impulses from the eye to the brain

organ of Corti
part of the cochlea which contains sensory neurons that make up the auditory nerve and is responsible for sending impulses to the brain

organ systems
systems made of two or more organs that work together to perform a specific function, such as the skeletal system, nervous system, and circulatory system

organic chemistry
the study of the chemistry of carbon compounds

origin
the point of muscle attachment to the bone that does not move

osmosis
a special form of diffusion that allows larger substances to pass through the pores in the cellular membrane

ossification
the process of bone formation, where by cartilage is changed to bone.

osteichthyes
the class of bony fish which includes most of the common fish that people are most familiar with. These exist in a variety of shapes and sizes and include the goldfish, tropical fish, and those that are most often found in rivers, ponds, lakes and oceans. The bony fish range in size from the tiny guppy to the 900 pound tuna. Bony fish are adapted to a wide range of environments.

ovaries (human)
the reproductive glands in females. They produce estrogen and progesterone which are responsible for female sexual development, and maintaining pregnancy.

ovaries (plant)
the organ that contains the ovule; it is the organ in which fertilization occurs, and the seed is produced. The ovary matures around a seed to form the fruit.

oviparous
the producing of young by means of eggs that are hatched after they have been laid by the parent

ovoviviparous
the producing of young by means of eggs that are hatched within the body of the parent, as in some snakes

ovum
egg cell

oxytocin
a hormone that converts glycogen into glucose within the liver and muscles. It does this when the concentration of glucose in the blood is low.

pacemakers
specialized cells in the heart that control the electrical system of the heart to cause it to beat

pancreas
an organ with two main functions. First, it produces hormones that regulate the amount of glucose (sugar) that is carried in the blood. Second, the pancreas produces pancreatic juice, which passes into the duodenum through the pancreatic duct.

paramecium
an animal-like protist that is covered in short hair-like organelles called cilia which are used for movement, for attachment to surfaces and other organisms, for feeding, and for sensation. Paramecia are found in all water habitats and soils.

parasympathetic nervous system
part of the autonomic nervous system. After an emergency, the parasympathetic system returns the body to its normal state. It is also known as the "rest and digest" system.

parathyroid glands
four small glands located on the posterior (back) surface of the thyroid gland. These glands are only eight millimeters long and are the smallest endocrine glands, they are about the size of a single grain of rice. The parathyroid glands produce parathyroid hormone, which regulates the levels of calcium and phosphate in the blood.

passive transport
the movement of materials through the cell membrane without energy being used

patella
kneecap

pectoral girdle (frog)
a group of bones that almost surround the thorax and attach front limbs to the body

pellicle
the outer protein layer of euglena. Euglena lack cell walls but have spiral strips of protein on the outer surface of their cells. The protein forms a flexible pellicle, which gives the cell its shape.

pelvis
a ring of bones at the bottom of the tailbone

pepsin
the principal protease found in gastric juice

pericardium
a protective membrane around the heart

periodontal membrane
connective tissue that serves to anchor the roots of each tooth to the jawbones. Also called the periodontal ligament.

peripheral nervous system
system comprised of the sensory organs and nerves

periosteum
a tough outer membrane that surrounds and protects all bones. It is the location where muscles attach to bone.

peristalsis
wave-like contractions of the intestines that mix the chyme (mass of partially-digested food) with more digestive enzymes as the process moves along the length of the intestine

phagocytes
a type of white blood cell. When the white blood cells in the phagocyte group come in contact with foreign material, such as disease-causing microorganisms, the blood cells phagocytize (digest) the material. In this manner, foreign bodies are eliminated from the body.

phalanges
the fourteen little finger bones of the hand, three per finger and two on the thumb

pharynx
or throat, a hollow section of the body that connects the oral cavity to the esophagus

phloem
a vascular tissue in plants composed of elongated cells that connect with each other. Its job is to transport starches and sugars from one part of the plant to another.

phospholipids
a type of molecule that is a part of the plasma membrane. A plasma membrane is not a fixed structure but is fluid-like and is composed of two layers of molecules called phospholipids. Each phospholipid molecule is made up of one head and one tail. The head is composed of a phosphate group, and the tail consists of a lipid or fatty acid.

phototropism
the movement or growth of part of an organism towards a source of light, without the overall movement of the whole organism

phycoerythrim
a red photosynthetic pigment found in cyanophytes

pineal gland
a gland located within the brain. It is important in regulating the sleep cycle through a hormone called melatonin.

pituitary gland
a tiny gland that sits just under the hypothalamus in the brain. The pituitary gland is made of an anterior lobe or rounded projection and a posterior lobe.

pivot joint
a joint like that between the two bones of the forearm and allows the forearm to rotate so that the palm can face up or down. A pivot joint in the neck allows turning of the head from left to right.

placenta
the organ that attaches the unborn child to the uterine wall of the mother. The placenta has many functions and serves as a liver, kidney, and lung for the embryo. It is responsible for nourishment and the production of hormones.

placoid scales
small, spiny projections that cover the skin of sharks. These scales are often referred to as dermal denticles and are similar in structure to teeth. They give shark skin its gritty sandpaper-like texture.

plasma
a clear yellow fluid in the blood. Plasma carries molecules in solution. Some of the molecules in plasma are glucose, amino acids, carbonic acid, and urea.

platelets
small cell fragments responsible for blood clotting to protect the body from excessive blood loss

pleura
a thin tissue layer that protects the lungs

polysaccharides
complex carbohydrates made by joining more than two monosaccharides together. Starch is a polysaccharide made from hundreds or even thousands of glucose molecules that are linked together.

posterior pituitary
part of the pituitary gland that stores and releases two hormones that are produced in the hypothalamus. One of these hormones is called oxytocin, and the other is called antidiuretic hormone (ADH), which is also known as vasopressin.

precocial
types of birds that are very advanced in their development when they hatch. They have down feathers, can feed themselves, and can hop and swim.

prokaryotes
single-celled organisms such as bacteria that do not contain a nucleus. In these organisms, the chromatin is located in an area called the nucleoid region. In addition, the organelles of prokaryotes are not covered with membranes as are the eukaryotes.

prophase
part of the process of mitosis. During this stage, the nucleolus and nuclear membrane dissolve. Structures called centrioles begin to appear on the outside of the nuclear membrane.

prostaglandins
hormones that also play important roles as mediators. They are made from a fatty acid known as arachidonic acid and are produced by all cells in the body, except lymphocytes. There are ten unique receptors for prostaglandins in the body. Although they have a variety of functions, prostaglandins are known for their responses to injuries, wounds, allergens, and pain.

proteins
among the most important kinds of molecules in the chemistry of life. They regulate all chemical reactions within the cell.

pulmonary artery
the artery that takes deoxygenated blood from the right ventricle and brings it to the lungs so that it can be oxygenated

pulmonary circulation
circulation that brings deoxygenated blood to the lungs to discard carbon dioxide and to pick up oxygen. The right side of the heart is responsible for pulmonary circulation.

pulmonary veins
the veins by which oxygenated blood from the lungs returns to the heart

pupil
the black opening in the center of the iris. The adjustment of the muscles of the iris will change the size of the pupil. In strong light, such as on a bright sunny day, the pupil will become smaller, so that less light is admitted into the eye. In dim light, such as at night, the pupil becomes larger, so that the maximum amount of light can enter the eye.

radius
one of two bones which constitute the forearm or forelimb of vertebrates. The radius articulates with the humerus at the elbow, and with the carpal bones of the hand at the wrist.

rectum
the last 20-30 centimeters of the large intestine

reflex
an automatic response to a stimulus, without the person consciously thinking about the response. The brain does not control many simple reflex responses.

renal artery
an artery that has branched off of the abdominal aorta to supply the kidney with blood. There is one renal artery for each kidney.

renal cortex
the outermost layer of the kidney

renal medulla
the part of the kidney located under the renal cortex. The filtration of blood and the formation of urine take place in the cortex and medulla.

renal pelvis
a large cavity that acts like a funnel to bring urine into the ureter

renal veins
veins that drain the kidney. Renal veins connect the kidney to the inferior vena cava. They carry the blood purified by the kidney.

replication
the process where an exact copy of the DNA is produced. During replication, the left and right sides of the DNA ladder are "unzipped" by an enzyme known as helicase. This process results in two strands; the nucleotides of the left strand will pair up with the nucleotides of a newly synthesized right strand, following the base-pairing rule. The identical process occurs for the right strand.

retina
a thin membrane on the back of the eye that contains light-sensitive receptors. These receptors are called cones and rods.

rhizome
an underground stem

ribose
a five-carbon sugar contained in RNA. The difference between ribose and deoxyribose is that ribose has one more oxygen atom.

right ventricle
the ventricle of the heart that pumps blood exclusively to the lungs

schistosoma
shoulder blade

sclera
a tough outer layer of tissue that covers the eye. The front portion of the sclera is called the cornea.

semicircular canals
canals in the inner ear that are involved in the senses of balance and motion

sensation
a physical feeling or perception resulting from something that happens to or comes into contact with the body

small intestine
part of the digestive system. If the small intestine were straightened out instead of coiled up inside the body, the length would be about six meters, or about twenty feet long. Most chemical digestion and absorption of food molecules occurs in the small intestine.

smooth muscles
involuntary muscles controlled without conscious thought. Smooth muscle is found mostly in internal organs, such as in the digestive system and blood vessel walls.

spina bifida
a defect of the spinal cord where part of the spinal cord may stick out through an opening in the vertebra

spinal cord
a cord of nervous tissue that extends downward from the medulla oblongata. The spinal cord passes sensory and motor information between the brain and other parts of the body.

stomach
a large, J-shaped organ at the end of the esophagus, on the left side of the body. It has three main functions: (1) breaking down food into smaller pieces, thus creating a larger surface area for easier digestion, (2) holding food and releasing it at a constant rate, and (3) killing most harmful bacteria that are ingested.

superior vena cava
a large vein that drains the blood from the upper part of the body into the right atrium

swim (air) bladder
an air sac, in the body cavity of many fish. It originates in the same way as the lungs of air-breathing vertebrates, and in the adult may retain a tubular connection with the pharynx or esophagus.

sympathetic nervous system
part of the autonomic nervous system. It dominates in times of great stress. It is also known as the "fight or flight" system.

systemic circulation
circulation that brings oxygenated blood from the heart to the rest of the body. The left side of the heart is responsible for systemic circulation.

systole
the force exerted by the contraction of the ventricles

tarsals
ankle bones

telophase
part of the process of mitosis. Telophase begins when the chromosomes have finished moving to the opposite poles of the nucleus. The chromosomes bunch up on each end of the nucleus, and two new nuclear membranes and nucleoli form. The spindle disappears. The interphase stage can begin again, and the entire process of mitosis can repeat itself.

tendon
a very strong band of fibrous tissue that connects a muscle to a bone

thallus
the multicellular body of the non-vascular plant

thymosin
a hormone that converts glycogen into glucose within the liver and muscles. It does this when the concentration of glucose in the blood is low.

thymus gland
a gland that lies in the upper part of the chest and directly behind the sternum, or breastbone. It provides an area for lymphocytes to mature and does this through the action of the hormone thymosin. The thymus gland is important in protecting the body against disease and in protecting it against attacking itself through autoimmune diseases.

thyroid gland
a gland that lies over the top part of the trachea in the neck and is shaped like an "H." It produces a hormone called thyroxine that regulates the metabolism of the body.

thyroxine
a hormone produced by the thyroid gland that regulates the metabolism of the body

tibia
shinbone

trachea
also called the windpipe, the trachea extends downward from the pharynx. When you swallow, the epiglottis prevents food and liquid from entering the trachea.

transduction
the process by which a virus carries bacterial genes from one bacterium to another in a lysogenic phage

trichocyst
stinging organelles on the paramecium, which are thread-like structures that are barbed at the end. These trichocysts are use for hunting and for self-defense.

tricuspid valve
the valve that separates the right atrium from the right ventricle on the "lung" side of the heart

trinchina
a type of roundworm parasite

tropic hormones
hormones that specifically target other endocrine glands, the thyroid, the adrenal cortex, and the gonads

ulna
one of the lower arm bones

ulna (birds)
a bone in the middle of a bird's wing, and are necessary to supply "lift." They are also used in courtship displays.

Ungulata
a major division of Mammalia comprising of all hoofed, plant-eating mammals; divided into the orders Perissodactyla(odd-toed ungulates) and Artiodactyla (even-toed ungulates)

uterus
also called the womb; where a fertilized egg attaches and develops until the unborn baby is ready to be born

veins
the vessels that transport blood back to the heart

vitamins
organic compounds that function as coenzymes, which are a necessary part of many chemical reactions that take place in the body

volvox
a colonial green alga

xylem
a vascular tissue in plants composed of elongated cells that connect with one another. Xylem tissue dies immediately after cell formation.

Z-bands
dark lines in muscle fiber that separate each myofibril into many identical-looking units known as sarcomeres

zygote (human)
the name given to the fertilized cells that duplicate from about twelve hours after fertilization until about the third day

Index

absorption, 40, 81, 98, 181, 187-189, 202-205, 259, 284, 323, 345, 368, 411

acetylcholine, 136, 153

acetylcholinesterase, 136, 153

actin, 170-171, 174

activation energy, 59-60, 97, 120-122

Adam, 3, 432

adenine, 9, 61, 211-214, 222

ADP, 36, 63-64, 121-122

adrenal cortex, 98, 103-105, 111

adrenal gland, 98

adrenaline, 98, 106

adrenal medulla, 98, 106, 111

adrenocorticotropic hormone, 103-105

African sleeping sickness, 254

agglutination, 77

Agnatha, 362-364

air bladders, 279

alimentary canal, 182-188, 191, 194, 205

allantois, 21, 391, 414

alleles, 225-226

altricial, 414, 433

alveoli, 26, 118-119, 128

amino acid, 58-61, 212-216, 222-223

amylase, 183-184, 188, 202

amyloplasts, 47

anaphase, 219-221

angiosperm, 299

anterior pituitary, 103-106, 111

antheridium, 278, 281-282

antibodies, 18, 76-80, 83

anticodons, 216, 223, 228

antidiuretic hormone, 103

antigens, 77-78, 83

Anura, 375-376

anus, 182, 191, 304, 316-320, 323, 328, 340, 351, 368, 371

aorta, 70-73, 88-89, 352, 369, 379, 412

aortic semilunar valve, 70-72

Apoda, 362, 375, 382-383

appendix, 190

Arachnida, 334-335, 338-339, 342

arachnids, 334, 339, 342, 358

archegonium, 281-282

arteries, 17, 21-25, 68-75, 85-86, 89-91, 124, 345, 369, 379

arteriosclerosis, 85-86, 91

arthropod, 334-338, 358

Artiodactyla, 426-427

asexual, 242, 261, 268, 271-272, 276-278, 282, 298, 307, 330

asexual reproduction, 261, 268, 271, 278, 282, 298, 307

atlas, 163

ATP, 36, 63-65, 116, 120-122, 170-173, 177, 193, 283, 439

atria, 70, 75, 411

automatic response, 139

awareness, 147, 440-447, 451

axis, 159, 163, 305, 367

axon, 134-136, 153

axoneme, 255

bacteria, 48, 76, 80-82, 117, 185, 190-191, 209, 232-251, 254-256, 264-265, 279, 303

bacteriophage, 249, 264

ball-and-socket joint, 162

basal body, 46

basal disk, 309

beak, 404-407, 411, 432

bending, 20, 59, 162-164, 172, 337, 443

bicuspid valve, 70-71

bilateral symmetry, 305, 311, 323

bile, 123, 188-191, 368, 378, 394

bile duct, 188, 378

bilirubin, 90, 123

binary fission, 218

Biogenesis, 439

biology, 234, 315, 335, 363, 437

bioluminescence, 253, 374

biosynthesis, 40

bivalves, 302, 325, 331

bladder, 19, 76, 123-128, 182, 188, 315, 340, 363, 370, 379, 384

blastocyst, 6-7, 14-17

Blastomere, 9

blastula, 304, 308

bond, 51-52, 58-60, 63, 150, 212

book lungs, 340

Bowman's capsule, 124-125

brain, 18-24, 27-29, 75, 78, 86-88, 95-98, 101-104, 120, 127, 132-133, 136-148, 153, 163, 170, 173-174, 199, 285, 292, 312, 321, 326, 338-340, 345, 350-351, 363, 370-371, 377, 380, 384, 393-394, 412, 415-417, 444-445, 449-450

brain stem, 137-139

breathing center, 120, 127

breathing rate, 114, 120, 129

bronchioles, 118

bronchus, 411

Brown algae, 268, 279-280

bryophytes, 280, 298

bud, 143

burrowing, 325, 383, 418-419

cacti, 295

Calcitonin, 100

capillaries, 17, 28, 73-78, 81, 91, 117-118, 125, 128, 315, 321, 369, 379-380

carbon, 23, 36, 50-58, 65, 70, 87, 90, 116-123, 127, 194-196, 211, 222, 240, 312, 317, 320, 323-325, 343-345, 366, 369, 380, 391, 411-413, 420, 437-439, 442

carbon dioxide, 23, 50, 70, 87, 90, 116-123, 127, 240, 312, 317, 320, 323-325, 343-345, 366, 369, 380, 391, 411-413, 420, 438-439, 442

cardiac muscle, 141, 156, 169-170, 173-174

carpals, 165

cartilage, 26, 156-158, 161-164, 168, 362, 365, 368-370

catalyst, 59-60, 121

cell layer, 190

cellular respiration, 40, 45, 54, 63-64, 81, 100, 114-116, 119-123, 129, 171, 241, 283, 369

cellulose, 42, 55, 193-194, 240, 252-254, 260, 270, 308, 337, 427

cell wall, 36, 42, 65, 235, 246, 249, 252-254, 259, 270, 276-278, 307, 427

Cementum, 184

centipede, 346-347

central nervous system, 27, 105, 132-136, 141, 153, 380

centriole, 46, 220, 228

centromere, 220

cephalopods, 326, 331

cerebellum, 137-139, 153, 370, 380, 412, 415

cerebral cortex, 137-140, 148

cerebrum, 137-139, 153, 370, 380, 412, 415

chalaza, 413, 432

chelicerates, 334-335, 338-339

chelipeds, 344

Chiroptera, 422

chitin, 259-262, 336-338, 344-345, 348

Chlamydomonas, 276, 298

Chlorenchyma, 285

chlorophyll, 46-47, 240-241, 247, 251-252, 259, 270, 276, 279, 286-287

Chlorophyta, 268, 276

chloroplasts, 46-47, 240, 251-252, 270, 277, 285

Chondrichthyes, 362-367, 384

chorionic villi, 17, 21

chromatin, 43, 47-48, 65, 219

chromosome, 7, 43, 218-221, 226-228, 242-243

cilium, 255

circulation, 21-23, 33, 68-72, 76, 81, 84, 87-89, 258, 292, 302, 309, 320-323, 326-328, 334, 340, 345, 352, 362, 369, 379, 388, 394

cisternae, 45

Class, 61, 236-237, 299, 302, 311-315, 319-321, 324-327, 334-335, 339, 342-347, 362-365, 368, 388, 401-402, 415, 422, 433

classification scheme, 235-236, 264, 335, 363

clavicle, 165, 409

Cleavage, 10, 13-14, 33

closed circulatory system, 323, 326, 336, 369, 379

coccyx, 164-165

cochlea, 145

codon, 212-216, 222-223

coelenterates, 302, 306-308, 311

coelom, 318-319, 323, 326-328

Coenzymes, 61, 195

collagen, 159-161, 176, 304, 307

collecting duct, 124-126

Collenchyma, 285

combination, 85, 263

conceptus, 7

cones, 146, 274, 286, 289-290, 293-294, 299

conifer, 274, 293

conjugation, 243, 256, 264, 277

contractile vacuole, 252, 256

convolutions, 138, 415, 420

cornea, 146-147

coronary arteries, 72, 86

cortex, 98, 103-105, 111, 124, 137-140, 148, 444

corticoid hormones, 98, 105

corticoids, 103-105

cotyledon, 296

crab, 344

crayfish, 218, 315, 334, 344-346, 358

cretinism, 99, 111

crop, 244, 261, 320, 351, 410

crown, 69, 184, 284, 318

Crustacea, 334-335, 344

cuticle, 280, 284, 313, 317, 338

cycad, 293

cyclic AMP, 96, 111

cyclostomes, 364

cytokinesis, 219-221, 228

cytoplasm, 36, 41-45, 119-121, 134, 149, 174, 210, 214-223, 243-245, 249, 252-255, 259, 277, 284-285, 379

cytoplasmic bridge, 243

cytosine, 9, 61, 211-214, 222

cytoskeleton, 46, 65

Daddy longlegs, 342

decomposer, 264

deoxygenated blood, 17, 23, 70-73, 89-91, 118, 379, 394, 411

deoxyribose, 63, 211, 214, 222

diabetes, 100-103, 110-111, 168, 200-201

diabetes mellitus, 100, 110-111

diaphragm, 24, 70, 118-120, 128-129, 416

diastole, 74

differentiation, 11, 33, 208, 221, 228, 433, 438, 442

digestion, 17, 50, 100, 181-191, 202, 205, 255, 302, 309, 320, 323, 328, 334, 340, 345, 350, 358, 362, 368, 378, 388, 393, 396, 410, 427

digestive glands, 328, 345

diploid, 226, 254, 272-274, 278, 281-282, 289-291, 294, 298

Diplopoda, 334-335, 346

distal tubule, 124-126

DNA, 7-12, 33, 43, 63-65, 174, 208-224, 228, 234, 240-243, 248-250, 270

Domain, 234-236, 311, 335, 363

dominant, 17, 138, 224-229

dorsal aorta, 369

double helix, 9, 208-214, 222

duct, 124-126, 187-188, 378

duodenum, 182, 187-189, 378

earthworm, 319-323, 330

ectoderm, 18, 304-305, 309-311

Edentata, 428

Biology For Life 489

egg, 6-14, 19-21, 42, 57, 213, 226-227, 245, 272-274, 277-282, 287, 291-294, 307, 318, 322, 325, 329, 346, 349, 352, 355, 367, 381, 390-394, 413-414, 418, 432, 438, 444

elimination, 81

embryo, 6-7, 11-22, 25, 33, 57, 161, 274, 281-282, 287, 290-291, 296, 305, 330, 381, 388-391, 413-420, 433, 438

embryonic leaves, 296

emotion, 79, 445-448

endoderm, 18, 304-306, 309-311, 316

endoplasmic reticulum, 45-46, 65

enzyme, 60-61, 120-122, 136, 183-184, 188-189, 202, 214, 217, 241, 249

epidermis, 175, 280, 284, 298, 304-306, 309, 317, 320-321, 327

epinephrine, 98

epithelium, 189, 304

erythrocytes, 77, 91, 256

esophagus, 180-185, 191, 202-203, 320, 323, 328, 340, 345, 351, 368, 378, 394-396, 410

estimative sense, 445

estrogens, 102

Euglena, 251-252

eukaryote, 238, 253

Excretion, 40, 114, 123, 126, 190, 302, 320, 325, 334, 340, 345, 350, 362, 368, 379, 388, 393

exoskeleton, 336-338, 344, 348-351

expiration, 119

extension, 172, 252

eyespot, 252

facultative anaerobe, 264

fallopian tube, 14-16, 20

family, 85-86, 99, 152, 161, 166, 201, 236-237, 241, 260, 297, 311, 335, 348, 363

fatty acids, 40, 45, 56-57, 65, 183, 188-191, 201

feathers, 388, 401-414, 432

feces, 191, 194, 315

femur, 166

fern, 271, 287-289, 299

fertilization, 6, 9-18, 33, 257, 271-274, 277-278, 284, 287, 291-294, 299, 308-309, 321-322, 325, 329, 352-355, 367, 371, 374, 381, 390, 395, 413, 416-419, 442

fetus, 4-8, 13, 18, 22-29, 33, 88

fibrin, 76

fibula, 166

filaments, 170-171, 247, 277-278, 298

flagellum, 252-254, 264

flatworm, 312, 315

flexion, 172, 337

flower, 271, 274, 295, 298-299, 305, 407

fluke, 313-315

Fragmentation, 272

Fr. Mendel, 225

fructose, 54-55, 122, 193, 201

fucoxanthin, 279

Fungi, 39-41, 209, 231-237, 240, 258-262, 279, 287-288, 303

gall, 19, 123, 182, 188

gallbladder, 191, 378

gamete, 254, 270

gametophyte, 273-274, 281-282, 287-291, 294-296, 299

ganglia, 140-141, 325, 351

gastrodermis, 309

gastrovascular cavity, 309

gemmae, 272, 282

gene, 7-9, 33, 174, 213, 222-228, 250

genetics, 2, 7-8, 11-12, 33, 208, 224, 228

genotype, 224-228

Genus, 236-237, 246, 289, 311, 335, 363

gill arches, 370

gill rakers, 370

gills, 262, 323-325, 328, 344-345, 363-376, 383-384

gland, 19, 29, 94, 98-111, 134-135, 176, 183, 323, 342, 358, 378-380, 393, 413

gliding joint, 163

glomerulus, 124-125

glucagon, 19, 100

glucose, 40-42, 47, 54-55, 59-60, 63-64, 76, 95, 98-101, 116, 120-122, 125, 151, 171, 183, 187-193, 202, 240, 286, 355, 378-379, 438-439

glycerol, 56, 183, 188-191

glycogen, 15, 55, 98-100, 171, 188, 191, 194, 202, 379

gnetophytes, 268, 290-292

Golgi body, 46

grafting, 271-272, 298

grass, 240, 271, 311, 315, 343, 419

gravitropism, 443

gray matter, 27, 137, 140

Gregor Mendel, 224, 228

growth hormone, 103-106

guanine, 9, 61, 211-214, 222

gullet, 251-252, 256, 378-379

gut, 254, 312

gymnosperm, 291, 299

haploid, 254, 272-274, 277-278, 281, 289-293, 298

hardening, 85

head, 19-20, 24-25, 28-30, 42, 88-89, 140, 146, 149, 162-166, 170, 213, 281, 312, 315-316, 324-326, 329, 338-341, 344-352, 367-370, 373, 377, 382, 388, 392-393, 399, 402, 406-409, 432

hearing, 22, 26, 132, 141, 144, 149-151, 174, 350, 371, 377, 393, 407, 448

heart, 18-19, 23-24, 27, 30, 38, 49, 52, 68-75, 79-81, 86-91, 95, 98-99, 118-119, 128, 141, 170-177, 200, 220, 285, 320, 323, 336, 340, 345, 352, 369, 379-380, 394, 400, 411-415

helix, 8-9, 208-215, 222, 324

hemispheres, 138-139

hemoglobin, 77, 119, 320, 323, 379, 403

hemolymph, 336, 340, 351-352

heredity, 7-9

heterozygous, 226-229

holdfast, 278

holiness, 31, 181

homozygous, 225, 228-229

hookworm, 317, 330

hormone, 16, 94-96, 99-106, 110-111, 167, 289

horse, 289, 426-427, 445

humerus, 165, 409

hybrid, 225-226, 229

hydrocarbons, 51-52, 56

hypertension, 68, 86, 91, 168

hyperthyroidism, 99, 111

hyphae, 259-262, 287

hypoglycemia, 202

hypothalamus, 29, 94, 103-106, 110-111, 139, 153

hypothyroidism, 99, 111

immaterial, 440, 449-450

implantation, 14-17, 33, 87

impulse, 96, 106, 132-136, 140, 144, 147, 153

incus, 144

inferior vena cava, 70, 73, 88-89

inhibit, 104-105, 148, 448

Insectivora, 428

insertion, 172

inspiration, 119

insulation, 56, 134, 411

insulin, 19, 100-101

intellect, 3, 39, 61, 278, 439, 450-452

Interferon, 82

internal respiration, 116, 119, 129

interneuron, 140

interphase, 219-221

intestine, 180-182, 185-191, 194, 202-205, 240, 314-317, 320, 323, 368, 378-379, 410

iodine, 99-100, 383

iris, 146

Islets of Langerhans, 94, 100, 111

isogamy, 277

Jawless Fish, 362-364

joint, 83, 156, 162-163, 166-169, 172, 177, 289, 337

jointed limbs, 338

Joules, 193

kidney, 16, 98, 103, 124-128, 316, 368

kilocalories, 192-193

Kingdom, 31, 37-39, 157, 232-238, 245, 248, 251, 258-259, 264, 267-270, 302-304, 311, 335, 363, 437, 441

koala, 235, 419

Lagomorpha, 421

lanugo, 24, 27-28

large intestine, 180-182, 185-187, 190-191, 203-205, 379

larynx, 118, 163, 411

leaf, 283, 318

left atrium, 70-71, 88-89, 379, 412

left ventricle, 70-73, 88, 412

legless amphibians, 375, 382-383

legumes, 198, 241

lens, 28, 146-147, 341

leukocytes, 78, 91

lichen, 263

ligaments, 17, 158, 162-163, 172, 177

light, 28, 32, 49, 122, 141, 146-149, 175-176, 240, 248, 252-253, 270, 276, 279, 312, 321, 325, 340-341, 348, 374, 404, 409-412, 441-447

lipids, 15, 23, 36, 45-46, 56, 65, 182, 188-190

liver, 16-19, 49, 55, 87-90, 98-100, 123, 171, 180-182, 187-191, 194, 202-205, 256-257, 314-315, 368, 378-379, 394

liverworts, 268, 276, 280-282, 298

lobe, 380

locomotion, 256, 306, 313, 327, 335-338, 349, 445

loop of Henle, 124

lumen, 185, 189-191

lung, 16-17, 70, 89, 117-118, 128, 324, 340, 374, 379, 394, 411

luteinizing hormone, 103

lymph, 81-83

lymphatic system, 49, 68, 76, 81, 91, 189-190, 317

lymphocytes, 78-82, 98, 102

lysogenic phage, 250

lysosomes, 46, 65

malaria, 256-258

malpighian tubules, 351, 358

maltose, 55, 60, 184, 190, 193

mandible, 184

mantle cavity, 323-326

marsupials, 388, 416-420, 433

meconium, 29

medulla oblongata, 139, 153, 415

megasporangia, 293

meiosis, 208, 218, 226-228, 242-243, 254, 272-274, 277-278, 293, 307

meiospore, 277

Melatonin, 102

membrane, 17, 21, 25, 41-46, 65, 70, 96, 118, 135-136, 144-146, 151-153, 158-159, 169, 177, 184, 189, 210, 219-221, 235, 240-242, 246, 249-252, 255-256, 259, 319, 348-350, 377, 380, 390-393, 413-414

meristematic, 284, 298

mesoderm, 18, 304-306, 311-313, 316-319

mesoglea, 309

messenger RNA, 63, 214-215, 228

metacarpals, 163-165

metamorphosis, 347-352, 358, 375, 382-383

metaphase, 219-221, 228

metatarsals, 166

microsporangia, 293

middle lamella, 42

millipede, 346

minerals, 15, 40, 49, 160-161, 180, 192, 195-200, 203-205, 241, 262, 272, 283-286, 344, 439, 442

mites, 246, 335, 338-339, 342-343, 358

mitochondrion, 45, 64, 121

mitosis, 10, 13, 175, 208, 218-221, 226-228, 242, 251-256

mitotic spindle, 220

Mollusks, 302, 306, 322-326, 331

monosaccharide, 54

morula, 6-7, 13-14

moss, 281, 288

mouth, 19, 23, 27, 49, 117, 128, 148, 151, 167, 180-188, 191, 213, 304, 308-309, 312, 316-320, 323-325, 328, 338-340, 344, 350, 364-371, 374, 377-380, 392-396, 400, 407

mucosa, 185, 189

mucus, 82, 117-118, 129, 143, 184-185, 313, 325, 368, 376-377, 384

muscle, 11, 26, 49, 55, 72, 75, 86, 100-102, 119, 128, 134-136, 140-141, 146, 156, 169-174, 177, 185, 200-201, 304, 309, 313, 317-321, 336-337, 351, 370, 380, 394, 415-416, 438

mycelium, 259-262

myelin, 134, 142, 149

myofibril, 170, 177

myosin, 170-171, 174

nasal cavity, 117

nematode, 317

nephridia, 320, 323-325

nephron, 114, 124-125, 129

nerve cord, 321, 336, 340, 351

nervous system, 18, 27-28, 50, 96, 104-106, 128, 131-136, 140-141, 147-149, 153, 166, 304, 312, 325-326, 329, 334, 338-340, 345, 351, 362, 370, 380, 385, 388, 394, 412

neuron, 133-137, 140-142

neurotransmitter, 135-136, 153

nitrogen, 52-53, 61-63, 118, 123, 129, 194, 211-212, 222, 241, 247, 264, 320, 340, 345, 351, 368-369, 391, 420, 437

nitrogen fixation, 241, 247, 264

nitrogenous base, 9, 211

node, 75, 83, 271

norepinephrine, 98

nostrils, 3, 370, 377, 392, 396, 399, 406, 411

nuclear membrane, 43-45, 219, 235, 251

nuclei, 10, 218, 251, 293, 379

nucleolus, 43, 219, 251-252

nucleotides, 8, 61, 211-214, 217, 222, 249

nucleus, 8-10, 36, 41-43, 47-48, 51, 61, 65, 77, 97, 134, 174, 214, 217-221, 235, 238, 252-256, 270, 276, 379

nutrition, 14, 17, 23, 40, 180, 192, 201, 205, 232, 237, 240, 259, 309, 312, 416, 437-442, 450

obligate anaerobe, 264

Oedogonium, 278, 298

olfaction, 143-144

olfactory receptors, 141-144, 444

operculum, 369

opossum, 418-419

optic nerve, 148, 444

oral groove, 256

organelle, 46

organic chemistry, 51

organization, 50, 65, 440

organ of Corti, 145

origin, 11, 31, 172, 330, 401

ossification, 26, 159

Osteichthyes, 362, 368, 384

osteoblasts, 158

osteoclasts, 158

osteocytes, 159-161

ostia, 336, 345

oval window, 145

ovary, 341, 346, 352, 381, 413

oviparous, 395, 432

ovoviviparous, 395, 432

oxygen, 7, 17, 24-27, 30, 50-56, 64, 70-73, 77, 87-90, 115-122, 127-128, 159, 171-172, 176, 193-194, 214, 240-241, 252-253, 312, 317, 320-326, 336, 340, 345, 349, 352, 366, 369, 379-382, 391, 396, 403, 411-413, 420, 423, 437

oxytocin, 103-104

pacemaker, 75, 86

pancreas, 18-19, 100-101, 180-182, 187-188, 191, 204-205, 378

paramecium, 235, 255-256, 438

parasite, 248, 254, 257-258, 313-317

parasympathetic nervous system, 141

parenchyma, 276, 284-285, 298

patella, 166

pea, 224-226, 229

peanut, 57

pectoral girdle, 165, 377, 409

pedipalps, 339-342

Pellicle, 252, 255-256

pelvic girdle, 165, 377, 393, 396, 409

pelvis, 124-126, 164-166

pentamerous radial symmetry, 326

pericardium, 70

Periodontal Membrane, 184

periosteum, 158-159

peripheral, 132-136, 140-141, 149, 153, 380

Perissodactyla, 426-427

peritoneum, 319

phagocytes, 78, 82-83

phalanges, 163-166

pharynx, 117-118, 144, 184-185, 312, 317, 320, 340, 410-411

phenotype, 213, 224-228

pheromones, 354-355

phloem, 49, 276, 285, 299

phosphate, 8-9, 42, 56, 61-64, 100, 121, 159, 211, 222

phospholipids, 42-43, 56

Biology For Life

photosynthesis, 46-47, 54, 241, 247, 252, 270, 276-279, 285-287, 291-292, 443

phototropism, 441-443

phycoerythrin, 247, 279

Phylum, 232, 236-237, 245-246, 251-256, 261-262, 302-303, 306-308, 311-312, 316-318, 322, 326, 330, 334-335, 338, 363

pine, 268, 293, 299

pineal gland, 94, 102

pituitary gland, 29, 94, 103-106, 109, 380

pivot joint, 163

placenta, 8, 14-25, 28-30, 33, 87-89, 395, 420, 433

placoid scales, 365

plasma, 42-46, 56, 65, 68, 76-77, 81, 91, 125, 189, 202, 210, 221

plasma membrane, 42-46, 65, 189, 210, 221

Plasmodium, 256

Plasmodium vivax, 256

plastids, 36, 46-47, 65

platelets, 76

pleura, 118

pollen, 270-271, 274, 291-295, 355, 422

polypeptides, 58, 186, 189

polysaccharides, 55, 65, 183, 190, 193-194, 202

posterior pituitary, 103, 106, 109-111

pregnancy, 7, 15-22, 25-28, 78, 100, 150, 165-166

prokaryotes, 48, 65, 213, 218, 232, 238-242, 264

prolactin, 104-106

prophase, 219-220, 228

prostaglandins, 97-98, 111

protease, 186-188

protein, 8-9, 18, 36, 57-65, 76-77, 82, 96, 99, 119-122, 159, 170, 177, 186, 191, 194-195, 198-199, 208-210, 213-218, 221-223, 228, 248-252, 304, 307, 320, 389, 413, 422

proteins, 8-9, 15, 23, 36, 40, 43-46, 49, 57-61, 65, 76-77, 80-83, 96-98, 120, 123, 174, 177, 180-183, 186-195, 199-202, 205, 208-217, 220-222, 241, 249, 258, 338, 379, 412-413

protonema, 281

pseudopods, 255, 307

pulmonary artery, 70, 73, 89, 379

pulmonary circulation, 72

pulmonary semilunar valve, 70-71

pulmonary surfactant, 26

pulmonary vein, 379

pulp cavity, 184

pupil, 28, 146

purebred, 225, 229

pyruvate, 64, 121, 171

quickening, 24

radial symmetry, 305, 308, 326

radius, 165

radula, 323-324

rakers, 370

rational, 436, 440-441, 448-450

Rational Soul, 436, 440, 448

recessive, 224-228

rectum, 182, 191, 315, 351, 410

Red Algae, 268, 279

reflex, 139-140, 149-150, 153

renal artery, 124

renal pelvis, 124-126

replication, 182, 217-218, 226-228

reproduction, 7, 40, 207-208, 217-219, 232, 241-242, 249, 256-257, 260-261, 264, 268-284, 290, 296-299, 302-307, 313, 318, 321, 329-330, 334, 341, 346-349, 352-354, 358-359, 362-363, 367, 371, 380, 384, 388-390, 395, 413, 416, 437-443, 450

reptiles, 375, 382, 388-395, 399, 403, 408, 411, 420, 432

reservoir, 251-252

respiratory system, 50, 114-116, 129, 323, 388, 411

retina, 146-147

Rh factor, 78, 91

Rhinoceros, 426-427

rhizoids, 262, 280-282, 287

rhizomes, 271, 289, 299

ribose, 63, 214

ribosomal RNA, 63, 214-215

ribosome, 215-216, 223

right atrium, 70-75, 88-89, 379, 411

right ventricle, 70-73, 88-89, 411

RNA, 63-65, 208-210, 213-216, 222-223, 228, 235, 248-250

rods, 146-147, 161

root, 272, 283-284

roundworm, 317

Runner, 271

sacrum, 164-165

saddle joint, 163

salts, 46, 123, 126-128, 187, 191, 279, 372

saprophytes, 259-261

sarcomere, 170-171, 177

scapula, 165, 409

Schistosoma, 315

science, 11-12, 29, 46, 51, 79, 170, 228, 233, 437

sclera, 146

scorpion, 342

sebum, 24

secretion, 45, 94-96, 103-106

segmented body, 336, 348

self-conscious, 441

semicircular canals, 145-146, 346, 371

semilunar valve, 70-72

seminal receptacle, 341, 352

sensation, 24, 133, 139-142, 147, 151, 444-445

sensitive, 49, 141-142, 146, 151, 176, 252, 312, 341, 345, 371, 436, 441, 444-445, 450

Sensitive Soul, 436, 444-445

shark, 362-367, 384

shrub, 272, 292

sight, 142, 146-149, 312, 399, 407, 444, 450

siphon, 326

skin, 10, 18, 24-28, 50, 82, 90, 102, 127, 133, 140-143, 147, 153, 156-157, 169, 175-177, 185, 202, 213, 221, 285, 315-317, 327, 338, 353, 362-365, 376-377, 380-382, 389, 392-394, 413, 416, 438, 445

sloth, 429

small intestine, 180-181, 185-191, 205, 378-379

smell, 132, 141-144, 148-153, 312, 339, 346, 351, 370-371, 377, 380, 394, 407, 412, 428, 451

smooth muscles, 170, 174, 177

somatic nervous system, 141

somatotropin, 103

soul, 3-4, 7, 31, 37, 115, 157, 210, 303, 436-445, 448-450, 453

sperm, 6-10, 13, 19-21, 103, 213, 218, 226-227, 243, 270-274, 277-282, 287, 293-294, 307-309, 313, 316-318, 321, 325, 329, 341, 346, 352, 355, 367, 371, 380-381, 390, 393-395, 413, 442

spermatozoa, 324

Spherical Symmetry, 305

sphincter, 185-187, 202, 205

spider, 249, 339-342, 358

Spina bifida, 19, 166

spinal cord, 18-19, 132-141, 153, 166, 363, 370-371, 377, 380, 384, 413

spinal nerves, 140, 371, 380

spiny anteater, 416-418

spiritual, 36-37, 74, 90, 115, 181, 210, 437-439, 449

Spirogyra, 277-278, 298

spleen, 123, 258

sponges, 302-308, 311, 330-331, 365

spores, 242, 245, 256, 260-262, 271-277, 281-291

stem, 137-139, 284-286, 289, 296, 304, 355, 443

stem-cell, 4

sternum, 26, 102, 128, 164, 409

steroid hormones, 97

stipe, 279

stomach, 19, 49, 180-188, 191, 202-205, 244, 248, 323, 328, 340, 345, 351, 368, 378, 394, 410, 427

striated, 120, 141, 170, 177

sucrose, 55, 193

sugar, 8-9, 54-55, 61-64, 98-100, 125-127, 141, 148, 183, 187-188, 193, 199-202, 211-214, 222, 368

sunflowers, 296

superior vena cava, 70, 73, 89

swim bladder, 363, 370, 384

swimming, 70, 162, 309, 315, 335, 344-345, 365, 375, 408, 421

symbiotic, 240, 263-264, 287-288

symmetry, 302, 305, 308, 311, 323, 326, 330, 358

sympathetic nervous system, 141

Biology For Life

synapse, 134-136

synthesis, 45, 58, 63, 77, 177, 208-210, 213-215, 221-222, 228, 241, 270

systemic circulation, 72

systole, 74

systolic pressure, 74

taste, 132, 141-143, 148-153, 184, 339, 350, 371, 380, 407

T-cells, 83

telophase, 219-221

telson, 345

tendons, 17, 158, 172, 177

termination, 213-215

thalamus, 139, 153

thallus, 276, 279-282

thorax, 164, 338-339, 345, 348-351, 377

thymine, 9, 61-63, 211-214, 222

thymosin, 102

thyroid, 18-19, 94, 99-100, 103, 110-111

Thyrotropic hormone, 103

thyroxine, 99, 103

tibia, 166

tick, 246, 343

touch, 22-24, 132-134, 141-142, 148-151, 176, 271, 277, 309, 312, 321, 325, 338-339, 345, 365, 368, 445

trachea, 99, 118, 394, 411

trait, 7, 222-226

transcription, 208, 214-215, 223, 228

transduction, 250

transfer RNA, 63, 214-215, 223, 228

transformation, 243, 264

translation, 208, 214-216, 223, 228

trichina, 317

tricuspid valve, 70-71, 89

trilobites, 335

triplet, 215-216

trophoblast, 14, 17, 21

tropic hormone, 105, 111

Tuber, 271

tympanic membrane, 144-145, 151, 380

ulna, 165, 405

umbilical cord, 17-25, 30, 420

unicellular, 36, 39, 49, 65, 251, 276, 318

unity, 157, 438-439

uracil, 61-63, 214

urea, 76, 123-126, 412

ureter, 124-126, 340, 379-381, 394

urethra, 124, 340

urine, 90, 101-103, 109, 124-128, 315, 340, 368, 372, 379, 394, 403, 412

Urodela, 362, 375, 382

uropods, 345

uterus, 6, 14-17, 29-30, 352, 381, 416-422

vacuoles, 45-46, 65, 255-256, 285, 312

valves, 70, 74, 81, 325

vas deferens, 413

vasopressin, 103

vegetative reproduction, 271-272

vegetative soul, 436, 440-442

veins, 70-75, 81, 91, 124, 296, 353, 369, 379

ventral nerve cord, 340

ventricles, 70, 74-75, 400, 411

venules, 73

vermis, 139

vertebra, 19, 162-164

vertebral column, 118, 124, 139, 162-166, 364, 370, 379

vesicles, 45-46

vibrations, 118, 145, 341-345, 393, 396

viroids, 232, 250, 264

viruses, 76, 82-83, 210, 232, 237, 248-250, 264

vitamins, 19, 166, 180, 192-200, 203-205

viviparous, 395

voluntary, 120, 141, 170, 181, 451

Volvox, 277, 298, 305

water soluble, 52

whisk fern, 287-288, 299

xylem, 49, 276, 283-285, 299

Z-bands, 170-171

zoospore, 278

zygospore, 278

zygote, 6-13, 254, 257, 270-274, 277-278, 294, 304, 308, 413, 438